I10643330

*Emerging Illnesses
and Society*

YMARBIL SUGMAD HTUOC .D.C .T

# Emerging Illnesses
# and Society

## Negotiating
## the Public Health
## Agenda

Edited by

RANDALL M. PACKARD

PETER J. BROWN

RUTH L. BERKELMAN

HOWARD FRUMKIN

The Johns Hopkins
University Press

*Baltimore & London*

L.C.C. SOUTH CAMPUS LIBRARY

RA
418
' E353
2004

© 2004 The Johns Hopkins University Press
All rights reserved. Published 2004
Printed in the United States of America on acid-free paper

9 8 7 6 5 4 3 2 1

The Johns Hopkins University Press
2715 North Charles Street
Baltimore, Maryland 21218-4363
www.press.jhu.edu

Library of Congress Cataloging-in-Publication Data
Emerging illnesses and society: negotiating the public health agenda / edited
by Randall M. Packard, Peter J. Brown, Ruth L. Berkelman and Howard Frumkin.
    p. cm.
Proceedings of a conference held in February 2000 at Emory University.
  ISBN 0-8018-7942-6 (hardcover: alk. paper)
  1. Social medicine—Congresses. 2. Public health—Congresses. 3. Health—
Congresses. I. Packard, Randall M., 1945–
RA418.E353 2004
362.1—dc22    2003023210

A catalog record for this book is available from the British Library.

MAY 0 1 2006

# CONTENTS

## PREFACE

From 1998 to 2000, the Center for the Study of Health, Culture, and Society (CSHCS) at Emory University organized three semester-long interdisciplinary faculty seminars to explore three themes: emerging illness and communities of suffering; environmental hazards, community activism, and the public health; and emerging illnesses and institutional responses. Most of the case studies in this volume were first presented in one of these seminars. In addition, the CSHCS conducted a symposium, "Emerging Illnesses and the Media," in March 1999.[1] Finally, in February 2000, the seminar participants gathered at Emory for a conference that focused on the similarities and differences among all the case studies. The CSHCS seminars brought together historians, sociologists, and anthropologists who had studied the emergence of health problems within different communities and cultural settings, with physicians, public health researchers, practitioners, and policy analysts. The seminars also included activists who were involved in efforts to bring specific ailments and disabilities to the consciousness of both health professionals and their own local communities. The linking of scholarly work in the social sciences and humanities with the work of health practitioners and community activists created a rich environment for exploring a set of problems that have historical and contemporary importance.

Funding for these seminars was generously provided through a two-year grant from the Mellon Foundation's Sawyer Seminar Program. In all these activities, the CSHCS was very fortunate to have the active participation of scientists associated with the Centers for Disease Control and Prevention (the main campus of which is conveniently located next door). This is a major reason that this volume's focus on "institutional responses" often emphasizes the role of national institutions rather than local and state public health authorities.

Although it was the intention of the seminar organizers to include studies from a range of cultural settings, this volume focuses largely on the U.S. experience. Future work will, we hope, provide more comparative perspectives on the social negotiation of public health agendas in international contexts. Indeed, the role of global authorities and institutions, like the World Bank and the World Health Organization, will be an important dimension of those analyses.

The style and content of the chapters that follow reflect the diversity of the participants. As editors, we spent considerable time discussing how, and to what extent, we should attempt to encourage the authors to conform to a common stylistic template in order to create a coherent volume. In the end, we felt that we should impose certain guidelines, for such things as citations and formatting, and should encourage the contributors to be aware of the potentially diverse backgrounds of the volume's readers. However, we would not attempt to make health activists write with the analytical distance and "objectivity" of a historian, or the sociologist with the passion and conviction of the activist. We would not attempt to homogenize the volume. We believe that exposing readers to a diversity of approaches and perspectives, as well as to a range of case material, adds to this book's value.

We have organized the essays into two groups. The first section, "Making Illnesses Visible," focuses attention on the efforts of illness support groups and organizations—the communities of suffering (COS)—to gain public health recognition for their specific health problems. The second section, "Institutional Responses to Emerging Illnesses," focuses on how and why the institutions of public health and medicine respond to different types of emerging illnesses. In Chapter 1, the introductory essay, we draw from the materials in these essays to analyze the broader social processes through which emerging illnesses do or do not gain a place on public health agendas.

NOTE

1. Support for this symposium was provided by a grant from the Rockefeller Foundations Humanities Program.

AAMC  Association of American Medical Colleges
APHIS  Animal and Plant Health Inspection Service
ASM  American Society for Microbiology
BSC  Board of Scientific Counselors
CDC  Centers for Disease Control and Prevention
CISET  Committee on International Science, Engineering, and Technology
CSTE  Council of State and Territorial Epidemiologists
DHHS  Department of Health and Human Services
EPA  Environmental Protection Agency
FDA  Food and Drug Administration
IOM  Institute of Medicine
NCID  National Center for Infectious Diseases
NIAID  National Institute of Allergy and Infectious Diseases
NIH  National Institutes of Health
NINDB  National Institute for Neurological Diseases and Blindness
OMB  Office of Management and Budget
USDA  United States Department of Agriculture
USPHS  U.S. Public Health Service
WHO  World Health Organization

*Emerging Illnesses
and Society*

# CHAPTER 1

# Introduction

## Emerging Illness as Social Process

RANDALL M. PACKARD, PETER J. BROWN,

RUTH L. BERKELMAN, AND HOWARD FRUMKIN

A series of publications in the early 1990s, including the Institute of Medicine's *Emerging Infections: Microbial Threats to Health in the United States* and Laurie Garrett's *The Coming Plague,* drew public attention to the rising threat of newly emerging diseases. These and other studies noted that over the past twenty years, human populations in various parts of the globe have experienced illnesses resulting from new infectious agents, such as HIV/AIDS, *E. coli* 0157:H7, and hepatitis C (HCV), as well as from existing pathogens that have become resistant to biomedical treatment, such as new strains of tubercle bacilli. Emergent diseases have also resulted from environmental changes that have increased contact between humans and existing pathogens, such as hantavirus and brucellosis. Exposure to a range of toxic substances produced by industrial production or built environments has produced another group of emerging illnesses. Finally, there exists a set of widely experienced but poorly understood emerging illnesses, such as chronic fatigue syndrome (CFS),[1] multiple chemical sensitivity, and fibromyalgia.

The threat posed by emerging diseases took on a new reality after September 11, 2001. The weaponization of anthrax and growing fears over the possible use of other biological agents, including smallpox, as terrorist weapons greatly increased the government's awareness of the need for public health preparedness. This concern resulted in efforts to increase national and international surveillance capacities in order to ensure the early detection of emerging diseases. These efforts certainly aided the rapid identification and initial control of the severe acute respiratory syndrome (SARS) pandemic, which began in 2003.

Increased surveillance capacities may detect acute emerging infections. However, it is important to recognize that emerging health problems do not always surface as acute disease outbreaks. They may, like HIV/AIDS, appear in a less dramatic fashion. In many cases an emerging illness is experienced by individuals and their caretakers long before public health authorities identify it.[2] Moreover, recognition by public health authorities may not lead to broader societal recognition of the importance of a particular health problem. Improving a society's ability to respond effectively to emerging health problems requires more than simply increasing existing public health surveillance mechanisms. It also requires recognition of the complex and often contested processes through which emerging health problems either become visible and the target of interventions or remain hidden and neglected.

This book is a collection of case studies about the social processes by which new illnesses or diseases get placed on the public health agenda and trigger responses by public health institutions. These historical and contemporary case studies examine how illnesses "emerge" from the suffering of individual patients to become medically recognized problems and public health issues requiring programs for research, control, and prevention. On a social level, some new diseases or illnesses provoke fear—even panic—and immediate response, whereas other illnesses are met with skepticism or apparent apathy. The central question is: How do medical, social, cultural, and political-economic factors define public health agendas?

The cases in this book demonstrate that the "emergence" of a disease or illness is both an epidemiological phenomenon and a social process. Of the variety of actors involved in the social process of emergence, two kinds of social groups are the most salient. On the one hand, representatives of the communities of suffering (COS) advocate for recognition of the new health problem in part by garnering broad social attention to the issue.[3] On the other hand, representatives of the institutions of public health and medical research are obligated—ideally—to use the tools of epidemiological and biomedical science to investigate legitimate claims about a new health problem. In some instances, health authorities must act as gatekeepers debunking illegitimate claims; in others they must work to increase public attention regarding problems that are being ignored. For leaders of public health institutions, the choice of which role to play is often difficult and shaped not only by epidemiological evidence but also by a range of cultural, political, and economic factors.

The interaction of the COS and the powerful institutions of medicine and public health is the crux of the social negotiation of the public health agenda. This negotiation is, of course, affected by important variables, including the epidemiological characteristics of the problem, the social class of sufferers, the interest of the media, and the political savvy of activists. The case studies in this volume are interesting because of their epidemiological and social particularities and also because they reflect variations of a general process of how the illnesses and disabilities of individuals emerge and evolve into public health problems. The essays also explore the struggles to define public health agendas from the perspective of both COS and the institutions responsible for ensuring the public's health.

## Emerging Diseases versus Emerging Illnesses

To understand the processes through which illnesses emerge or reemerge, it is important to widen the purview of the discussions beyond the realm of emerging infectious diseases with recognized etiologies. In this regard, the social scientific distinction between disease and illness is critical. *Disease* refers to an accepted biomedical category, often based on laboratory tests or precise clinical definitions.[4] Biomedical practitioners generally believe that disease categories are "objective" facts, even though historians of science have repeatedly demonstrated that disease categories are also social constructions of a culturally specific cognitive system of medicine.[5] In contrast, *illness* refers to the subjective experience of a person suffering from symptoms that are salient within his or her cultural context. People attempt to understand their symptoms and suffering through "explanatory models" that provide answers about what the illness is, why they got it, and what the prognosis is.[6] A group of people may share a set of explanatory models of illnesses in an ethnomedical or folk medical system. It is easy to see that these ethnomedical systems are also socially constructed systems of belief. It is less obvious that our scientific biomedical system is also a socially situated ethnomedical system, albeit a powerful and efficacious one. In the context of the present volume, if the case studies were limited to only emerging *diseases,* then the analyses would be confined to examples that had already been legitimated by biomedicine, and critical elements of the social negotiation process would have been missed.

Important social processes come to light by examining *contested* cases, such as CFS, for which there is no recognized etiology. The history of medicine is full of cases in which "legitimate" recognition of a new nosological category was hotly debated before it came to be an accepted fact. The recent history of HIV/AIDS is also a good example of such a contest—one in which the community of suffering and the public health authorities found themselves in a very public struggle. Uncertainties concerning etiology and treatment create challenges for individuals trying to achieve a positive resolution to their illness, as well as for the institutions charged with protecting the public's health. In the face of uncertainty, individual sufferers struggle to understand their conditions and gain public recognition of them. Health institutions also struggle to know how to respond to emerging illnesses or, in some cases, whether to respond to them at all.

### Histories of Emergence: From "Conditions" to National Public Health Concerns

Popular media portrayals of emerging infectious diseases, as seen in the movie *Outbreak* or on a recent episode of *ER,* have highlighted the dramatic events associated with viral or bacterial disease such as illness caused by *E. coli* 0157:H7, West Nile virus, hantavirus, Ebola virus, or anthrax. Illnesses such as these are marked by sudden and dramatic onset and lead to rapid recognition and action by both sufferers and public health authorities.

From a historical perspective, however, many emerging illnesses begin as "conditions," a personal set of symptoms whose significance is neither understood nor appreciated. These conditions may be viewed as normal life circumstances, like chronic pain, the flu, perimenopausal symptoms, or "impotence" in older men. Alternatively, they may be seen as the inevitable corollaries of particular circumstances of individual sufferers, such as shortness of breath in aging coal miners, the physical symptoms of office worker stress, or the mental slowness of children raised in slum housing. Yet at some point, those suffering from these conditions, or their doctors, or medical researchers, or public health authorities, may come to see these "conditions" as illnesses with organic causes.

How does this transformation come about? There are social processes through which a new illness category becomes accepted by some sectors of society. Chronic pain becomes fibromyalgia. Frequent flu symptoms and fatigue become CFS. Mineworkers' shortness of breath becomes pneumoconiosis. The effects of office worker stress become linked to sick building syndrome or multiple chemical sensitivity. The mental slowness of slum children becomes the effect of lead poisoning. The acceptance of new medicalized labels is an essential first step in the process by which emerging illnesses gain public recognition. When a group of people are united by a common illness category and are motivated to garner more attention to and recognition of that illness category, then a community of suffering can become an actor in the social process of emergence.

The chapters in this volume explore the conditions that have led to the emergence of particular illnesses, as well as barriers that have retarded recognition in some cases. Comparison of the case studies in this volume reveals that there are some fundamental social processes through which a health "condition" becomes a recognized and legitimated issue that demands funding and action by public health authorities. These processes are shaped by the particular characteristics and sometimes idiosyncratic interaction of

- the epidemiology of the disease or illness;
- the community of suffering's communication and leadership;
- the responsible public health institution's culture and constraints; and
- the response of the media.

## Epidemiological Characteristics of Emerging Illnesses

The features of an illness itself may hasten its recognition and the response it triggers in public health institutions. A disease with a high case-fatality rate and dramatic presentation of symptoms, as in the case of Ebola, is likely to gain attention and provoke fear. High levels of public attention occur when people are anxious that they are personally at risk of contracting the disease. This can happen when the epidemiology is unknown and still under investigation. It can also happen when the route of transmission appears to have a random quality because it is spread by an insect vector, like Lyme disease or West Nile virus, or through the mail, as recently occurred with anthrax. On the other hand, if people believe that they are somehow protected from getting the illness because it is an illness of an isolated or stigmatized group— "those people"—then there is less public interest and potentially a less dramatic reaction by health authorities. Finally, certain illnesses achieve particular *symbolic* salience in a society, often for historical reasons—like plague or leprosy—and therefore provoke immediate attention.

There is generally, though not always, as we shall see, a correlation between the magnitude of a public health problem as reflected by its prevalence and the intensity of the public health response. It is for this reason that communities of suffering seek to establish the epidemiological importance of their illness. In some cases this may lead such communities to overestimate the prevalence of their illness. Colin Talley's essay (Chapter 2) on the history of multiple sclerosis (MS) suggests that the National Multiple Sclerosis Society inflated its estimates of the number of cases of MS in the 1950s in order to attract attention to the problem. Howard Kushner (Chapter 3) suggests that the leaders of the Tourette Syndrome of America may have used the same strategy. Yet in some cases, like HCV in the 1990s, prevalence alone was insufficient to garner attention. This was partly because the incidence of new cases of HCV did not appear to be increasing. Many public health officials argued that the majority of those infected had been exposed to the illness in the 1950s, 1960s, and 1970s.

An increasing incidence rate of a disease indicates that the disease may become an even bigger problem in the future, especially if public health measures are not instituted. Yet the significance of incidence data may be related to the social and economic composition of those at risk. In the case of HCV in the 1990s, the incidence of disease was growing but appeared to be limited primarily to intravenous drug users. For health activists, such as Larry Mass (see Chapter 14), this marginalization of the problem hid the fact that a much wider population was at risk of infection.

In general, the visibility of an emerging illness is very much dependent on whom the disease affects. It is a long recognized, though still unfortunate, reality that emerging illnesses that affect wealthy and educated people garner more media attention, public interest, and public health response than illnesses that affect the poor. There is a striking contrast between the history of the public health response to Lyme disease (for which five-acre zoning is a significant risk factor) and the histories of the responses to diseases that characteristically affect the poor (like tuberculosis). The epidemiological definition of *epidemic* centers on a *greater than expected* incidence of disease.[7] New diseases, therefore, can be considered an epidemic by definition. However, the *expected* incidence of a disease or illness is necessarily a subjective issue that raises the question, Expected by whom? Excess disease burden on the poor tends to receive less public attention largely because society expects the poor to be sicker than the rest of society. Indeed, a consistent observation of the seminars organized by Emory University's Center for the Study of Health, Culture, and Society (CSHCS; see the Preface) concerned the way that diseases "settle in on the poor," a poignant phrase used by Paul Farmer. When this "settling in" takes place, the problem consequently loses media attention and its position on the public health agenda. Conversely, when diseases of the poor (or migrants) are perceived as threats to the health of the dominant society, they reemerge as public health concerns.[8]

From an epidemiological standpoint, the investigation of an outbreak is more likely to be significant if the disease lends itself to a clear case definition. This in part reflects the culture of public health institutions, described below. The investigation is epidemiologically stronger if the case definition is based on laboratory tests rather than clinical signs. If no laboratory test is available, as is often the case with new diseases, it is better if the clinical signs are specific, rare, and intensely focused on a particular social group. As James Curran, who directed the early efforts of the Centers for Disease Control and Prevention to deal with HIV/AIDS, pointed out in seminar discussions, the early epidemiological investigations of the disease through the diagnosis of Kaposi's sarcoma (KS) were enhanced by the specificity of KS pathology and its low incidence in the United States, especially among young men. Epidemiological features that make it easier to identify a "cluster" make recognition and reaction by public health institutions more likely.[9] This is an important reason why *infectious* diseases have an advantage in the social process of emergence as explored in this book. It is also the reason that communities of suffering, like those that coalesced around CFS, strongly push for research that will identify an infectious etiology. Public health institutions like the CDC were historically designed around the goal of controlling communicable infectious diseases. Illnesses resulting from other etiologies—environmental, nutritional, iatrogenic, psychogenic, or others that are unknown—make medical identification and public health response more difficult. These health problems are often the purview of other institutions, like the Environmental Protection Agency or the Food and Drug Administration, that have different missions and cultures.

Environmentally caused illnesses are particularly difficult to investigate and demonstrate from an epidemiological standpoint. Symptoms of toxic exposure may be relatively common as "background" illness and therefore not specific enough to demonstrate causation. For example, the background prevalence of cancer in a community may be high enough that an epidemiological investigation of a "cancer cluster" cannot statistically demonstrate a clear cause-effect relationship between environmental hazards and neoplasms.[10] Moreover, the difficulty of demonstrating the health effects of pollution reflects the epidemiological complexity involved in correlating a variety of chemical pollutants with a large assortment of symptoms, such as occurred in the case of the African American community of Newtown in Gainesville, Georgia, described by Ellen Spears (Chapter 7). In contrast, the historical case of lead poisoning, described by Christian Warren (Chapter 9), had the significant epidemiological benefit of a single environmental pollutant and available laboratory tests for measuring exposure to lead.

Case studies of emerging illnesses with environmental etiologies often involve issues of social and political power, whereby the least powerful are victims of the illness. This makes demonstrating causation doubly difficult. The case of sick building syndrome, described by Michelle Murphy (Chapter 8), is not simply an epidemiological quagmire of multiple stressors and symptoms—including psychological stress in the workplace. It also becomes an issue of labor contestation whereby unsympathetic management minimizes employees' health complaints as a type of mass hysteria. The socio-economic disparities at play in Spears's discussion of Gainesville, Georgia, were even more striking.

## Characteristics of the COS

Some communities of suffering have demonstrated remarkable organizational and political ability in gaining media attention or directly influencing the activities of public health institutions. From a recent historical standpoint, the most salient example of an organized and energetic community of suffering has been the gay community and its response to HIV/AIDS. The public demonstrations of organizations like ACT UP powerfully influenced the public health community's response to the AIDS epidemic and led to remarkable levels of HIV/AIDS funding at the CDC and National Institutes of Health. Steven Epstein's case study (Chapter 4) on the development of lay expertise in the gay community and its influence on the design of clinical trials for AIDS treatment demonstrates a historically remarkable development in which AIDS activists actually earned a "seat at the table" in setting research agendas. Other illness communities that have developed around Lyme disease and CFS have learned from AIDS activism and adopted similar tactics and strategies.

COS leadership can arise in several ways. In many cases, a single individual comes forward as an energetic leader to move a health cause along the pathway of "emergence." Polly Murray played this role in the emergence of Lyme disease. Lois Gibbs, founder of the Love Canal Homeowners Association, led the initial organized efforts to draw public health attention to the toxic waste problem in the Niagara, New York, area. The motivations of these individual leaders are often the result of a family member or friend having been stricken by the disease. One important result of such individual efforts has been the establishment of a wide range of permanent bureaucracies and nongovernmental organizations (NGO) organized around a particular illness.

When disease problems occur, or have "settled in," among the poor and powerless, it is less likely that leadership will emerge from the COS—in part because members of the COS have more immediate concerns of survival. However, the history of the Newtown Florist Club, described by Ellen Spears, serves as a warning to those who would assume that these communities cannot mobilize around their own health problems. The history of Ken Saro-Wiwa and the Movement for the Survival of the Ogoni People's battle against Shell Oil in Nigeria, presented to the Mellon Foundation's Sawyer seminar "Environmental Illness, Activism and the Public Health" by his brother, Dr. Owen Wiwa, is another case of leadership emerging from among the poor.

Leadership also may emerge from outside the community of suffering. In some cases, a particular interest group may use health issues to further its own political interests. In this regard, it is important to see AIDS activism within the larger agenda of the gay liberation movement, at least in the beginning of the epidemic. Similarly, the sick building syndrome described by Murphy was taken up by the organization 9to5 as part of its efforts to build an office workers' labor movement. Lead poisoning was taken up by Democratic political leaders in the 1960s as part of the larger War on Poverty. There are cases in which surrogate leadership from an NGO—for example, from Partners in Health (PIH) working with persons with multiple-drug-resistant tuberculosis (MDRTB) in Peru, in the case described by Sandy Smith-Nonini (Chapter 10)—can provide a powerful force for change.

The early participation of medical and public health researchers in the leadership of COS can greatly facilitate the speed with which an illness emerges on the public health agenda. The early involvement of medical researchers in the emergence of Lyme disease resulted in part from the fact that a number of them lived or vacationed in infected areas. Other medical researchers were attracted by challenging research questions posed by the emerging illness. Moreover, a new disease can provide a valuable niche for the establishment of an individual scientist's research career. Within medical research and public health institutions, there are political struggles for different research agendas. To gain advantage in such a competitive setting, an alliance between scientists and the COS or other champions of the health cause can be very useful. The alliance between the Multiple Sclerosis Association of America and neuroscientists and that between the Fibromyalgia Association of America and rheumatologists are cases in which the needs of communities of sufferers for medical leadership coincided with the needs of medical scientists in underfunded disciplines for a disease that would attract research funding.

It is interesting to compare the advantages and disadvantages of different forms of leadership in terms of the success of COS in gaining a place on the public health agenda. As in other social movements, indigenous leaders are often better able to understand the needs of the community they serve and to articulate these needs in terms that are understandable to the members of that community. On the other hand, this interior view may make it difficult for the indigenous leaders to express the community's needs in terms that are understood by the public health community from which they seek support and recognition. The importance that leaders of the Newtown Florist

Club and CFS and fibromyalgia support groups place on local knowledge—or in Spears's words, "situated knowledge"—versus the clinical or scientific knowledge of health experts often creates tensions between these leaders and the public health and medical authorities with whom they need to work. External leaders, although better equipped to make connections with the public health community, may find it more difficult to understand the perspectives of the constituency they serve. Moreover, Epstein's discussion of the AIDS activists who became experts in the medical science of AIDS and gained a "seat at the [research] table" suggests that COS leaders who become effective at working within the biomedical community may be perceived as having lost their connection to the movement.

The leaders of an effective community of suffering often have links with the leadership of other communities of suffering. COS leaders pay attention to other such groups and emulate their successful strategies and symbols. The most obvious example is the way in which the activities of HIV/AIDS groups have been taken up by other COS. The red lapel ribbon associated with AIDS, which symbolizes the need to "cut red tape" during an epidemic emergency, has been used by other causes, substituting ribbons of other colors. The tactics of ACT UP have been borrowed by CFS advocates. Some communities of suffering have had circulating and overlapping leadership. Thomas Sheridan, one of the leading AIDS activists, later helped organize and coordinate CFS patient support groups. Fibromyalgia and CFS support groups have developed cooperative strategic alliances for political action.

Communities of suffering, through their communications with one another, transform individual illness experiences into recognition of a wider collective health problem and thus initiate the process of illness emergence. An increasingly common vehicle for such communication is the Internet. Thousands of chat rooms and illness support groups currently exist on the Internet. As Diane Goldstein (Chapter 5) demonstrates, such sites provide opportunities for illness sufferers to share their subjective illness experiences and develop consensus regarding the nature of these experiences. Andrew Spielman, Peter Krause, and Sam Telford (Chapter 11) describe how elite social networks—involving the residents of wealthy vacation areas, physicians, researchers, and politicians—facilitated the recognition of Lyme disease. Murphy's chapter examines how the women's office worker movement linked office workers experiencing work-related discomforts and redefined the office workplace as a site of cumulative physical assaults from copier fumes, computer screens, and the chemicals found in various building materials. Deborah Barrett (Chapter 6) describes how communications among ad hoc support groups involving sufferers of chronic pain and interested rheumatologists transformed an ill-defined condition into fibromyalgia. Communication within the local community of African American women described in Spears's chapter constructed a pattern of disability and death within the community and worked to give it a collective identity by linking the physical symptoms to toxic exposures associated with local industries.

Although the Internet has facilitated the rapid coalescence of COS, it has also created problems for these communities. The Internet provides a medium through which information concerning a particular form of illness can circulate widely. Yet the information that circulates is often not subjected to any kind of biomedical screening. As Goldstein notes, the Internet community gives primacy to the subjective experiences of illness sufferers. The openness and subjectivity of these communications can make it difficult for COS to define the boundaries of their illness. In addition, as open as the Internet is to the participation of a wide range of illness sufferers, it remains a medium that has an economic threshold of participation, since many poorer communities may not have access to the Internet. This type of forum may therefore preclude the participation of a larger population of sufferers. A clear example of this exclusion is the formation of CFS networks, which at first were composed overwhelmingly of middle-class white women. Later epidemiological surveys by the CDC indicated that large numbers of minority women suffered from similar sets of symptoms.

Not all the illnesses in this volume have produced a community of suffering. Indeed, a major point of Mass's essay on HCV is the difficulty of getting media attention for a health problem for which there is no community, no community leadership, and no biomedical leadership. This case is interesting precisely because the slowness with which HCV gained a place on the public health agenda reflected the absence until recently of an organized social group able to demand public attention.

## Characteristics of the Responding Institutions

Variations in the culture and mission of public health institutions influence their ability to respond to newly emerging illnesses and the manner of their responses. One of the most valuable aspects of the CSHCS seminars was the active participation of public health officials and the regular inclusion of the viewpoint of public health institutions, particularly the CDC. There are many cases in which the rhetoric of a community of suffering can demonize the people working in public health institutions. Viewing such institutions as characteristically uncaring and oppositional to COS concerns is both simplistic and wrong.

To understand how public health institutions respond to the emergence of an illness in society, it is necessary to recognize the mission of the institution as well as the programmatic and fiscal restraints on the institution. Most public health institutions are government agencies whose funding is controlled by political forces. Protection of the health of the population has been the general mandate of public health institutions, but the historical decline of public resources allocated to these institutions, both nationally and internationally, should be a matter of concern.[11] Newly emergent illnesses may lead to the creation of NGOs, as in the case of the National Multiple Sclerosis Society and the Lyme Disease Foundation. But NGOs related to emergent illnesses do not have the same responsibilities, diversity of mission, or political vulnerability of funding as government agencies. In fact, the goal of such NGOs may be primarily to influence the response of governmental health agencies.

In the United States, there is neither a coherent and integrated public health system nor an agreement of what "public health" entails.[12] A very wide range of institutions and agencies at the federal, state, and local levels have missions with some impact on the public's health. The character of such organizations, whether they are regulatory agencies, providers of clinical care, or research institutions, has a great deal to do with how they respond to emerging illnesses.

The Department of Health and Human Services is the umbrella organization designated to protect the health of all Americans and to provide essential human services. The department, in turn, designates the CDC as the lead federal agency for protecting the health of the American public. The CDC figures prominently in the response to emerging illnesses in the case studies throughout this volume. For more than half a century, the CDC has been the nation's front-line defense for the prevention and control of infectious diseases. In recent decades, CDC has broadened its role to include environmental health issues, the prevention of chronic diseases, the promotion of healthy lifestyles, and injury prevention.[13] Local and state public health agencies work most directly at the national level with the CDC in response to an emergent public health problem, but these agencies work with many other institutions in providing and financing clinical care.

The National Institutes of Health is the premier U.S. institution for biomedical research. Early NIH research focused on finding solutions to practical health problems. In recent decades, however, the NIH has increased its emphasis on the study of human biology: immunology, genetics, and cellular and molecular biology.[14] Biomedical research, conducted within the NIH or by academic researchers funded by the NIH, is often essential for establishing the etiology of emerging illnesses. NIH funding is also critical to the development of treatments for emerging illnesses. Generating NIH funding for an emerging illness is thus an important goal of illness activists.

Unlike the CDC and NIH, the Food and Drug Administration is a regulatory agency responsible for assuring the safety of foods and cosmetics and the safety and efficacy of pharmaceuticals, biological products, and medical devices. Agencies outside the Department of Health and Human Services may also play a prominent role in public health, but this is generally not their sole mission. The Environmental Protection Agency's mission is to protect human health and to safeguard the natural environment—air, water, and land.[15] The U.S. Department of Agriculture plays a role in assuring the safety of agricultural products for consumption, but its primary mission has been to assist farmers and ranchers. The U.S. Department of Transportation devotes resources to the safety of our highways, but its primary mission is to assure efficient transportation systems that meet the nation's needs.

The number of different agencies and institutions responsible for various aspects of public health sometimes makes it difficult for COS to know where to turn for support. Nonetheless, although all these large agencies have a stake in the public's health, the agency in charge is the CDC.

Historically, the CDC has had a mission of controlling and preventing disease, and it developed two distinct cultures: one based on the work of outbreak investigators—epidemiological "cowboys" who ride into a troubled town, quickly solve the epidemic, and leave[16]—and the other based on carrying out service functions, particularly the preventive services for priority areas such as childhood immunizations, tuberculosis, and sexually transmitted diseases, later expanded to include chronic diseases, injuries, and HIV. The two cultures are distinct, with relatively little overlap.

The first culture centers on the outbreak investigation of an infectious disease, in which Epidemic Intelligence Service (EIS) officers are called into a locality by state authorities. The EIS model can be remarkably efficient and effective for controlling outbreaks of infectious diseases. Foodborne diseases like *E. coli* 0157:H7 are good examples of controllable outbreaks because tools of epidemiology combined with laboratory methods, such as the tracing of the genetic fingerprint of the pathogen, can lead to the identification of a single source of contamination. The appropriate action for immediate control is therefore obvious and well established in the history of public health—removing the handle from the proverbial Broad Street pump. The rapid-response EIS model for dealing with infectious outbreaks has historically added to the prestige and credibility of the CDC as well as local health departments.

The CDC's culture of outbreak investigation has served the institution very well, but it has also been a source of tension when problems like chronic fatigue were identified and placed into this setting. Chronic fatigue did not have a clear epidemiology, and putting it into an area of the CDC traditionally strong in outbreak response created problems. Importantly, the CDC has less capacity to address and hence less interest in a health problem like chronic fatigue, in which the lack of existing scientific knowledge means that there is little or nothing the institution can do for prevention or cure. Neither the rapid-response model nor CDC's traditional approach to supporting prevention services worked well at the beginning of the HIV/AIDS epidemic. Although HIV/AIDS was an infectious disease, it had a long incubation period and no cure. Moreover, preventing AIDS, unlike stopping the spread of Legionnaires' disease, required the use of a broad set of social and behavioral approaches applied over an extended period of time. Coping with HIV/AIDS required a different level of commitment and sustained response than the CDC was prepared to deliver, and the CDC has had to adopt a series of new policies and approaches.

Although *E. coli* 0157:H7 outbreaks have been effectively brought under control by the CDC's rapid-response methods, these outbreaks also demonstrate the weaknesses of a public health response that is designed to put out fires. The first *E. coli* outbreaks, in the 1980s, sparked a rapid public health response. Yet once the outbreak was brought under control, little was done to prevent another outbreak from occurring. This was in part because of the fire department culture of the public health institutions. Yet it was also because the public health system was not designed to devote sustained attention to an acute infectious disease problem once the initial outbreak had been resolved, unless it fell into a major program of emphasis, such as childhood vaccine-preventable diseases or sexually transmitted diseases. The lack of a sustained response to *E. coli* O157:H7 in the 1980s and early 1990s was further complicated by the fact that it occurred at a time when resources for dealing with infectious diseases were being cut back. As Ruth Berkelman and Phyllis Freeman (Chapter 13) and Smith-Nonini point out in their contributions to this volume, the belief that infectious diseases were a thing of the past had led to a shift in public health priorities away from acute infectious disease outbreak investigation and control and toward AIDS prevention, chronic illnesses, and injury prevention, areas that did not rely so heavily on the "outbreak" model for their program activities. Consequently, the ability of the public health infrastructure to handle the threat of outbreaks—including the ability of physicians to diagnose the problem and the ability of laboratories to identify the pathogen—has been limited. One can speculate that many of the outbreaks that occurred in the late 1990s might have been missed or their recognition delayed if there had not been some improvements in local and state health departments' technical abilities to diagnose these epidemics, improvements resulting from the "Emerging Infectious Disease" initiative described by Berkelman and Freeman in this volume.[17]

The new public health threats created by the proliferation of antimicrobial-resistant infections have also required new approaches. In this case, the CDC has had to seek long-term solutions involving the participation of medical, legal, and political institutions together with pharmaceutical companies, in addition to more classic public health approaches in the form of surveillance and health education. Shifting approaches and perspectives have not come easily within the agency.

In the wake of the anthrax attacks in the United States in 2001, new resources have been invested in developing improved methods for identifying and responding to infectious agents purposefully introduced into human populations. This heightened concern for public health preparedness is viewed by many in the public health community as a positive development that will help strengthen the country's overall public health system. This may well be so. Yet it needs to be recognized that increased interest in public health preparedness may contribute to a skewing of public health resources toward institutions and approaches that reproduce a rapid-response approach to public health at the expense of efforts to find more effective long-term approaches to health problems.

An important variable affecting institutions is their fiscal situation at any given time, and this financial health is closely linked to politics. Public health agencies are dependent on public funding. To maintain this funding, they are also dependent on a certain degree of public interest. In times of economic difficulties, or if logical public health policy offends some politicians' moral sensibilities (e.g., safe-sex education or needle-exchange programs in HIV prevention), funding sources can be cut. Although public health scientists are dependent on public interests and government support, they are legally unable to lobby politicians directly. However, government-supported public health scientists can indirectly influence public opinion and political support. In this volume, Berkelman and Freeman's analysis of the strategic initiative to highlight emerging infectious diseases as microbial threats to the health of the United States demonstrates a working alliance between concerned independent scientists, the media, congressional staff, and representatives of the CDC. On the other hand, it is clear that this was an exceptional case and that the CDC's unwillingness to become engaged in lobbying efforts generally has placed it at a disadvantage compared with the various institutes of the NIH in soliciting funds for public health.

Another important variable is that of leadership within public health agencies. Emerging illnesses often require creative problem solving to answer new questions. For a variety of reasons that have been analyzed by sociologists since the nineteenth century, creative adaptation to novelty is never a typical characteristic of bureaucratic organization. Within public health agencies there are often political struggles for expansion of budget, space, and personnel. At the same time, agency personnel are often reticent to become involved in political action outside the agency. Indeed, such action can be a poor career strategy or, in some cases, even illegal. More important, within an agency there can be a cultural belief in the preeminence of science and scientific investigation, even when this runs counter to broad public concerns. The case of maternal HIV testing, established by the CDC as an anonymous surveillance system, described by Lydia Ogden (Chapter 12), is a good example of the way in which the CDC's institutional culture prevented it from responding in a timely manner to the changing political environment of AIDS.

## The Response of the Media

The process by which emerging illnesses move from individual conditions to public health issues is shaped at all stages by how the illnesses are covered by the media. An epidemic, particularly of a new and unknown pathogen, is newsworthy. This is true regardless of how widespread or epidemiologically important the epidemic is. The outbreak of disease caused by the hantavirus in the United States, beginning in 1994, received widespread coverage in newspapers and newsmagazines and on the nightly news even though it involved only a small number of cases. The exotic nature of the infection and its relatively high case mortality rate made it news.

For illnesses with less dramatic presentation, media coverage is often slow to develop and much less extensive. Slow-developing chronic illnesses like AIDS are, at least initially, less newsworthy. Even illnesses that affect large numbers of people, such as HCV, may not draw rapid or extensive news coverage. The news reporters, for the most part, take their lead in reporting on emerging illnesses from public health authorities. Thus illnesses that have yet to gain public health recognition often receive limited coverage. Barrett notes in her chapter on fibromyalgia and CFS that there was not a single news story on fibromyalgia until the American College of Rheumatology issued its diagnostic criteria. At the same time, the history of AIDS suggests that even when public health authorities were becoming concerned about the growing incidence of AIDS, there was little reporting of the problem on the nightly news. As Cook and Colby have argued, network news program directors were reluctant to present a health story that focused on a stigmatized minority, especially during the dinner hour.[18] Mass's chapter in this volume argues that newspapers and newsmagazines resisted printing stories about HCV for similar reasons.

In some cases, however, the news media may be ahead of public health concern. One reason communities of suffering seek media attention for their cause is that they recognize that media coverage is a critical instrument for gaining public acceptance, political interest, and research funding. The chapters by Kushner and Talley in this volume describe how patient advocacy groups formed around MS and Tourette syndrome used the media in this way. Both AIDS and CFS activists have used the media to announce research findings prematurely as a way of promoting further research. News writers may also be first in identifying an emerging health problem or break-through in the hope of getting a "scoop."

Public health authorities also recognize the power of the media to shape public awareness of health problems, as well as to influence legislators to fund measures to prevent emerging health problems. Berkelman's chapter on the CDC's efforts to place emerging infections on the public health agenda describes how the National Center for Infectious Diseases' staff took steps to inform popular scientific writers Richard Preston (*The Hot Zone*) and Laurie Garrett (*The Coming Plague*) about the CDC's plan to deal with emerging infections. The NCID also trained a number of scientists to respond to media requests for information on emerging infections. At the same time, public health authorities want the media to report accurately and responsibly on emerging health problems in a way that does not prematurely generate public concern or panic, as occurred following the publication of an article on CFS in *Rolling Stone* magazine in 1987.[19] Health authorities also see the media as a legitimate outlet of health education messages, even when those messages are not "news" at all.

### Social Emergence: A Processual Model

Although COS and public health institutions are the central focus of the studies in this volume, the social process of illness emergence potentially involves at least four sets of actors in addition to the community of suffering, the responding institutions, and the media: medical practitioners, the public, the political authorities, and members of the legal system. The social relationships among these diverse groups vary a great deal, both over time and within different illness experiences.

In the following section we provide a general model (fig. 1.1) for understanding the interaction of these actors within the process of illness emergence. This simplified model does not fit all the case studies. Yet the overall process of emergence—through which illness changes from being an individual experience, to becoming publicly recognized, to meriting social action and resources as part of the public health agenda—shares important similarities across these case studies. In this figure, the primary flow of historical events occurs in three stages, indicated by rectangles. The large arrows refer to the sequential process; the thin arrows point out potential complicating factors. The first stage is the occurrence of sickness on an individual level. The second stage is a period of advocacy when efforts to understand the illness clinically and epidemiologically occur. The final stage is placing the health problem on the agenda of public health activities with sufficient priority to garner some resource allocation. Linking these three stages are two critical points of transformation: initial recognition, which often happens within the lay community, and "official" biomedical recognition. These two periods of transformation can be strongly influenced by the four previously described variables (epidemiological, COS, institutional, media), as well as by political and legal opposition or support, all indicated by ovals.

In general, "new" illnesses must proceed through the entire chain of events in order to earn a place on the public health agenda. HIV/AIDS is a good example of this, although the process took several years. Other historical cases like MS may be also traced through a similar progression of recognition and advocacy. Infectious diseases can pass through the process extremely quickly, especially when the disease falls within the purview of an institution with a culture of responsiveness. For example, reaction to the outbreak of West Nile virus in New York in September 1999 was extremely rapid—the gap between the first reports of dead crows and human encephalitis and the initiation of mosquito spraying activities was only eight days.[20] Social factors, especially the power and influence of the community of suffering or people at risk, act as catalysts to the overall process.

The case studies in this volume include examples of illnesses that have not completed the entire process and are unlikely to ever get on the public health agenda. The dual role of advocacy—to gain biomedical recognition

*Facing page*

Fig. 1.1. A processual model of social emergence. Rectangles are historical stages; diamonds are points of transformation; ovals are external variables.

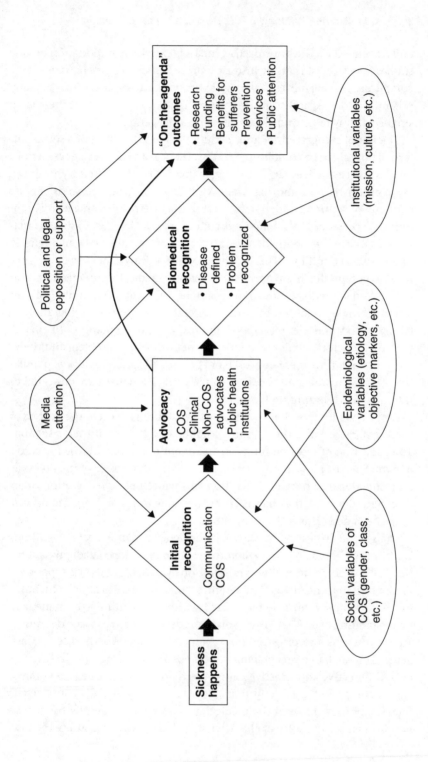

**"On-the-agenda" outcomes**
- Research funding
- Benefits for sufferers
- Prevention services
- Public attention

Political and legal opposition or support

Institutional variables (mission, culture, etc.)

**Biomedical recognition**
- Disease defined
- Problem recognized

Media attention

Epidemiological variables (etiology, objective markers, etc.)

**Advocacy**
- COS
- Clinical
- Non-COS advocates
- Public health institutions

**Initial recognition**

Communication COS

Social variables of COS (gender, class, etc.)

**Sickness happens**

and placement on the prioritized agenda—is extremely important. The case studies here suggest that the historical process of emergence becomes more complicated as one moves to the right portion of this model. When so many potential variables and contingencies are in play, it is clear that there is no single pathway to getting placed on the health agenda.

The initial recognition of an individual sickness as a part of a social pattern of illness can be complicated by the absence of wide prevalence, serious medical consequences, and objectively demonstrable laboratory findings. Recognition of common experiences of suffering often depends on the existence of communication networks like the Internet. In other words, conditions experienced by isolated individuals, who lack access to support networks, labor movements, or the Internet, may remain epidemiologically unidentified conditions. This lack of recognition can occur even in the context of potentially high prevalence, serious medical consequences, and objective demonstrable tests. For example, HCV may constitute such an untransformed condition for many sufferers. Despite the proliferation of CFS networks and patient support groups, a large population of sufferers (and their doctors) do not recognize the nature of their problem and therefore remain outside these networks and groups. Additionally, a large population of low-income and minority people with CFS remain unconnected to Internet-based communities of suffering.

For illness sufferers, there can be much at stake in regard to the process of social emergence and the gaining of public recognition of their situation. The allocation of research funds for treatment development, the initiation of public health prevention campaigns, and the acceptance of the illness as an insurable condition all depend on this wider public recognition of an emergent illness. As the chapters in this volume indicate, many factors can facilitate or deter this wider recognition.

The second critical point of social emergence is biomedical recognition. Within the United States, the dominance of biomedical paradigms assures that the movement from illness recognition by patients and their supporters to wider public health action requires "medical certification." In other words, the illness must be recognized within the confines of biomedical understanding of disease. This involves the development of a case definition and a consensus regarding the etiology and other aspects of the disease. Case definitions are based on epidemiological observation. They are designed to be both sensitive and selective in order to help in the identification of new and existing cases. Case definitions, however, are also socially constructed. The disciplinary biases of the medical researchers who identify the illness may influence the choice of characteristics that come to constitute the case definition of an emerging illness, as happened in the case of Lyme disease[21] and Tourette syndrome described by Kushner in Chapter 3. Political processes may also shape case definitions. The history of AIDS was marked at several points by political debates over its case definition. For example, protests over the inclusion of illness markers experienced by women in the early 1990s led the CDC to revise its definitions, despite the strong belief of a number of CDC researchers that the inclusion was not epidemiologically indicated.[22]

Medical certification, although important, does not ensure that an illness will gain necessary funding for research, prevention, and treatment. Moreover, as the cases of fibromyalgia and CFS suggest, some degree of certification may occur without a definitive understanding or consensus on etiology. Nonetheless, without certification, there seems little chance that an illness will be placed on the public health agenda. An important exception to this generalization is the history of the Gulf War syndrome, which was forced onto the public health agenda because of direct political decisions, pushed strongly by COS and veterans' groups, despite some protests by public health authorities.[23]

It is also important to note that the cultural dominance of biomedical paradigms makes medical certification a powerful psychological goal for sufferers of emergent illnesses. Barrett's chapter indicates that the incorporation of fibromyalgia within the International Classification of Diseases (ICD-10) in 1993 was viewed by fibromyalgia sufferers as a major achievement that validated the legitimacy of their suffering. On a personal level, patients want their suffering to be considered with respect. Given the power of the biomedical paradigm in our society, merely the implication that an illness may have a psychogenic etiology or that symptoms are psychosomatic is considered by the sufferers as disrespectful and condescending. The insistence on the predominance of a biomedical paradigm by the COS can sometimes result in the denial of alternative research hypotheses and treatment modalities. The quest for biomedical legitimacy certainly has practical material benefits for members of the COS. Only with medical certification are some financial resources available through health insurance and government programs.

### Factors Influencing Institutional Responses: A Comparative Model

The weakness in the processual model of figure 1.1 is that not every illness follows a single pathway. Moreover, that model is biased toward the point of view of the COS and incorrectly implies that both biomedical and public health institutions are simply adversarial gatekeepers of the public health agenda. Another type of model is presented in figure 1.2. The figure relates the rapidity with which a number of the illnesses discussed in this volume obtained public health recognition (the response gradient) to five other variables: the epidemiological importance of the illness; the availability of an unequivocal diagnostic test; the social class of sufferers; the level of COS activism; and the level of coverage by the news media. These variables cover a spectrum from high to low, and the illnesses discussed in this volume are placed along the spectrum to reflect their position vis-à-vis other illnesses. This type of model permits interesting cross-illness comparisons.

There are two major difficulties in placing these illnesses along the six gradients. The first has to do with changes over time. In some cases, the position of an illness on a gradient has shifted over time, and this is indicated with the narrow arrows within the gradient box. HIV/AIDS, for example, began as an illness with limited morbidity/mortality but rapidly shifted to the left on the epidemiological gradient. In addition, there is evidence that although the illness began in the United States among a socially privileged, if stigmatized, population of middle-class gay males, it moved to being viewed as an illness that affected a wider socioeconomic section of the population. (as indicated with arrows in fig. 1.2). By the 1990s, its demographics had changed again, as it increasingly became a disease of the poor and women. Finally, in the last few years, there is evidence of a resurgence of HIV infection among young gay men who, in the wake of the development of new antiviral treatments, have come to see AIDS as a chronic disease rather than as life threatening. There is thus a potential shift back to the left of the social gradient.

The second source of difficulty in placing illnesses on the gradients is simply a lack of appropriate information. The epidemiological prevalences of many of the illnesses on the gradients, including HCV, CFS, MS, and fibromyalgia, are uncertain. Part of the uncertainty with illnesses such as CFS and fibromyalgia has to do with the difficulty in knowing what constitutes a case in the absence of a clear diagnostic test.

The comparisons displayed in figure 1.2 reveal three interesting patterns of emergence. First, certain illnesses, such as Lyme disease and illnesses caused by *E. coli* 0157:H7, hantavirus, and West Nile virus, seem to have a low epidemiological salience but nevertheless receive a great deal of public

*Facing page*

---

**Fig. 1.2.** Factors influencing the timing of public health institution responses to emerging illnesses.

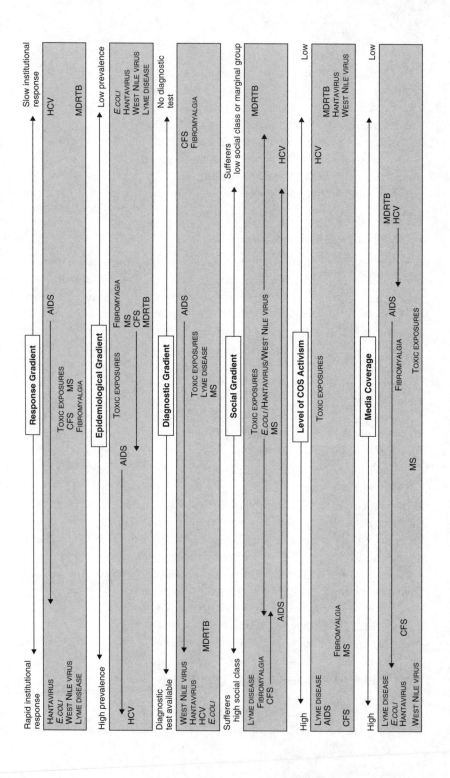

health attention. Second, other illnesses, for example, HCV and increasingly MDRTB, are epidemiologically important yet do not receive timely public health response. Finally, a third group of illnesses, including CFS, fibromyalgia, and sick building syndrome (a type of toxic exposure), appear to become mired in contestation, both legal and medical, that results in delays and uncertainty about appropriate public health action. Examining the placement of these illnesses on the gradients in figure 1.2 helps to explain these three patterns of emergence and illuminate the ways in which various factors combine either to facilitate or to inhibit the process of illness emergence.

Timely Institutional Response to Emerging Illnesses

Why have certain illnesses of limited epidemiological importance received rapid public health attention? Looking at *E. coli,* hantavirus, and West Nile virus across the various gradients in figure 1.2 gives some answers. First, each has produced clusters of cases with dramatic symptoms, in some cases leading to deaths. This is precisely the kind of pattern that public health authorities are trained to investigate. Second, each of these illnesses is located on the high end of the diagnostic gradient, as each can be identified through a laboratory test. Third, all three occupy a wide middle position on the social gradient. The arrows pointing to the left and right of these illnesses are meant to indicate the wide social spectrum of those at risk of infection. These are illnesses that appear to strike at random, without regard for social class (perhaps more so for West Nile and *E. coli* than for hantavirus). The potential randomness of infection gives these illnesses a greater emotional salience with the public than illnesses like MDRTB or HIV/AIDS in the 1980s, which were seen as limited to particular subgroups of the population. Anthrax poisoning shares this emotional salience. Finally, and perhaps because of their place on the epidemiological, social, and diagnostic gradients, these illnesses have received widespread media attention.

Lyme disease is somewhat different from the other three illnesses that have received rapid public health attention. To begin with, there is no diagnostic procedure for definitively identifying whether a person has been infected with the pathogen (*B. burgdorferi*), which is the cause of Lyme disease. Identification of previous infection becomes increasingly difficult in the later phases of the illness, marked by chronic pain and fatigue, when many potential cases come to light. Lyme disease differs in two other ways from the other illnesses with low epidemiological importance but high public health response. First, Lyme disease is situated on the high end of both the social gradient and the gradient for COS activism. This combination has been important in Lyme's emergence. Whereas *E. coli,* hantavirus, and West Nile virus have benefited from being perceived as "equal opportunity illnesses," Lyme has benefited from being located on the high end of the social gradient. As the chapter by Spielman, Krause, and Telford makes clear, the social position and influence of many of those who early on either suffered from Lyme disease or felt threatened by it played a central role in moving the illness onto both the local and national public health agendas. Their social position also allowed them to organize an effective community of sufferers that had the resources and political influence to push their case for public health funding of Lyme disease research. Finally, the combination of high social class and COS activism accounts for the high level of media attention that Lyme disease has attracted.

## Delayed Institutional Responses to Emerging Illnesses

At the other end of the response gradient is hepatitis C, an illness that affects millions of people but has received relatively little public health attention. Again, looking at HCV's position on the various gradients in figure 1.2 helps us understand this anomaly. First, although HCV is located on the high end of the epidemiological gradient, with an estimated prevalence of 3.9 million HCV-infected people in the United States and as many as 100 million worldwide, the *incidence* of the disease appears to be declining. Since the introduction of blood screening in the early 1990s, the number of new infections has dropped dramatically. The CDC estimated that there were 230,000 new infections per year in the 1980s.[24] By 1999 this number had dropped to 35,000.[25] In fact, most medical authorities believe that the real epidemic occurred in the 1960s and 1970s in association with drug culture and in the absence of blood screening. These figures, unfortunately, may hide the true epidemiological significance of HCV. Given the long latency period between initial infection and its more severe health effects, many of those infected at an earlier time are just now developing symptoms. About 20 percent of those infected will develop serious liver disease as a consequence of HCV. Roughly a quarter of those who develop symptoms will die from the disease and its complications. Although the number of HCV deaths in the United States is currently between 8,000 and 10,000, the CDC estimates that by 2017 that number will reach 24,000 per year as those infected earlier develop the disease. The epidemiological importance of HCV therefore lies in the large number of cases of liver disease emerging among those already infected and the possibility that the large, but as yet undiagnosed, population infected with HCV may contribute to the further dissemination of the virus if prevention measures are not put into place.[26]

A second reason that there has been a public health delay in responding to HCV lies in its position on the social gradient. Because many people were infected before there was a test for HCV, the social spectrum of those infected is wide. However, the current epidemic of new infection is centered primarily among intravenous drug users. It is estimated that the prevalence of HCV infection among intravenous drug users in the United States is between 60 and 90 percent after five years' usage.[27] Prison populations also have high rates of infection. It is estimated that 41 percent of the prison population of California is infected.[28] HCV is thus perceived as an illness that is limited to marginal populations. As indicated earlier, this is a false perception.

HCV is also located on the low end of the COS activism gradient. Until recently there has been no effective COS activism with regard to HCV. That a large population of middle-stream Americans are unaware that they may have been infected during an earlier period has inhibited the development of an effective activist movement. One population that may be at risk of infection and that has demonstrated the ability to mobilize effective COS activism in the past is gay men. However, as Mass argues in his chapter on HCV and the news media, the actual risk of HCV transmission among gays has until recently been considerably greater than the perceived risk.

Finally, media coverage of the HCV epidemic has been sparse. Few articles appeared in U.S. newspapers between 1993 and 1997, in part because of the epidemiological estimates of HCV incidence that were being developed at this time. It was not until *Newsweek*'s cover story on the "silent epidemic" in 2002 that HCV became a serious focus of media attention. Mass's article provides a personal account of the difficulties he encountered in trying to get newspapers and journals to publish stories on the HCV epidemic.[29] The public acknowledgment by several celebrities that they were infected with HCV has contributed to recent media coverage in the same way that the revelations of Rock Hudson and Magic Johnson generated media interest in HIV/AIDS.[30]

## Contested Emergent Illnesses

A number of illnesses discussed in this volume have been slow to emerge because recognition of them has met considerable resistance from medical and public health authorities as well from the wider public. Included in this category are CFS, fibromyalgia, and illnesses related to various types of environmental exposures, including those as specific as lead poisoning and as amorphous as sick building syndrome. Examining the positions that these illnesses occupy along the various gradients in figure 1.2 can help illuminate the various factors that have delayed the emergence of these illnesses.

Resistance to the recognition of these illnesses is due to in part to uncertainties associated with their positions on the epidemiological and diagnostic gradients. As indicated earlier in this chapter, it is difficult to know how
many people suffer from these illnesses and thus how important they are
from an epidemiological perspective. Lack of a clear diagnostic test also
leads many physicians to question the legitimacy of these illnesses or to see
them as essentially psychogenic.[31]

The placement of these illnesses on the social gradient is complex and the
illnesses themselves difficult to categorize. Some illness activists would
argue that all these illnesses potentially affect a wide spectrum of the U.S.
population. Yet at various times, and particularly during the early history of
their emergence, most of the illnesses in this category have been associated
with particular subgroups. This has limited their visibility.

Early cases of CFS, fibromyalgia, and sick building syndrome appeared to
involve primarily middle- and upper-middle-class white professionals, particularly women. This association contributed to the view that these illnesses were a form of stress disorder associated with the social conditions experienced by these sufferers.[32] This association in turn limited the extent to
which these illnesses were given serious public health attention.

Lead poisoning during the 1970s was viewed as an environmental illness
in which atmospheric pollution from gasoline played a major role. This view
placed everyone at risk and made lead poisoning a major public health issue.
Yet for most of its history, lead poisoning has been viewed as limited to particular classes of workers and the poor. Limited in this way, lead poisoning
received attention only through the concerted efforts of health activists and
when linked to broader political agendas, as described by Warren.

Illnesses associated with toxic waste dumping have occasionally affected
mainstream populations, such as occurred in the Love Canal area of New
York. Toxic waste is thus a potential health hazard for a wide spectrum of the
U.S. population. Yet historically the populations most at risk have been
poorer, often minority populations forced to live in close proximity to polluting industries, such as the African Americans in Gainesville, Georgia, discussed by Spears.

The low position that illnesses associated with exposure to lead and other
toxic substances often occupy on the social gradient is particularly problematic in that those who are often responsible for these exposures occupy a
higher socioeconomic position. This differentiation, along with the ability
of polluters to use the courts to deflect both responsibility for and attention
from the health effects of toxic waste dumping, has limited the emergence
of the illnesses associated with toxic waste.

Given the uncertainties concerning the epidemiology and diagnosis of contested illnesses, the limited numbers of people affected by them, the often marginal status of those most at risk, and the economic interests that may oppose their emergence, effective COS organization and leadership are particularly important in determining whether contested illnesses gain public health attention. COS groups associated with CFS and fibromyalgia have been successful in overcoming the challenges to illness emergence described earlier in this chapter. This activism accounts for much of the public health attention that these illnesses have gained.

The level of activism regarding CFS has been particularly high. By 1990 there were four national CFS patient organizations and some four hundred local support groups. CFS sufferers are predominantly female, and many are upper-middle-class professionals, even though recent prevalence surveys for CFS in Chicago showed the illness to be widespread in low-income and minority communities.[33] The professional status of many CFS activists has facilitated their ability to organize and educate themselves about their illnesses and various medical responses.[34] CFS activists have also benefited from the lessons of AIDS activists, who, as Epstein's chapter in this volume suggests, created a new model for medical activism in this country.[35]

The level of COS activism with regard to toxic exposures has been much more variable. When the communities at risk have been relatively well off, as in the case of Love Canal or the women office workers described by Murphy, communities of suffering have enjoyed some success. When the communities at risk have been poor and marginalized, as in the case of lead poisoning in the sixties and toxic dumping in the South, the level of activism has been limited and reliant on external leadership. The activities of various environmental justice organizations have helped raise the consciousness of poor communities exposed to toxic dumping and assisted them in fighting for compensation and the establishment of environmental protection measures. Still, these communities have historically found it difficult to gain sustained public health attention for their illnesses.

Regardless of the social position of the populations at risk of toxic exposures, their struggles for public health attention have tended to be local in nature. There is no environmental illness equivalent of the national patient support organizations that have been so successful in pushing forward the interests of sufferers of CFS, HIV/AIDS, and fibromyalgia. The local character of the communities of suffering that form around toxic exposures has restricted the visibility of these illnesses.

Another problem facing the communities of suffering that form around toxic exposures is the difficulty of knowing to whom to turn. Government responsibility for the environment and environmental hazards is divided among half a dozen federal, state, and local agencies. These include the U.S. Environmental Protection Agency, the National Institute for Occupational Safety and Health, the National Center for Environmental Health, the Agency for Toxic Substances and Disease Registry, the U.S. Department of Housing, the Department of Agriculture, and the Food and Drug Administration, as well as state and local departments of health. These agencies have overlapping responsibilities and represent different sets of interests, governmental missions, and regulatory mandates. As Spears points out in her analysis of Newtown, Georgia, working through this array of agencies can be a daunting experience for those seeking redress or even just explanations for illnesses associated with environmental exposures.

Finally, gaining sustained media attention for contested illnesses has been difficult. Media coverage of toxic exposures has waxed and waned over the past thirty years. Incidents like Love Canal and Three Mile Island in the 1970s attracted a great deal of media attention, and more recently, films like *A Civil Action* and *Erin Brockovich* have focused media attention on the health consequences of industrial dumping. Yet, in between these events, many other cases, especially ones that occurred in minority communities like that described in Spears's chapter, have received little or no media attention. Media and public health attention to lead poisoning, described by Warren, has also fluctuated over the past forty years.

The extent of media coverage of contested illnesses is directly related to the effectiveness of COS leaders in publicizing the plight of their communities. CFS leadership has been exceptionally effective in this area. In the 1990s, roughly fourteen hundred articles in major English-speaking newspapers either mentioned or discussed CFS. This coverage came even though the medical field had made few advances in understanding this illness during this same period. CFS leadership has also pushed hard to shape the kind of coverage that the illness has received, encouraging stories focused on the biomedical causes of the illness. Edward Shorter argues that media emphasis on the biomedical causes of CFS, consequently discounting its psychogenic origins, has played a key role in the biomedicalization of CFS.[36] This in turn has helped keep CFS on the public health agenda.

The various positions of illnesses on the six gradients indicate that the process of illness emergence is complicated. No single factor determines whether an emerging illness gains public health attention. Rather, the extent to which an illness gains visibility is determined by the combination of factors shown in figure 1.2.

## Lessons Learned

The case studies in this volume highlight factors that have facilitated or hindered public health recognition of emerging illnesses. These case studies reveal epidemiological and historical peculiarities as well as a general process. The interaction of the communities of suffering and the powerful institutions of medicine and public health is the crux of the social negotiation of the public health agenda. This negotiation is influenced by variables like the epidemiological character of the illness, the social class of sufferers, the interest of the media, and the organizational abilities of the activists. An awareness of these factors suggests the need for some new types of public health policies or organizational structures.

The institutions designated to respond to emerging illnesses may not be well designed to accomplish this task. This is not simply a problem of increasing the surveillance capacity of local and state health agencies. The institutional culture of public health institutions may impede their ability to respond to certain types of illnesses. Noninfectious illnesses that do not appear to represent an immediate public health threat, illnesses that cannot be linked to specific exposures, or illnesses that primarily affect marginalized populations may continue to go unrecognized despite increases in surveillance.

In order to make public health institutions more responsive to emerging illnesses, and not just to acute episodes of infectious diseases, there needs to be greater coordination among the state and local agencies responsible for monitoring and responding to health issues related to environmental exposures. In addition, there needs to be a rethinking of the "one illness, one exposure" paradigm that drives occupational and environmental health. Greater recognition needs to be given to sufferers from illnesses with complex etiologies or those whose illnesses may be the product of accumulated exposures. The legal implications of such exposures and the evidentiary demands of the legal system make this shift in approach difficult but should not preempt efforts to do so.

Also needed is a greater recognition of the extent to which political and economic forces may prevent agencies from effectively fulfilling their mandate. This is particularly true for environmental illnesses, but political interests may also impede the emergence of other forms of illness. For example, in the early 1990s, health officials concerned about the emergence of MDRTB among farm and poultry workers in Georgia found their efforts to screen these workers blocked by politicians representing the interests of farmers and poultry plant operators.[37] Defining the public health is at heart a political process, but the process needs to be more transparent if we are to develop public health policies that are responsive to the health needs of the entire population.

Communities of suffering (and those at risk) deserve to have a "seat at the table" of research and public health policy decision making. There needs to be greater sensitivity to the etiological understandings of COS. On the other hand, communities of suffering also need to recognize that their etiological understandings may not be totally correct. Adversarial actions taken primarily for media attention can impede the process of getting new health problem on the agenda. Until we have all the answers, each side needs to be willing to listen and adapt, and research agendas need to be broadly based and not prematurely narrowed to meet the interests of either party. Public health and medical research institutions can benefit in their mission of improving society's health from collaboration. Medical researchers, public health officials, and communities of suffering need to be partners.

Finally, the importance of COS in negotiating the public health requires that attention be given to providing poor and disenfranchised populations with the resources that will allow them to mobilize and effectively advocate for consideration of their health problems. Much of the work in this area is currently conducted by NGOs. Finding ways to make government agencies more proactive in uncovering emerging illnesses among the poor would help make these illnesses visible. We must also seek to end the process by which health problems "settle in" on the poor and lose their priority on the public health agenda.

These types of changes are necessary to increase the probability that illnesses of actual or potential epidemiological importance become visible in a timely fashion and to attract necessary resources for prevention and treatment. It is a certainty that new illnesses and public health challenges will appear in the future. Understanding the social processes of illness emergence will help us be bettered prepared to meet those challenges.

### NOTES

1. "Chronic fatigue syndrome" is the current name employed by the Center for Disease Control. This name is hotly contested by some patient support groups who view the name as demeaning and prefer the name "chronic fatigue immunity deficiency syndrome." This name implies a physiological cause for the cluster of symptoms associated with the illness.

2. In this regard it is worth noting that 9/11 also produced a cluster of cases involving pulmonary problems of unknown etiology among firemen and others who were involved in the Twin Towers disaster.

3. The term "communities of suffering" was first used by anthropologist Victor Turner, and later by Arthur Kleinman, to describe groups of people coming together to help one another deal with their illnesses and other afflictions. It is employed here to convey not simply the shared suffering experienced by those afflicted with various illnesses but also their collective reactions and resistances to misfortune.

4. Robert Hahn, *Sickness and Healing: An Anthropological Perspective* (New Haven: Yale University Press, 1995).

5. Robert A. Aronowitz, *Making Sense of Illness: Science, Society and Disease* (Cambridge: Cambridge University Press, 1998); Charles E. Rosenberg and Janet Golden, *Framing Disease* (New Brunswick, N.J.: Rutgers University Press, 1997); Charles E. Rosenberg, *The Cholera Years: The United States in 1832, 1849, and 1866* (Chicago: University of Chicago Press, 1962).

6. Arthur Kleinman, *The Illness Narratives: Suffering, Healing and the Human Condition* (New York: Basic Books, 1988).

7. Leon Gordis, *Epidemiology*, 2nd ed. (Philadelphia: W. B. Saunders, 2000).

8. A striking analogy could be drawn between, on the one hand, the public response to emerging illnesses that strike mainstream society compared with the response to illnesses that affect only the poor and, on the other hand, the public response to the sniper shootings in the Washington-Baltimore metropolitan area in the fall of 2002 compared with the response to the more than 250 homicides that occur annually in Baltimore.

9. HIV/AIDS needed to cross an epidemiological threshold before gaining recognition. Retrospective analysis showed that HIV was being transmitted in Africa, in relatively low numbers, for three decades before its identification in the United States. See Edward Hooper, *The River: A Journey to the Source of HIV and AIDS* (Boston: Little, Brown, 1999).

10. CDC, *MMWR*, "Guidelines for Investigating Clusters of Health Events," July 27, 1990, 1–16, www.cdc.gov/epo/mmwr/preview/mmwrhtml/00001797.html.

11. See Laurie Garrett, *Betrayal of Trust: The Collapse of Global Public Health* (New York: Hyperion, 2000), for a useful historical overview of the international setting of declining public health infrastructure.

12. Ibid., 6–8.

13. The CDC "is recognized as the lead federal agency for protecting the health and safety of people—at home and abroad, providing credible information to enhance health decisions. . . . CDC serves as the national focus for developing and applying disease prevention and control, environmental health, and health promotion and education activities designed to improve the health of the people of the United States." www.cdc.gov/aboutcdc.htm.

14. Images from the History of the Public Health Service, photographic exhibit, U.S. Department of Health and Human Services, Washington, D.C., 1994.

15. www.epa.gov/history/org/origins/mission.html.

16. Elizabeth W. Etheridge, *Sentinel for Health: A History of the Centers for Disease Control* (Berkeley: University of California Press, 1992).

17. The case of *E. coli* 0157:H7 reveals another weakness of the public health response. One of the major causes of *E. coli* 0157:H7 outbreaks is the absence of adequate safety measures in the preparation of food. This is a problem over which local, state, or even national public health institutions have only limited control. Local health authorities can police local restaurants, yet *E. coli* 0157:H7 and other intestinal pathogens often enter the food supply through the food industry (e.g., meatpacking plants), whose activities are monitored by the USDA or FDA. Once these pathogens are in the food supply, even if food recalls are issued, it is often up to local consumers to safeguard themselves and others from infection. One of the remarkable aspects of the Georgia Department of Health's response to the Whitewater recreational water outbreak was the way in which it used the news media to hammer home the message that contaminated kiddie pools were only a small part of the problem. The public health authorities used the crisis as an opportunity for health education, with the central message that food consumers played a critical role in protecting themselves from infection in their own homes.

18. Timothy E. Cook and David C. Colby, "The Mass Mediated Epidemic: The Politics of AIDS and the Network Nightly News," in Elizabeth Fee and Daniel M. Fox, eds., *AIDS: The Making of a Chronic Disease* (Berkeley: University of California Press, 1992), 89–90.

19. Edward Shorter, *From Paralysis to Fatigue: A History of Psychosomatic Illness in the Modern Era* (New York: Free Press, 1992), 315.

20. Robert Shope lecture, "Emerging Viruses' Big Questions," Emory University, September 11, 2000.

21. Aronowitz, *Making Sense of Illness,* 68

22. Communication by Dr. James Curran during Sawyer seminar "Emerging Illnesses and Institutional Responses."

23. H. Gavaghan, "NIH Panel Rejects Persian Gulf Syndrome," *Nature* 369 (1994): 8; Jeffrey Sartin, "Gulf War Illnesses: Causes and Controversies," *Mayo Clinic Proceedings* 75 (2000): 811–819.

24. CDC, *MMWR,* "Recommendations for Prevention and Control of Hepatitis C Virus (HCV) Infection and HCV-Related Chronic Disease," October 16, 1998, 1.

25. CDC, *MMWR,* "Summary of Notifiable Diseases, United States 1999," April 6, 2001, 1–104.

26. CDC, *MMWR,* "Recommendations for Prevention and Control of Hepatitis C Virus (HCV) Infection and HCV-Related Chronic Disease, " 6.

27. Ibid.

28. *Los Angeles Times,* May 30, 2000, Metro, B:6, Editorial Writers Desk.

29. In 1994, only 6 of the 189 articles on HCV were published in U.S. newspapers. British and Canadian papers, in contrast, gave a great deal of attention to HCV. Media attention in the United States rose in 1997 and 1998 with the development of new therapies. The correlation between increased media coverage and the development of a more effective treatment for HCV (Rebetron) may reflect the media's need for what Mass terms "a hook," a piece of new information that makes an ongoing story partic-

ularly newsworthy. The rise in media coverage in association with the development of Rebetron may reflect more than media interest in the new drug combination, however. Schering-Plough aggressively marketed its new drug and has been accused by some public health authorities of overstating the risk of contracting HCV in order to increase sales. One Schering-Plough ad stated, "To put it bluntly, every living breathing human being can get hepatitis C—even you." Denise Grady, "Hepatitis C: How Widespread a Threat?" *New York Times,* December 15, 1998, F:3.

30. Perhaps most influential in this regard was the experience of popular country music singer Naomi Judd, who was forced to withdraw from public performances because of the effects of HCV. Judd received treatment for her infection and has apparently become virus-free. Judd has become an HCV activist, pushing for early screening and treatment. More recently, Pamela Anderson, of the TV show *Baywatch* fame, has gone public with her HCV infection. The changing awareness of risk brought about by the media and the marketing activities of Schering-Plough may have led Congress to allot $13 million to CDC for HCV prevention.

31. Fibromyalgia activists working with rheumatologists have attempted to counter this problem by developing a set of diagnostic criteria based on pain points.

32. Evelyn Kim, "A Brief History of Chronic Fatigue Syndrome," *JAMA* 272, no. 34 (1994): 1070–1071.

33. L. Steele, J. G. Dobbins, K. Fukuda, M. Reyes, B. Randall, M. Koppelman, and W. C. Reeves, "The Epidemiology of Chronic Fatigue in San Francisco," *American Journal of Medicine* 105:3A (1998): 83S–90S; Dorothy Wall, "New Research Debunks Chronic Fatigue Syndrome Myth," *San Francisco Chronicle,* May 12, 2000, A27.

34. Juanne Clarke, "The Search for Legitimacy and the 'Expertization' of the Lay Person: The Case of Chronic Fatigue Syndrome," *Social Work in Health Care* 30, no. 3 (2000): 72–94.

35. CFS patient support groups have played an important role in directing CFS research within the NIH and CDC. For example, in February 2000 the National Institutes of Health held an internal meeting, "State of the Science Consultation," on CFS in Bethesda, Maryland. This meeting generated considerable anger among members of the CFS Coordinating Committee (CFSCC) and patient advocates because the CFSCC was not involved in the planning, the CFS community was not invited to observe, and no CFS clinicians were asked to participate. In addition, the NIH-chosen CFS experts were three psychiatrists who hold controversial views about the diagnosis and treatment of CFS. Because of the uproar, the NIH expanded the participant list, and no observer was turned away from the meeting. In addition, a true "State of the Science" meeting, with full participation by the CFS community, was held in November 2000. Dr. Anthony Komaroff, a CFSCC member and longtime CFS researcher from Harvard, led the planning committee, which included members of the medical and patient communities.

CDC researchers also meet regularly with representatives of the CFSCC and several advocacy organizations for patients to review research. CFS groups have emphasized the need for the CDC to focus research on biomedical sources of their illness. Notable in this regard is the location of CFS research in the National Center for Infectious Diseases. The NCID first became involved in CFS research while investigating the possible links between the illness and Epstein-Barr virus. Even though these and other investigations produced no conclusive evidence that CFS is caused by an infectious agent and the International Classification of Diseases classifies CFS as a neurological disorder, CFS research remains located in the NCID. It is unclear to what extent the decision to keep CFS in the NCID was influenced by CFS activists, but it certainly conforms to their efforts to view the illness as having an organic cause.

36. Shorter, *From Paralysis to Fatigue,* 314–320.

37. Interviews conducted by Drs. Packard and Naomi Block with state public health workers in Athens and Waycross in the summer of 1995.

# PART I // MAKING ILLNESSES VISIBLE

# CHAPTER 2

# The Combined Efforts
# of Community and Science

*American Culture, Patient Activism,*

*and the Multiple Sclerosis*

*Movement in the United States*

COLIN TALLEY

The chapters in this section focus on the efforts of illness support groups, or communities of suffering, to make their illnesses visible by gaining medical and public health attention and support. Multiple sclerosis (MS), a disease that affects the nervous system of young adults, "emerged" as a public health problem in the United States during the 1950s. As Colin Talley indicates, the emergence of MS resulted in large part from the activities of the National Multiple Sclerosis Society (NMSS). The NMSS represents a kind of transition in the history of patient activism. It linked the active volunteerism associated with earlier, medical crusade–style organizations, such as the American Cancer Society (1913), the American Heart Association (1915), and the National Foundation for Infantile Paralysis (1933), with the activism of the communities of suffering that emerged in the 1980s around HIV/AIDS, chronic fatigue syndrome, and Lyme disease.

Like the former groups, the NMSS focused much of its work on raising funds for biomedical research. Yet it shared the additional burden of later communities of suffering in having to promote support for an illness that was surrounded by much uncertainty. The NMSS was one of the first patient activist groups in the United States to mobilize around an illness for which there was neither a clear etiology, case definition, or diagnostic test. In this context, the NMSS took a more activist role in raising public consciousness of the disease and making it visible. Like later patient advocacy groups, the NMSS attracted attention for a relatively obscure disease by enlisting the support of prominent Americans who had some experience with MS sufferers. In addition, the NMSS linked itself to a medical subspecialty, in this case neurology, which was "in need" of a culturally salient disease with which it could increase research support.

The NMSS resembled more recent examples of patient activism in two other ways. First, it actively sought to gain and shape media coverage of the group's illness. The NMSS was highly successful in getting MS into the national consciousness through articles in popular media such as *Reader's Digest*. Second, it sought to encourage and shape the direction of biomedical research into the etiology and treatment of MS. This meant more than old-style fund-raising. It required the NMSS to "gain a seat at the table" where decisions concerning the allocation of research funding were made. The NMSS was highly successful in this activity, virtually dominating research on MS through its connections to the National Institute for Neurological Diseases and Blindness.

The NMSS differed from later examples of patient activism in one important respect. It was not confrontational or oppositional. It did not engage in "street politics"—it did not hold public rallies, sit-ins, or other forms of protest aimed at publicizing the cause and influencing key decision makers. In this regard, the NMSS was clearly an example of a pre-HIV/AIDS style of patient activism.

---

Large voluntary health associations emerged in the United States in the early twentieth century. The first was the National Tuberculosis Association (Christmas Seals), which was founded in 1904.[1] Other organizations included the American Society for the Control of Cancer, founded 1913 (reorganized as the American Cancer Society in 1945); the American Heart Association, founded in 1915; and the National Foundation for Infantile Paralysis (Easter Seals), founded 1933.[2] Voluntary organizing has dominated disease awareness education in U.S. history. These groups have engaged in national crusades to educate the population about disease, for disease prevention, and to raise huge sums of money for medical research. They have been major actors in drawing public attention to emerging diseases in American history as well as forceful constituent and pressure groups with the goal of expanding the role of the federal government in medical research.

These organizations and their crusades are a peculiarly American cultural phenomenon.[3] As Walter S. Ross wrote in *Crusade: The Official History of the American Cancer Society* (1987), "The powerful spirit of voluntarism that Tocqueville discerned in the United States in the 1830s has not only endured, but intensified. Nowhere else are peaceable voluntary associations so numerous, so widespread, or so varied as in the United States of America. And nowhere else on earth do so many people get together on their own to help one another."[4] For example, the American Cancer Society had 1.5 million volunteers in 1956 and 2.5 million volunteers in 1987.[5] Robert Hamlin, M.D., M.P.H., noted that by 1961 voluntary health agencies had become big businesses in the United States. For example, in 1959 dollars the National Foundation (March of Dimes, polio) had an income of $37,093,400. The other large voluntary associations had similar annual incomes in 1959: the American Cancer Society had an income of $35,379,900, the American Heart Association $24,917,200, and National Tuberculosis Association (later named the American Lung Association) $28,775,800.[6] Given the enormous impact of these massive organizations on the history of health, medicine, and disease in the United States, surprisingly little has been written by medical historians on the topic.

This is the story of the emergence of a new voluntary health association in the United States in the late 1940s, the National Multiple Sclerosis Society. The NMSS modeled itself on these previous organizations, especially the National Foundation for Infantile Paralysis. However, unlike the case with cancer or polio crusaders, the multiple sclerosis activists worked in a culture in which their disease was relatively unknown.

From the 1870s to about 1920, American neurologists considered multiple sclerosis a rare malady in the United States.[7] During the 1920s they began to diagnose more MS cases, and by the 1930s American neurologists considered MS a common neurological disease in the United States.[8] Nevertheless, MS was not an object of intense scientific scrutiny in the 1930s and early 1940s, though a few prominent American neurologists had studied the disease. For the lay public, MS remained a virtually unknown disease during this same time period. Suddenly, after 1946, scientific research on MS and the public's interest in it increased dramatically.

Neurologists demonstrated increased interest in MS during the 1930s and early 1940s at least as expressed in the numbers of articles published on the subject in American medical journals. Beginning in 1947, there was a sharp rise in the number of articles devoted to MS. This reflected an intensification of experimental research and clinical interest in the disease in the United States.[9]

This new concern with MS resulted in several scientific conferences. The NMSS sponsored and initiated all these meetings. In December 1948 the Association for Research in Nervous and Mental Diseases (ARNMD) devoted its annual conference exclusively to MS. The New York Academy of Sciences held a conference in April 1953 entitled "The Status of Multiple Sclerosis," which was chaired by Pearce Bailey, director of the National Institute for Neurological Diseases and Blindness.[10] In 1957 the NMSS co-sponsored a symposium entitled "New Research Techniques of Neuroanatomy."[11] In 1962, the NMSS co-sponsored a conference, "Mechanisms of Demyelination," at the University of California, Los Angeles."[12]

These efforts made MS a research priority in American neurology during the late 1940s and 1950s by creating economic incentives to do MS work and by increasing the interest of existing neurological organizations in the disease. Between 1946 and 1955 the fifty-four chapters of the NMSS raised a cumulative total of about $1.3 million for research.[13] By 1960 the NMSS had raised a total of $6,067,381 for research into MS[14] and was funding seventy-two research projects.

This expansion of research into MS cannot be simply ascribed to the general increase in biomedical research after World War II. Much of this general increase in research after 1945 came as a result of the expansion of the federal government into biomedical research. However, the federal government did not fund neurological research in any significant way until 1952. The expansion of research into MS predated the federal interest in the disease. An intensification of research in MS began in 1947 and was in full stride by 1951. Even after 1951 the NMSS could legitimately take credit for pushing the NINDB (funded for the first time in 1952) to investigate the disease.

Observers at the time noted the increased interest in the disease.[15] Vermont professor George Schumacher noted in 1952 that the "general awareness among medical scientists and practitioners of the great prevalence of multiple sclerosis has been a surprisingly recent development in the history of the study of this disease. Although isolated investigators in the past quarter century devoted much energy and time toward an understanding of multiple sclerosis, it remained an esoteric problem of medicine generally neglected by the medical scientist and clinician alike. . . . Impetus for the first great wave of concentrated effort and research on a wide scale came from a group of laymen who in 1946 founded the National Multiple Sclerosis Society."[16] Also in 1952, physician Cornelius Traeger noted the "renaissance of interest in the problem of multiple sclerosis and the demyelinating diseases and the widespread increased activity in this field of research."[17] In 1954 W. H. Sebrell Jr., director of the National Institutes of Health, wrote that "the research attack on multiple sclerosis has been greatly accelerated in the past few years."[18] Also in 1954, Pearce Bailey, director of the NINDB, noted the "gathering public and scientific interest in this disorder."[19] Concomitantly, awareness of MS increased among the lay public. The number of articles about MS in popular journals rapidly increased in and after 1946.[20]

What explains this sudden increase in medical activity concerning MS and the growing awareness of the disease by the lay public, especially considering there was no dramatic epidemic of MS and no evidence that it was an infectious condition?[21] The answer is that persons with MS, their families, and their partisans put MS on the cognitive map of American medicine and popular culture. In other words, lay activists made MS a research priority in American neurology and were directly responsible for the increase in scientific work on this malady.

## The National Multiple Sclerosis Society

The story of lay activism and MS began in 1945 with two New Yorkers in their twenties. Sylvia Lawry was becoming increasingly distraught over the decline of her brother, who was suffering from MS. Her brother's doctor would refuse to see him when a new symptom occurred unless it was catastrophic. This physician's advice, according to Lawry, was to go home and rest because there was nothing medicine could do for this disease. Because of this attitude Lawry began reading the medical literature herself, hoping to find some clue about treatment. She discovered that many people with MS experienced remissions, and she decided to investigate the factors that might precipitate a remission. To do this she placed a personal advertisement in the *New York Times* in May 1945 that read: "Multiple Sclerosis. Will anyone recovered from it please communicate with patient. T272 Times." Lawry received about fifty replies, mostly from people with the disease. She and the respondents continued to correspond about the problems surrounding MS. They then began to hold meetings at the New York Academy of Medicine and at the Red Cross headquarters in New York City. They decided to start a national organization dedicated to finding a cure for MS. The New York Academy of Medicine then donated office space to the fledgling group. In March 1946, people with MS, their families, and their partisans formed the Association for the Advancement of Research into Multiple Sclerosis (AARMS).[22] In 1948 the group reorganized and named itself the National Multiple Sclerosis Society. In 1946 AARMS gave its members, mostly people with the disease, stacks of cards to enroll their families and friends in the new organization; because of this, Lawry described these people with MS as the "prime movers" in the new organization.[23] They placed an advertisement in a Boston paper that attracted more potential members.[24] Membership quickly rose after 1946. In 1946 AARMS had 600 members, a figure that grew to 7,500 in 1948. In 1949 the NMSS membership had climbed to 15,000. By 1954 the organization had 33,000 members, and in 1958 the figure rose to 120,000.[25]

For AARMS, Lawry recruited a lay board of directors and a medical advisory board. For the lay board she wanted professionals and businesspeople whose prominence and influence were national in scope. Raymond Moley, contributing editor at *Newsweek*, became the first chairman of the Public Education Committee and chaired the first press conference of AARMS. Carl W. Owen of the law firm of Owen, Willkie, Otis, Farr and Gallagher became chairman of the board of directors.[26] Other prominent sponsors included Mrs. James S. Rockefeller; Mrs. Wendell Wilkie; William J. Norton, secretary of the Children's Fund of Michigan; and Senator Brien McMahon of Connecticut.[27]

To select a chair for the medical advisory board, Lawry read the medical literature to find the most prominent researcher in MS in the United States. Richard Brickner and Tracy Jackson Putnam were two of the most prominent researchers in MS in the United States during the 1930s and 1940s. Lawry chose Putnam to be the first chair of the medical advisory board because he was director of the Neurological Institute at Columbia University and professor of neurosurgery and neurology at Columbia University.[28] Other prominent members of the medical advisory board included Roger I. Lee, retiring president of the American Medical Association; Thomas M. Rivers, director of the Hospital of the Rockefeller Institute for Medical Research; Dr. Ernest L. Stebbins, dean of the School of Hygiene of the Johns Hopkins University; Henry Woltman of the Mayo Clinic; and neurologist Leo Alexander of Boston.[29] Lawry had rapidly put together lay and medical boards with very prominent members. What explains the suddenness of the emergence of the MS movement in 1946? Why was Lawry so successful so quickly considering the apparent obscurity of MS as compared with well-known diseases such as polio and cancer, diseases for which activists staged massive advertising campaigns from the mid-1940s through the 1950s?[30]

Part of the answer is that physicians increasingly diagnosed patients with MS during the 1930s and 1940s, compared with previous decades. In effect, this increase in diagnoses created a significantly larger population with this incurable chronic disease than had existed in the first three decades of the twentieth century. By increasingly naming the disease, physicians, especially neurologists, created the potential for a patient social movement centered on MS. Lawry found fertile ground on which to build a new organization because of several factors: many people with MS were young, little could be done for them, medicine seemed to be ignoring their plight, and those with the disease were desperate. Moreover, because the life expectancy of persons with MS was only slightly less than average, the increasing tendency of neurologists to diagnose people with MS from the early 1930s onward meant that there was a snowball effect in terms of numbers of people with the disease, which reached a critical mass by 1946. In short, there were simply more people with MS. Lawry's efforts helped overcome the isolation of many of them and tapped into their deeply felt need for an organization to address what they saw as a lacuna in modern medical research and treatment. In addition, by 1946 many prominent figures in American society knew someone with MS. Raymond Moley, the editor at *Newsweek* mentioned earlier in this chapter, had two promising students who became ill with MS while he was teaching at Barnard College. Senator Charles Tobey of New Hampshire had a daughter with the disease. Henry Kaiser Jr., son of the industrialist, also suffered from MS.[31]

This newly created pool of people with MS in the 1940s coincided with a political and cultural climate highly favorable to disease crusades and the private and public outlay of funds for biomedical research. Moreover, American neurologists, whose specialty was relatively low status and poorly funded in the United States in the mid-1940s, quickly realized the benefits a vigorous patient movement could mean for their field.[32]

These conditions enabled people with MS, their families, their partisans, and their neurological allies through the NMSS to make MS a research priority for American neurology in the 1950s. They did this by raising large sums of money for research directly and indirectly by persuading the federal government to spend money on MS research. The NMSS carried the campaign for dollars, public awareness, and the attention of the medical profession on two main fronts. On the first front the NMSS ran a media campaign that attempted to raise public consciousness about MS and that helped raise money for the NMSS to distribute directly to neurological workers. On the second front the NMSS vigorously lobbied the federal government to fund and carry out research on MS.

### The First Front: Advertising Disease

Part of the problem the MS activists faced in the late 1940s was public ignorance about the disease, unlike the situation with cancer or polio.[33] As an AARMS pamphlet stated in 1946: "To the man in the street, the term multiple sclerosis is strange and unfamiliar. The time has come when the world should know about multiple sclerosis for this widely prevalent nerve disease has become an acute social problem. Only the combined efforts of the community and science can some day hope to solve the mystery of multiple sclerosis."[34] Howard Rusk wrote in the *American Mercury* in 1947 that "for the man on the street it [MS] is merely a strange and unfamiliar medical term."[35] In short, the NMSS sought to "make the public realize by a broad campaign of education that this is not a rare or mysterious disease."[36]

To do this the NMSS fed stories and press releases to science writers and journalists at popular magazines such as *Time, Newsweek, Saturday Evening Post, Look, Coronet, Parents, Survey, Cosmopolitan, American Mercury, Reader's Digest, Today's Health, Science Newsletter, Science Digest,* and *Business Week* and at major newspapers like the *New York Times.*[37] The successful placement of articles in popular magazines did not depend on luck. *Newsweek* editor Raymond Moley served on the board of directors of the NMSS, and Howard A. Rusk, a contributing medical editor at the *New York Times,* served on its medical advisory board.[38] These contacts gave the NMSS significant access to popular media.

The NMSS was able to garner more media attention and dollars by enlisting the rich, famous, and powerful in its fund-raising and organizational campaigns.[39] Senator Charles Tobey of New Hampshire became active on behalf of the NMSS in 1948 and 1949 because his daughter, Louise Tobey Dean, had the disease.[40] Henry Kaiser Jr., who also had MS, campaigned on behalf of the NMSS in 1949.[41] The NMSS enlisted Eleanor Gehrig, widow of the former Yankee baseball star, in its fund-raising drives beginning in 1949.[42] Ralph I. Straus, a director of R. H. Macy and Company, helped lead a fund-raising campaign in 1950, as did Mrs. John D. Rockefeller Jr. in 1952.[43] In 1953 the NMSS "announced the election of Edward Locke Williams as president of the Society and Oliver E. Buckley as Chairman of the Society's Board of Directors." Williams was a Long Island attorney and former president of the Insurance Executives Association. Buckley was former chairman of the Bell Telephone Laboratories.[44] Also in 1953, "with an assist from actress Shirley Temple, whose brother George has M.S., . . . the Los Angeles chapter staged a successful fund-raising telethon."[45] Mamie Eisenhower agreed to be the "honorary chairman of the 1954 appeal" of the NMSS for funds and was honorary chairman of the 1957 "Hope Chest" campaign.[46] In 1954 Ralph C. Block, vice president of the Bank of New York, became president of the NMSS.[47] In 1955 Robert W. Sarnoff, executive vice president of the National Broadcasting Company, volunteered to chair an NMSS fund-raising drive.[48] In 1956 the actress Grace Kelly chaired the women's activities section of the NMSS fund-raising drive.[49] In 1957 the NMSS appointed retired Vice Admiral H. R. Thurber as national chairman of its "Hope Chest" campaign.[50] In 1958 Senator John F. Kennedy of Massachusetts headed the "fund-raising appeal" of the NMSS.[51] In 1959 Alfred N. Steele, chairman of the board of the Pepsi-Cola Company, agreed to chair the 1959 NMSS fund-raising campaign. His wife, actress Joan Crawford, had served as chair of the women's activities in the 1957 and 1958 NMSS drives.[52] In April 1959 Steele died, and Crawford took over as chair of the 1959 NMSS fund drive.[53] These prominent people attracted significant media attention, and their elite status also aided in fund-raising.

Grassroots MS activists waged the public awareness and fund-raising campaigns at the local level as well. For example, in Washington, D.C., the local NMSS chapter asked Surgeon General Leonard Scheele to speak at a Shriners' luncheon "to inform leading citizens of the community . . . about multiple sclerosis." This meeting was "timed to coincide with other events connected with the M.S. campaign, such as the declaration of Multiple Sclerosis Week by the Commissioners of the District of Columbia."[54] The Washington, D.C., chapter of the NMSS held a benefit dance "in the Terrace Room of the Shoreham Hotel on March 6, 1953" that "netted . . . approximately $2,500." The Washington radio and TV stations gave "frequent spot announcements for M.S." Dr. Walter Freeman "was on WTOP TV from 7 to 7:15 the evening of March 17th " speaking about MS. The local chapter placed "car cards telling the M.S. story" in four hundred buses and streetcars. "The March 2, 1953 issue of the 'Transit News' carried an article on Multiple Sclerosis." The General Services Administration of the federal government distributed "M.S. cannisters in the government cafeterias for M.S. Week." The Washington Society of the Blind cooperated with the local MS chapter by placing M.S. cannisters "on the blind stands throughout the Washington area." The public libraries in Washington put up MS posters, and several of the city's hotels permitted the Multiple Sclerosis Society to set up public information tables in their lobbies.[55]

In New York, Governor Thomas Dewey proclaimed April 5–11, 1953, "Multiple Sclerosis Week and asked New Yorkers to help in the fight against one of the leading disabling diseases."[56] The city of New York declared April 4–27, 1954, "MS Week" in conjunction with an NMSS fund drive.[57]

Turning now to the content of these local and national multimedia campaigns, one sees that the NMSS waged its campaign for public awareness partly by comparing the incidence and consequences of MS with those of other diseases. In 1947 Marshall Hornblower, chairman of the NMSS, writing in the *New York Times,* pointed out that "at a time when so much publicity is being given to cancer and infantile paralysis campaigns, it is important that new and unheralded organizations for pushing back other frontiers of medicine not be crowded off the pages of the press."[58] Six years later, in a letter to Surgeon General Scheele, Hornblower still perceived competition with other diseases for attention and dollars as a key element of the NMSS crusade: "As you know, multiple sclerosis, despite its prevalence, is comparatively unknown to the general public. One of our big jobs in this area is to convince people who are already well aware of poliomyelitis, cancer, heart, etc. that multiple sclerosis is also a critically serious health problem which has been, comparatively speaking, neglected."[59]

One way the NMSS attempted to overcome public ignorance about MS was by comparing it with polio, perhaps the highest-profile malady of the late 1940s and 1950s before the Salk vaccine was developed.[60] *Time* magazine reported in October 1946 that "neurologists estimate that multiple sclerosis is more prevalent than infantile paralysis."[61] The *Science Newsletter* went further, arguing that "the disease is believed to be more than twice as common as infantile paralysis."[62] The *American Journal of Public Health* was more reticent than popular publications, stating in 1946, "It is thought possible that sclerosis victims may out number infantile paralysis sufferers."[63] Speaking at an AARMS dinner meeting on February 21, 1947, Tracy Putnam—director of the Neurological Institute at Columbia University, Columbia University professor, and first chair of the AARMS (early name of the NMSS) medical advisory board—announced that "in the period from 1931 to 1935 Boston City Hospital admitted twice as many people with multiple sclerosis as with infantile paralysis."[64] Howard A. Rusk announced in the *New York Times* in April 1947 that "the patients with multiple sclerosis outnumbered those with infantile paralysis by more than two to one."[65] In October 1947 Rusk repeated the claim in the *American Mercury.*[66] The *New York Times* reiterated this claim in 1948.[67] Senator Charles Tobey testified in Senate hearings in 1949 that MS was more prevalent than polio.[68] Not only did the NMSS and authors report that MS was more common than polio, but they argued that MS was more devastating.[69] Robert Grant Jr., who had MS, wrote in the *Saturday Evening Post* in 1953 that MS was "worse than polio, which does all its damage at one fell swoop."[70]

The NMSS attempted to get federal backing for its claims that MS occurred more frequently and was more devastating than polio. In 1953 Alice Friedman, director of public relations for the NMSS, asked Harold Tager Jr., information officer for the NINDB, for support of the NMSS position being promulgated in popular publications. Friedman wrote: "I am writing to ask if you would do a little digging for me on a subject which somewhat confuses me. An article in *Newsweek,* January 14, 1952, by Dr. Pearce Bailey, estimates that 300,000 people are afflicted with MS as compared to 250,000 with the after effects of polio. (I think the 300,000 figure is conservative, considering our diagnosis problem, etc.) These figures are substantially the same as the 1952 report of the National Committee for Research in Neurological Disorders which gives figures of 300,000 for MS and the demyelinating diseases as compared with 225,000 for chronic poliomyelitis. My question is—can we say from these figures—assuming they are correct—that MS is more prevalent than polio? This would depend on the definition of 'the after effects of polio.' But, I would like an interpretation from the National Institute."[71]

Tager responded cautiously: "There is a Government taboo on making comparisons of diseases, suggesting that one is more or less serious than another, and for various reasons avoidance of this kind of overt competition seems right to me. What obtains for us, however, is not necessarily so for you; but this is to say anyhow we wouldn't go on official record on statistics which are no more than educated guesses on both sides."[72]

After 1955, when the Salk vaccine was proved effective, direct comparisons of MS with polio in popular magazines and journals declined dramatically. MS then emerged as the neurological disease that took the place of polio in public consciousness. It became what Thomas M. Rivers, medical director of the National Foundation for Infantile Paralysis, called in 1958 the "the foremost neurological problem of our time."[73] This was an important pronouncement considering Rives directed the nation's largest polio foundation.

The politics of numbers as played out in popular journals held another key claim by the NMSS: that there were 250,000 people with multiple sclerosis in the United States. This assertion became a mantra in popular articles about the disease.[74] *Today's Health* reported in 1950 that "in the United States alone it is estimated that more than a quarter-million people have m.s. (as patients dub it), and the figure is probably much higher because of the difficulty of diagnosing it in the early stages."[75] In addition, *Today's Health, Science Digest,* the *New York Times, Time, Newsweek,* and the *Science Newsletter* all reported that there were 250,000 persons with MS in the United States from 1950 to 1957.[76] Occasionally there would be claims of even higher MS incidence. *Look* stated in 1954 that there were 300,000 cases of MS in the United States.[77] In 1956 the *New York Times* reported 500,000 Americans with MS,[78] but in 1957 it stated that the number was 300,000.[79] In 1957 *Newsweek* upped its estimate to 300,000.[80] These numbers were plastic estimations and not based on solid epidemiological evidence.

In fact, epidemiological studies sponsored by the NMSS and the U.S. Public Health Service from 1948 through 1951 had come up with much lower figures. Charles C. Limburg's study had found that there were between

50,000 and 150,000 cases in the United States.[81] Dr. Leonard T. Kurland's survey had found an incidence of 70,000 to 80,000.[82] The NMSS and the NINDB ignored these studies, which they had funded, and continued to claim that there were between 250,000 and 300,000 cases of MS in the United States. Even in the same publication one could find these competing estimates. In the 1954 publication of the proceedings of the 1953 conference entitled "The Status of Multiple Sclerosis," O. E. Buckley, chairman of the board of directors of the NMSS, maintained that there were "from 200,000 to 300,000 cases of" MS in the United States. Contradictorily, Leonard T. Kurland of the NINDB and Knut B. Westlund of the Johns Hopkins University estimated that there were "about 70,000" cases of MS in the United States.[83] Of course it would have been in the interest of the NINDB and the NMSS to hold publicly to the higher numbers, which they could always support through justified claims about the difficulty of diagnosing MS. However, considering that in 1995 the estimate of MS cases in the United States remained 250,000 to 350,000, the two organizations probably deployed inflated numbers in the 1950s, given the large population increase in the United States since the 1950s.[84]

These prevalence claims did not go uncontested. In April 1954 Lawrence C. Kolb, M.D., of the Mayo Clinic in Rochester, Minnesota, wrote the medical director of the NMSS, Harold R. Wainerdi. Kolb complained that he "was rather distressed, in reading the annual report of the National Multiple Sclerosis Society, to notice under the section dealing with prevalence the statement that 'the National Institute of Neurological Diseases and Blindness estimates that there are approximately 300,000 persons in the United States today with chronic, progressive, multiple sclerosis and related demyelinating diseases.' I do not know who is responsible for giving you this figure, but it is entirely out of line with the work that was done by your own statistical committee under the directorship of Dr. Leonard Kurland who, of course, is now attached to the National Institute of Neurological Diseases and Blindness. The figure is more like 70,000 or 80,000."[85]

Another element of the public relations campaign waged by the NMSS had to do with the cultural syntax writers used to represent persons with MS in the popular literature. Authors designed their magazine articles not only to increase public awareness of the malady but also to instill hope in persons suffering from MS. This was vital because of the despair many of them felt upon diagnosis. I found many cases of suicide attempts, depression, and despair in patient records.[86] A doctor from the University of California, Los Angeles Hospital, described a patient's depression this way in his "Physician's Notes": "Her depression is mostly from mistaken ideas of M.S. and her prognosis which she makes unrealistically morbid. Is currently sitting around home not working and only brooding because she felt working would make M.S. worse. Parents also are over concerned and cautious and slowly driving pt into nervous exhaustion from too much misguided attention. Have urged her to return to work and straightened out some of her misconceptions."[87] One way popular writers instilled hope was to point out that the average life span of persons with MS was only slightly less than normal. Another way was to proclaim that in the absence of an absolute physiological cure, healing could occur if the individual enacted the American myth of self-transformation.[88] To be healed of MS meant to transform oneself through individual effort, to resurrect oneself.[89]

For example, Paul de Kruif, in a 1948 *Reader's Digest* article entitled "The Patient Is the Hero," recounted the following tales of remission/resurrection: "In 1941, the *Washington Evening Star* carried a headline: HOPELESS CRIPPLE CONFOUNDS DOCTOR IN TRICYCLE TRAVELS. On his bed Wilford Wright had begun feebly but systematically to move his stiff, wasted muscles. At last he struggled up onto an adult's tricycle. Since then he's driven his tricycle from Florida to Nova Scotia and even to the West Coast. He is not completely rehabilitated, but his improvement is remarkable."[90] "In Cleveland, a young woman, Betty Bard, lay paralyzed from multiple sclerosis. On her own she began giving herself weak but infinitely determined exercises. She is medically famous as an advanced case now free from incapacity." "Twenty years afflicted, five years bedfast, Mrs. Henrietta Apatta was learning to walk, alone and unsupported."[91]

The emphasis was on individual effort and self-transformation through the use of "heavy resistance exercises." These labors when carried on with devotion might "mean resurrection to active life."[92] As *Today's Health* reported in 1950: "Patients must have the will to work."[93] Commenting on her own recovery in 1950, also in *Today's Health,* Jean Griffith Benge reported that "it took character and persistence, for the amount of work to make any progress is prodigious."[94] Benge witnessed to other sufferers: "My experience brings a message of hope to parents whose children have had polio, to those afflicted with m.s. and to sufferers from other paralytic conditions."[95]

*Cosmopolitan* magazine featured Joan McCarthy's "My Victory over MS" in 1960. In the article McCarthy, who had MS, advised others with the disease to "make your own miracle." She preached to those who wanted recovery: "You've got to make it happen yourself."[96] McCarthy recounted her own experience of "soul-searching." She overcame despair and hopelessness in a critical turning point in 1954, which she described in the syntax of a conversion experience. She then knew "that it was up to me to recover. Nobody could do it for me."[97] Later, others with MS asked McCarthy how she had recovered some of her lost abilities like walking and driving. She "told them it was nothing but work, and that work was everything."[98]

In 1953 the NMSS, in a pamphlet entitled *Self Help,* advised that until there was a cure "the multiple sclerosis patient will fundamentally make his adjustments through his own resources—through courage, persistence and self-discipline. More specifically, the multiple sclerotic should be aware of the dangers of depending too greatly on his family, desirable though this course may seem to be. Such dependence may aggravate the sufferer's condition; and in any case, the dependent patient rarely receives enough sympathy to satisfy him, and he becomes more and more of a burden to himself and his family. Only by maintaining a sure independence, within medical limits set by a physician, can the patient live a full creative life."[99]

Similarly, in 1956 *Coronet* magazine prefaced an article by Jane Sterling, who had MS, with: "This is the philosophy of a dauntless woman who found the power within herself to conquer her crippling malady."[100] Sterling taught that "self-discipline is of utmost importance if one is to cope with the disease successfully."[101]

Not only could persons with MS transform themselves through individual effort, but they could experience healing through fellowship with others who had the disease. In 1948 the NMSS framed the plight of a person with MS this way: "The Multiple Sclerosis victim has been peculiarly isolated . . . in the Biblical phrase, 'a man sitting in darkness.'"[102] To overcome this isolation the NMSS encouraged people with MS to engage in collective self-help. By 1953 the NMSS reported that many local chapters had "established patients' clubs and some patients' clubs have been founded locally without affiliation with any organization. Such clubs, apart from the therapeutic facilities they may afford, provide a useful social outlet for multiple sclerosis patients; they afford the ease of friendship and the common, sympathetic understanding of those who share the same difficulties. Meeting together, multiple sclerosis patients also have the opportunity of sharing information about hobbies and business or employment opportunities."[103]

Other collective experiences could be found in clinics, funded by the NMSS, devoted to MS research and treatment in key cities around the country. In 1948 the NMSS funded MS clinics in Boston at Beth Israel Hospital, Boston State Hospital, and Massachusetts General Hospital. Also in 1948 the NMSS funded MS clinics at Tulane University School of Medicine in New Orleans, Cedars of Lebanon Hospital in Los Angeles, the New York University School of Medicine and Montefiore Hospital in New York City, and the Albany Hospital in Albany, New York.[104] In 1954 the NMSS and its Washington chapter sponsored a new clinic at George Washington University Hospital in Washington, D.C.[105] The stated purpose of these clinics was to do research. However, these clinics, along with the patient clubs, also served as sites around which a movement culture could develop and sites where ongoing fund-raising activities could occur. As one lay director of the Washington, D.C., chapter of the NMSS put it in 1953: "Something concrete must be offered to the patients. . . . The chapters were authorized to keep 60% of their funds in the local area . . . this was a wise plan because without an information center, a clinic and efforts at rehabilitation, it would not be possible to raise any money for research." Pearce Bailey concurred, saying that "unless a clinic is set up the Chapter will not survive. Patients need a program." One physician commented that the clinics acted "as psycho-therapy. The patients are happier if they have somewhere to go."[106]

In addition to metaphors of self-transformation and collective self-help, people with MS and lay activists also deployed military metaphors in framing MS in popular journals. One person with MS wrote in 1950 that "muscle reeducation like Pavaroff's is a calculated campaign, plotted as precisely as any of the world's great battles."[107] Another described his disease in 1954 as a "phantom sniper" and that "mystery" was its "Iron Curtain."[108] *Newsweek* editor Raymond Moley, in "Weapons against a Pitiless Enemy," compared MS to a "guerrilla" attack: "MS attacks here while it retreats there."[109] Although it might be tempting to attribute this military framing of MS as due to World War II or the Cold War, military metaphors were not an unusual formulation for diseases; nor were they unique to the 1940s and 1950s. Nevertheless, the military jargon was an important element of the popular construction of the disease and the campaign against it. The martial language also helped to create a sense of mission in the popular crusade.

Healing, then, as the NMSS represented it, meant participating in a crusade against the disease itself—in other words, struggling against disability and death in a movement culture animated mainly by the American myth of self-transformation and collective self-help. Moreover, in terms of an ultimate cure, the participation of those who were ill was key as well.

## The Second Front: The NMSS and the NINDB

The second major push of the NMSS in the late 1940s and 1950s consisted of lobbying the federal government to fund research on MS. The initial stages of this campaign show the extent to which the lobbying effort was embedded in a new postwar cultural consensus concerning the relationship between disease, medical research, and the federal government.[110] The "wartime parade of miracle drugs," especially penicillin, led American society to call "for more and more medical research," for which it was prepared to "contribute handsomely," as medical workers at the time knew.[111] Cornelius H. Traeger, M.D., expressed the new faith in science when in 1949 he preached that "if you get enough people and give them enough money you will get an atomic bomb. If you get enough people who are interested and have genius and give them the wherewithal you will get the answer."[112] As one ill person put it in 1954: "I know that, in this age of atomic energy, antibiotics and radioactive isotopes, a cure for my trouble is around the scientific corner."[113] This vaulting cultural faith in medical science led to a new consensus that, as Mayor Robert Wagner of New York City expressed it in 1956, "government must play its role, in sponsoring medical research."[114]

The federal government had been involved in medical research before World War II; however, the amount of money the government spent on research before 1942 and that spent after 1942 differed greatly.[115] In 1887 the Marine Hospital Service set up the Hygienic Laboratory to conduct bacteriological studies to help control epidemics. The Biologics Control Act of 1902 authorized the Hygienic Laboratory to test biologicals and to expand into zoology, chemistry, and pharmacology; yet its annual budget was less than $50,000. In 1912 Congress authorized the USPHS, formerly the Marine Hospital Service, to study chronic and infectious diseases.[116] Through the 1920s the USPHS budget was still only about $300,000 per year.[117] In 1930 Congress authorized the conversion of the Hygienic Laboratory of the USPHS into the National Institutes of Health and the spending of $750,000 for this purpose; however, in the 1930s Congress never appropriated full funding for the NIH because of the Great Depression. In 1937 Congress created the National Cancer Institute (NCI) and appropriated $400,000 for its first year; however, in 1945 the NCI budget was still only $500,000 per year.[118] The low NCI budget notwithstanding, the federal government vastly expanded medical research during World War II. The Committee on Medical Research in the Office of Scientific Research and Development (CMR/OSRD) spent $15 million during the war. The momentum for federal funding of medical research continued as the government closed the CMR/OSRD at the end of the war and transferred remaining funds to the NIH. The NIH budget quickly expanded; its research budget went from $180,000 in 1945 to $4 million in 1947. In 1945 Congress had appropriated $700,000 total for the NIH. By 1948 the government expanded NIH spending to $29 million, and by 1955 the NIH budget was $98 million. By 1965 the figure had reached $436 million.[119]

The federal government established other research institutes through the 1940s with these moneys. By 1950 the NIH included the National Cancer

Institute, Experimental Biology and Medicine Institute, Microbiological Institute, National Institute on Mental Health, National Heart Institute, and the National Institute for Dental Research.[120] In this cultural and political context the NMSS lobbied the federal government to start a research institute for MS like the other dedicated institutes. The following comments show the arguments that the MS activists and their allies used to get the federal government involved in MS research, but they also show the extent to which a change had occurred in American culture—that is, disease and medical research were legitimate and necessary activities of the federal government. Solving the problems of disease had come to be seen as just another public works project for which many people wrote their congressional representatives.[121] For example, one ill person wrote in 1948: "Dear Senator Wherry: . . . in your position in the U.S. Senate, it occurred to me that you might see the opportunity of the establishment of laboratories or hospitals which could be devoted to the effort to establish the cause of this most baffling and insidious disease, about which at this time practically nothing is known."[122] Here is another example of how health projects came to be viewed like bridges, roads, dams, and other public works projects. In 1948 U.S. representative Mike Mansfield of Montana forwarded a letter from a constituent to Surgeon General Leonard Scheele. Mansfield wrote: "Herewith is letter which I have received from Miss Jane Sullivan relative to sclerosis. It is the intention of a group in Butte to carry on some research on this disease and they were wondering whether or not any research could be done at the Laboratory at Hamilton, Montana."[123]

In addition to grassroots letter writing, the NMSS activists initiated legislative action in the Senate. On May 10, 1949, a hearing was held before the Subcommittee on Health of the Committee on Labor and Public Welfare of the United States Senate to consider S 102, the National Multiple Sclerosis Act. Republican senator Charles Tobey of New Hampshire, whose daughter had MS, sponsored the bill. Tobey declared, "We cannot take this thing lying down. There is money enough in this country to take care of this job. When we spend $5,000,000,000 for the Marshall plan—which I voted for—and when we spent $12,000,000,000 every 30 days in World War 2 to kill men and destroy capital property forever, we cannot for a moment sit back idly and say, we cannot appropriate whatever millions are necessary to find the cause, and the research to look into this hellish disease and to give men courage and faith to restore these things. We may not be successful, but God will hold us responsible unless we try to do something for them."[124] Not only had studying disease become a legitimate concern of the federal government, but it had become a moral imperative.

Ralph I. Straus, president of the NMSS, stated succinctly the postwar acceptance of the significantly increased role of the federal government in medical research: "It seems to me that consideration by a committee of the United States Senate of legislation to combat multiple sclerosis is an important social achievement. It indicates a general acceptance of the fact that disease is everybody's business, and that which is everybody's business is the business of government."[125]

Cornelius H. Traeger argued: "Now, as to the need for funds, they are first of all needed for basic research. Basic research means picking apart the little building blocks which make up the human organism. The men who are qualified to do that research are available; talent, genius, and interest are here. What we need is money to pay these people to do the job."[126] Tracy Putnam concurred: "It is my personal belief that the time has come when we must turn to the State and Federal Governments for aid in the struggle against a disease such as multiple sclerosis, which carries such tremendous misery with it and economic loss."[127]

One Illinois citizen echoed the doctors' sentiments in a letter to Senator Tobey: "It is my understanding there is a bill before Congress to furnish more aid to such institutions (Kaiser Kabat Foundation in Vallejo). I am a die-hard Republican and am against all forms of Government subsidy, but when I know that only 10 percent can afford this treatment and that 90 percent are dying a slow and sure death then I shall be happy to revamp my opinion."[128]

In these hearings and elsewhere, MS activists claimed that the NIH in particular and the medical profession in general had ignored their disease even though the pace of federal funding for medical research was increasing. In 1946 The New York Times hinted at negligence on the part of medical workers: "This ignorance of a disease which is a social problem is no credit to science. . . . The Association [AARMS] has engaged in work which should have been undertaken systematically long ago, and which deserves all the philanthropic support that it can enlist."[129]

Testimony at the 1949 Senate hearing demonstrated this sentiment of neglect as well: Senator Tobey complained, "We might as well be candid about it, practically nothing of medical value is presently known about the cause, control, or effective treatment of multiple sclerosis. For all intents and purposes our knowledge concerning the disease is practically the same it was 80 years ago." Eleanor Gehrig agreed: "It is a tragic fact that my testimony on this subject is almost as acceptable as that of any doctor in the land. This is not an indication of my erudition. It is an indication of how little is known concerning multiple sclerosis—even by the doctors who are most interested in it." Straus concurred: "In the 80 years that have elapsed since the eminent French neurologist, Jean Martin Charcot, first identified the disease now known as multiple sclerosis, little, if any important progress has been made in the field concerning its cause and effective treatment."[130]

These charges were not accurate, as neuroscientist Tracy Putnam pointed out in the 1949 Senate hearing. In fact, American neurologists had been investigating MS since the late nineteenth century and had explored numerous hypotheses concerning its etiology. The Commonwealth Fund financed MS studies at the New York Neurological Institute from the 1920s to the 1940s. Physicians working for the military during World War I had conducted the first proto-epidemiological study of MS in the United States based on the famous study "Defects in Drafted Men."[131] What *was* accurate was that the U.S. Public Health Service and the NIH had virtually ignored MS and, in fact, had mostly ignored neurological questions altogether.

In 1947 the study sections of the USPHS research grants programs included antibiotics and bacteriology, biochemistry and nutrition, cardiovascular, dental, gerontology, hematology, malaria, metabolism and endocrinology, pathology, pharmacology, physiology, public health methods, radiobiology, sanitation, surgery, syphilis, tropical diseases, tuberculosis, and virus and rickettsial diseases.[132] There was no neurology study section until 1953.[133]

In a 1948 letter to Cornelius Traeger of the NMSS, David E. Price, chief of the Division of Research Grants and Fellowships within the USPHS, discussed the USPHS's lack of interest in multiple sclerosis research: "The Public Health Service has received very few applications for research projects relating to the demyelinating diseases. I believe I am correct in saying that to date we have not received any requests for grants that relate directly to multiple sclerosis."[134] However, the USPHS through the National Institute of Mental Health had begun an epidemiological study of MS in 1947 in cooperation with the NMSS.[135]

Despite these pleas and the fact that the NIH was basically not much involved in neurological research, the NIH opposed the formation of a separate institute for MS because of the administrative burden of a separate institute for one particular disease. It also opposed Tobey's bill because it would set a bad precedent if every disease required its own institute. As a result the NMSS changed its tactics and lobbied for the founding of a neurological institute with funding for MS research.[136]

These lobbying efforts of the MS activists succeeded. On August 15, 1950, President Harry Truman signed Public Law 692, which authorized the surgeon general of the USPHS to set up the National Institute for Neurological Diseases and Blindness and the National Institute for Rheumatism and Metabolic Diseases. NIH officials began work to organize the neurological institute; however, Congress did not appropriate funds for it until 1952.[137]

With the founding of the NINDB, the NIH's interest in MS increased significantly.[138] Altogether, by August 1955 the NMSS had persuaded the federal government to spend $1,035,166 on MS research.[139] These dollars represented NINDB support of seventeen research projects in 1954. I have incomplete data for 1955, but by August 1955 the NINDB was supporting eighteen projects related to MS for the year.[140] By 1955 the NMSS had spent $1,355,642 on research, which meant that between 1946 and 1955 the two organizations had pumped $2,390,808 into MS research in the United States.[141]

The NINDB's mandate was to study epilepsy, cerebral palsy, MS, blindness, and other neurological diseases. However, the persons with MS seemed to have been the most active partisans of a particular disease in lobbying the NINDB.[142] The NMSS had incorporated amyotrophic lateral sclerosis (ALS) within its purview, which meant that the scleroses together accounted for 55 percent of all requests to the Clinical Center during this time period. In 1952 the NINDB spent as a percentage of the total money for research, for the two largest categories, 11.9 percent for epilepsy and 11.5 percent for MS.[143] This represented a significant accomplishment for the NMSS activists, since there were probably somewhere between eight and ten times more people with epilepsy than with MS in the United States.

Moreover, the NMSS and the NINDB acted as virtually one organization in the 1950s. In their relationship there was a blurry boundary between public and private spheres. They operated with interlocking directorates. They engaged in joint planning and budgeting and worked out an explicit division of labor.[144] They also created a single structure ruling neurological research on MS in the United States that disciplined the boundaries of legitimate research on the disease.[145] The important point for this essay is that the NMSS predated the NINDB and played a significant role in the latter's creation and in sustaining the institute's interest in MS. Once the NINDB became operational in 1952, the NMSS continued to exert a powerful influence on the NINDB.

This was not a unique situation. The American Cancer Society and the National Cancer Institute also had interlocking directorates in the 1950s. They had an explicit division of labor and engaged in close cooperation. As the American Cancer Society noted in 1955, "Over the years the Society has consistently urged Congress to increase appropriations for the research and educational programs of the National Cancer Institute, U.S. Public Health Service. So that efforts may be coordinated, liaison is continuous. Represented on the Society's Board of Directors are officials of the Institute and members of the Board regularly served on the Institute's National Advisory Council."[146]

Medical historians have noted the importance of the intervention of the federal government in medical research after 1945. They have neglected the critical and powerful role the national voluntary health associations had in that expansion and the voluntary associations' powerful impact on the experience of health, medicine, and disease in the twentieth-century United States.

### NOTES

1. Richard Harrison Shryock, *National Tuberculosis Association: A Study of the Voluntary Health Movement in the United States* (New York: National Tuberculosis Association, 1904). The National Tuberculosis Association became the American Lung Association in 1973.

2. Walter S. Ross, *Crusade: The Official History of the American Cancer Society* (New York: Arbor House, 1987). James T. Patterson, *The Dread Disease: Cancer and Modern American Culture* (Cambridge: Harvard University Press, 1987). William W. Moore, *Fighting for Life: The Story of the American Heart Association, 1911–1975* (Dallas: American Heart Association, 1983). Richard Carter, *The Gentle Legions: National Voluntary Health Organizations in America,* with a new introduction by Virginia A. Hodgkinson (New Brunswick, N.J.: Transaction Publishers, 1992), originally published in 1961.

3. To see how the presence of a powerful voluntary organization or its absence can affect the science of a disease and the illness experience of the person with the disease in the different cultures of the United States and France, see Howard I. Kushner, *A Cursing Brain? The Histories of Tourette Syndrome* (Cambridge: Harvard University Press, 1999).

4. Ross, *Crusade,* xiii. This was not a new observation among health workers. Robert Hamlin, M.D., M.P.H., noted in 1961 that "the well-nigh universal habit of giving both time and money to worthwhile causes is uniquely American" (Robert Hamlin, M.D., M.P.H., *Voluntary Health and Welfare Agencies in the United States* [New York: Schoolmasters' Press, 1961], v). See also Robert Hamlin, National Tuberculosis Association, *Bulletin* 47, no. 7 (July 1961): 4–7. Cherry Tsutsamida of the United States Public Health Service noted in 1967, "De Tocqueville, the famous French philosopher, said that one of the unique ways that the American public solves its problems is through the voluntary organizing, the voluntary getting together of people. And when we say we organize people, this does not mean that we hierarchically set them up on an up and down basis like a military chain of command. Rather, organizing means that channels of communications are established that work along crosswise, where many, many people can come together and establish some type of dialogue. In public health, we have many, many voluntary agencies which are getting together to cooperate in a way in which the public health problems can be solved in a progress more representa-

tive of the ways and mores of the American public" (*Critical Health Problems: Proceedings of the New York State Conference, Albany, NY November 29, 1967* [Albany: University of the State of New York and the New York State Department of Education, 1967], 21–22).

5. American Cancer Society, Annual Report, 1956. Ross, *Crusade*.

6. Hamlin, *Voluntary Health and Welfare Agencies*, 84–87.

7. John I. Cook, "Multiple Cerebro-Spinal Sclerosis," *Richmond and Louisville Medical Journal* 14 (1872): 76. George S. Gerhard, "Cases of Multilocular Cerebro-Spinal Sclerosis," *Philadelphia Medical Times* 7 (Nov. 11, 1876): 49–52. W. M. Butler, "Disseminated Sclerosis with Case," *Hahnemanian Monthly* 25 (1890): 148. Archibald Church and Frederick Peterson, *Nervous and Mental Diseases* (Philadelphia: W. B. Saunders, 1899), 433. Charles L. Dana, *Text-Book of Nervous Diseases Being a Compendium for the Use of Students and Practitioners of Medicine* (New York: William Wood and Co., 1892), 374. Bernard Sachs, "On Multiple Sclerosis, with Especial Reference to Its Clinical Symptoms, Its Etiology and Pathology," *Journal of Nervous and Mental Diseases* 25 (1898): 314. Charles L. Dana, "Discussion on the Absolute and Relative Frequency of Multiple Sclerosis," *Journal of Nervous and Mental Diseases* 2 (1902): 288, 290. William G. Spiller, "A Report of Two Cases of Multiple Sclerosis with Necropsy," *American Journal of Medical Sciences* 125 (1903): 61. Leo M. Crafts, "The Early Recognition of Multiple Sclerosis," *Journal of the American Medical Association* 69 (1917): 1130.

8. Israel Weschler, "Statistics of Multiple Sclerosis," in Association for Research in Nervous and Mental Disease, *Multiple Sclerosis [Disseminated Sclerosis]* (New York: Paul B. Hoeber, 1922), 2:31. E. W. Taylor, "Multiple Sclerosis: The Location of Lesions with Respect to Symptoms," *Archives of Neurology and Psychiatry* 7 (1922): 561–62. Albert G. Odell, "The Signs and Symptoms of Multiple Sclerosis with Particular Reference to Early Manifestations," *New York State Medical Journal* 31 (1931): 1018. Paul De Nicola, "Diagnosis and Treatment of Multiple Sclerosis," *New England Journal of Medicine* 209 (1933): 837. Milton Lozoff, "Multiple Sclerosis" (medical thesis, University of Wisconsin, 1938), 9. Sidney D. Wilgus and Egbert W. Felix, "Priapism as an Early Symptom in Multiple Sclerosis," *Archives of Neurology and Psychiatry* 25 (1931): 153–57. Hinton D. Jonez, "Diagnosis of Multiple Sclerosis," *Postgraduate Medicine* 14 (1953): 121. Tracy J. Putnam, "Sclerosis and Encephalomyelitis," *Bulletin of the New York Academy of Medicine* 19 (1943): 302.

9. Sources include *Quarterly Cumulative Index Medicus* (Chicago: American Medical Association, 1900–56); *Index-Catalogue of the Library of the Surgeon-General's Office* (Washington, D.C.: U.S. Dept. of Health, Education, and Welfare, Public Health Service, 1880–1961); *Current List of Medical Literature*, vols. 29–36 (Washington, D.C.: Armed Forces Medical Library, 1956–59).

10. Association for Research in Nervous and Mental Disease, *Multiple Sclerosis and the Demyelinating Diseases: Proceedings of the Association, December 10 and 11, 1948, New York* (Baltimore: Williams and Wilkins, 1950). National Multiple Sclerosis Society [hereafter NMSS], *Light on a Medical Mystery* (New York: NMSS, 1948), 10. New York Academy of Medicine [hereafter NYAM], *The Status of Multiple Sclerosis*, Pearce Bailey,

M.D., Conference Chairman (New York: New York Academy of Sciences, 1954). *Annals of the New York Academy of Sciences* 58 (1954).

11. William F. Windle, ed., *New Research Techniques of Neuroanatomy: A Symposium Sponsored by the National Multiple Sclerosis Society,* foreword by Frederick L. Stone (Springfield, Ill.: Thomas, 1957).

12. Augustus S. Rose, M.D., and Carl M. Pearson, M.D., eds., *Mechanisms of Demyelination* (New York: McGraw-Hill, 1963).

13. NMSS, *Light on a Medical Mystery,* 7–10. "A Challenge to Medicine," *New York Times* [hereafter *NYT*], May 3, 1948, 20. "Multiple Sclerosis," *NYT,* Mar. 8, 1953), IV:8. "Study Grant Made on Nerve Disease," *NYT,* Mar. 11, 1953, 60. Raymond Moley, "The Fight against MS," *Newsweek,* Apr. 13, 1953, 116. Murray Illson, "New Booklet Out on Sclerosis Care," *NYT,* Nov. 15, 1953, 37. Typescript, "Fact Sheet National Multiple Sclerosis Society," attached to letter, Ralph C. Glock, President, NMSS, NYC [hereafter NYC], to Pearce Bailey, M.D., National Institute for Neurological Diseases and Blindness [hereafter NINDB], January 7, 1955, National Archives and Record Administration, National Institutes of Health Records [hereafter NARA/NIH], Record Group [hereafter RG] 443, ser. 47, box 2, folder "Assoc. I NMSS, vol II, Jan 54–." "$102,000 Medical Gift," *NYT,* Apr. 20, 1955, 26. "Sclerosis Report In," *NYT,* Oct. 13, 1958, 31.

14. Information Resource Center, NMSS, NYC, to Colin Talley, San Francisco, Dec. 31, 1997, in possession of author.

15. Sylvia Lawry, "Fighting 'M.S.,'" *Today's Health* 33 (Jan. 1955): 13.

16. George A. Schumacher, "Foreword: Symposium on Multiple Sclerosis and Demyelinating Diseases," *American Journal of Medicine* 12 (1952): 499–500.

17. Cornelius H. Traeger, M.D., Medical Director, NMSS, NYC, to Members of the Medical Advisory Board, Dec. 15, 1952, NARA/NIH, RG 443, ser. 47, box 2, folder "NMSS vol. I."

18. W. H. Sebrell Jr., M.D., Director, NIH, Bethesda, Md., to Senator Thomas A. Burke, Washington, D.C., July 22, 1954, NARA/NIH, RG 443, ser. 47, box 15, folder "W."

19. Pearce Bailey, M.D., Director, NINDB, Bethesda, Md., to Sylvia Sokal, Secretary, Kings County Chapter, NMSS, Brooklyn, N.Y., Oct. 22, 1954, NARA/NIH, RG 443, entry 47, box 11, folder "9—Speeches, Lectures, and Statements."

20. *Readers' Guide to Periodical Literature* (Minneapolis, Minn.: H. W. Wilson, 1901–19). *New York Times Index* (New York: New York Times Co., 1930–59).

21. Infection was one of the many etiological theories considered at the time, though not one of the most important or interesting ones, according to the leading neurologists of the day.

22. Jane E. Brody, "New Leads in the Multiple Sclerosis Fight," *NYT,* May 3, 1955, B9. Sylvia Lawry, founder of the NMSS, interview by author, Apr. 29, 1994, NYC, notes in possession of author. Sylvia Lawry, interview by author, Nov. 10, 1994, San Francisco, tape in possession of author. "Study Grant Made on Nerve Disease," 3. NMSS, *Light on a Medical Mystery.*

23. Lawry interview, Nov. 10, 1994. "Sclerosis Society Opens Fund Drive," *NYT,* May 11, 1958, 57.

24. Moley, "Fight against MS," 116.

25. Data on the number of members of the NMSS come from the following: NMSS, *Light on a Medical Mystery,* 7–8. Paul de Kruif, "The Patient Is the Hero," *Reader's Digest* 52 (May 1948): 75. "Study Grant Made on Nerve Disease," 60. NMSS, *Self Help* (Bethesda, Md.: NMSS, NINDB, 1953), NYAM. Illson, "New Booklet Out on Sclerosis Care," 1. Robert Grant Jr., "I've Got the Most Mysterious Disease," *Saturday Evening Post,* May 22, 1954, 122. "Sclerosis Society Opens Fund Drive," 5.

26. "Organization for Multiple Sclerosis Formed," *American Journal of Public Health* 36 (1946): 1357.

27. Ibid., 1357. Grant, "I've Got the Most Mysterious Disease," 26–27, 121–26.

28. Lawry interview, Nov. 10, 1994. "The Mystery of Sclerosis," *Newsweek,* Oct. 14, 1946, 79. "Organization for Multiple Sclerosis Formed," 1357. NMSS, *Light on a Medical Mystery,* 7. Grant, "I've Got the Most Mysterious Disease," 26–27, 121–26.

29. "Organization for Multiple Sclerosis Formed," 1357.

30. Patterson, *Dread Disease,* 172–200.

31. Lawry interview, Nov. 10, 1994. De Kruif, "Patient Is the Hero," 71–75. U.S. Congress, Senate, Subcommittee on Health of the Committee on Labor and Public Welfare of the United States Senate, National Multiple Sclerosis Act, 81st Cong., 1st sess., May 10, 1949, 1–9 [hereafter National Multiple Sclerosis Act].

32. For the history of neurology in the United States, see Bonnie Ellen Blustein, "New York Neurologists and the Specialization of American Medicine," *Bulletin of the History of Medicine* 53 (1979): 170–83. Bonnie Ellen Blustein, "Percival Bailey and Neurology at the University of Chicago, 1928–1939," *Bulletin of the History of Medicine* 66 (1992): 90–113. Bonnie Ellen Blustein, *Preserve Your Love for Science: Life of William A. Hammond, American Neurologist* (New York: Cambridge University Press, 1991).

33. Patterson, *Dread Disease,* 171. Paul Starr, *The Social Transformation of American Medicine* (New York: Basic Books, 1982), 346. Naomi Rogers, *Dirt and Disease: Polio before FDR* (New Brunswick, N.J.: Rutgers University Press, 1992). Margaret L. Grimshaw, "Scientific Specialization and the Poliovirus Controversy in the Years before WW2," *Bulletin of the History of Medicine* 69 (1995): 44–65.

34. Association for the Advancement of Research into Multiple Sclerosis [hereafter AARMS], *Join AARMS* (New York: AARMS, 1946), NYAM.

35. Howard Rusk, M.D., "Incurable Multiple Sclerosis," *American Mercury* 65 (Oct. 1947): 450. See also Howard A. Rusk, M.D., "Research Seeks Way to Curb Common Crippling Disease," *NYT,* Apr. 20, 1947, 53. "Challenge to Medicine," 3.

36. Raymond Moley, "Weapons against a Pitiless Enemy," *Newsweek,* May 3, 1954, 100. Lawry interview, Nov. 10, 1994.

37. Alice Friedman, Director, Public Relations, NMSS, NYC, to Harold Tager Jr., Information Officer, NINDB, Bethesda, Md., Mar. 20, 1953, NARA/NIH, RG 443, ser. 47, box 2, folder "NMSS, vol. I." Pearce Bailey, M.D., NINDB, Bethesda, Md., to Director of NIH, memorandum, Oct. 12, 1954, NARA/NIH, RG 443, ser. 47, box 20, folder "Research 3-2-3 MS Isoniazid Project." Lawry interview, Nov. 10, 1994. Examples of these articles include "Mystery Crippler," *Time,* Oct. 14, 1946, 51. W. K., "Research

Begun on Multiple Sclerosis," *NYT*, Oct. 6, 1946, IV:9. "Mystery of Sclerosis," *Newsweek*, Oct. 14,1946, 79. "Organization to Help Victims of Nerve Disease," *Science Newsletter*, Oct. 26,1946, 260. "Sclerosis Malady Subject of Study," *NYT*, Feb. 22, 1947, 3. Rusk, "Research Seeks Way to Curb Common Crippling Disease," 4. "Frontier of Medicine," *Survey* 83 (May 1947): 83. Rusk, "Incurable Multiple Sclerosis," 450–54. "Fund Asked to Fight Multiple Sclerosis," *NYT*, Feb. 10, 1948, 14. "The Menace of Sclerosis," *Newsweek*, Dec. 20, 1948, 48. "Our No. 1 Neurological Problem," *NYT*, Dec. 12, 1948, IV:9. "Mrs. Gehrig Backs Sclerosis Aid Bill," *NYT*, May 11, 1949, 31. Howard A. Rusk, "Victims of Multiple Sclerosis Aided in New Clinic Services," *NYT*, Apr. 2, 1950, 83. "Sentence Commuted," *Today's Health* 28 (May 1950): 16–17, 66. Jean Griffith Benge, "I Escaped a Wheelchair," *Today's Health* 28 (Oct. 1950): 20–21, 63–65. "Doctor Offers Hope for Sclerosis Cases," *NYT*, Feb. 20, 1952, 27. "Multiple Sclerosis: What Is it?," *Science Digest* 31 (May 1952): 41–42. "Raise Blood Pressure to Help MS Patients," *Science Newsletter*, May 9, 1953, 293. "Restore Brain Chemical Process in MS Patients," *Science Newsletter*, Dec. 26, 1953, 402. "Sclerosis Fund Drive Opens," *NYT*, Apr. 21, 1954, 23. L. Galton, "New Advances against Multiple Sclerosis," *Cosmopolitan* 136 (June 1954): 16. "R. W. Sarnoff to Head Drive," *NYT*, Feb. 12, 1955, 13. "$2,500,000 Is Sought by Sclerosis Group," *NYT*, Feb. 10, 1956, 4. "Diet vs. a Crippler," *Newsweek*, Oct. 1, 1956, 86. Russell N. DeJong, M.D., "Multiple Sclerosis," *Today's Health* 34 (Dec. 1956): 26–28, 52. Jane Sterling, "Today Is What Counts," *Coronet* 41 (Dec. 1956): 64–68. "Defeatism Hampers Multiple Sclerosis Fight," *Science Newsletter*, Mar. 16, 1957, 168. "Admiral Heads Sclerosis Drive," *NYT*, Apr. 25, 1957, 27. "Mrs. Eisenhower Heads Drive," *NYT*, Apr. 29, 1957, 14. "Ray of Hope?," *Newsweek*, June 17, 1957, 99. "Business Man Will Head Sclerosis Drive," *NYT*, Nov. 7, 1958, 19. T. L. William, "Multiple Sclerosis," *Parents* 34 (Apr. 1959): 72. "Joan Crawford Heads Drive," *NYT*, May 10, 1959, 95. "Radiation Threat to Brain," *Business Week*, Sept. 17, 1960, 83–85.

38. Rusk was also chairman of the Department of Physical Medicine of the Bellevue Medical Center; "Doctor Offers Hope for Sclerosis Cases," 1.

39. Sylvia Lawry, Founder/Director, NMSS, NYC, to Dr. Oliver E. Buckley, NMSS, NYC, memorandum, Sept. 8, 1952, NARA/NIH, RG 443, ser. 47, box 2, folder "NMSS vol. I."

40. De Kruif, "Patient Is the Hero," 71–75. National Multiple Sclerosis Act, 7.

41. National Multiple Sclerosis Act, 7.

42. "Mrs. Gehrig Backs Sclerosis Aid Bill," 4.

43. "Sclerosis Citations Go to Straus, Owen," *NYT*, Dec. 6, 1950, 40. "Doctor Offers Hope for Sclerosis Cases," 1.

44. "Elected as the President of the Sclerosis Society," *NYT*, Apr. 24, 1953, 18.

45. Grant, "I've Got the Most Mysterious Disease," 122.

46. "Mrs. Eisenhower's Plea," *NYT*, Apr. 23, 1954, 29. "Mrs. Eisenhower Heads Drive," 3. Ralph C. Glock, President, NMSS, NYC, to Mrs. Dwight D. Eisenhower, Washington, D.C., Dec. 24, 1957, NARA/NIH, RG 443, ser. 48, box 15, folder "Organizations and Conferences NMSS, 1956–61." Homer D. Babbidge Jr., Assistant to the Secretary, Department of Health, Education, and Welfare, Washington, D.C., to Mary

Jane McCaffree, memorandum, Jan. 9, 1958, NARA/NIH, RG 443, ser. 48, box 15, folder "Organizations and Conferences NMSS, 1956-61."

47. "Heads National Sclerosis Unit," *NYT,* June 30,1954, 2.

48. "R. W. Sarnoff to Head Drive," 2.

49. "$2,500,000 Is Sought by Sclerosis Group," 6.

50. "Admiral Heads Sclerosis Drive," 5.

51. "Senator Heads Sclerosis Drive," *NYT,* Apr. 28, 1958, 24.

52. "Business Man Will Head Sclerosis Drive," 4.

53. "Joan Crawford Heads Drive," 4.

54. Marshall Hornblower, Chairman, NMSS, NYC, to Dr. Leonard Scheele, Surgeon General, Washington, D.C., Feb. 21, 1953, NARA/NIH, RG 443, ser. 47, box 1, folder "NMSS Washington D.C. Chapter."

55. *MS Newsletter,* Washington Multiple Sclerosis Society, Mar. 1953, NARA/NIH, RG 443, ser. 47, box 1, folder "NMSS Washington D.C. Chapter."

56. "Multiple Sclerosis Week Set," *NYT,* Apr. 3, 1953, 9.

57. "Sclerosis Fund Drive Opens," 4.

58. Marshall Hornblower, "Fight against Multiple Sclerosis," *NYT,* May 13, 1948, 24.

59. Marshall Hornblower, Chairman, NMSS, NYC, to Dr. Leonard Scheele, Surgeon General, Washington, D.C., Feb. 21, 1953, NARA/NIH, RG 443, ser. 47, box 1, folder "NMSS Washington D.C. Chapter."

60. Patterson, *Dread Disease,* 171. Starr, *Social Transformation of American Medicine,* 346. Rogers, *Dirt and Disease.* Grimshaw. "Scientific Specialization and the Poliovirus Controversy in the Years before WW2," 44-65.

61. "Mystery Crippler," 51.

62. "Organization to Help Victims of Nerve Disease," 260.

63. "Organization for Multiple Sclerosis Formed," 1357.

64. Putnam quoted in "Sclerosis Malady Subject of Study," 8.

65. Rusk,"Research Seeks Way to Curb Common Crippling Disease," 4.

66. Rusk, "Incurable Multiple Sclerosis," 452.

67. "Fund Asked to Fight Multiple Sclerosis," 2.

68. National Multiple Sclerosis Act, 12.

69. Benge, "I Escaped a Wheelchair," 20-21, 63-65.

70. Grant, "I've Got the Most Mysterious Disease," 122.

71. Alice Friedman, Director of Public Relations, NMSS, NYC, to Harold Tager Jr., Information Officer, NINDB, Bethesda, Md., July 8, 1953, NARA/NIH, RG 443, ser. 47, box 6, folder "6 Morbidity and Mortality (includes reports) (alpha by disease.)"

72. Harold Tager Jr., Information Officer, NINDB, Bethesda, Md., to Alice Friedman, Director of Public Relations, NMSS, NYC, July 8, 1953, NARA/NIH, RG 443, ser. 47, box 6, "Folder "6 Morbidity and Mortality (includes reports) (alpha by disease.)"

73. "For Multiple Sclerosis Research," *NYT,* May 25, 1958, IV:11.

74. "Multiple Sclerosis: What Is it?," 41-42.

75. Benge, "I Escaped a Wheelchair," 20-21.

76. "Multiple Sclerosis: What Is it?," 41-42. "Mysteries of Multiple Sclerosis," *NYT,*

Dec. 14, 1952, IV:9. "Drugs Are Found to Ease Sclerosis," *NYT*, Mar. 3, 1953, 24. Moley, "Fight against MS," 116. "Raise Blood Pressure to Help MS Patients," 293. Illson, "New Booklet Out on Sclerosis Care," 1. "Source of the Crippler?," *Newsweek*, Jan. 4, 1954, 37. "Many Scars," *NYT*, Apr. 25, 1954, 10. Moley, "Weapons against a Pitiless Enemy," 100. Lawry, "Fighting 'M.S.,'" 13. DeJong, "Multiple Sclerosis," 26–28, 52. "MS and Spirochete," *Time*, June 24, 1957, 82.

77. "Victory in a Wheel Chair," *Look*, May 18, 1954, 34.

78. "Sclerosis Society Gets $100,000 for Research in Anonymous Gifts," *NYT*, Oct. 13, 1956, 21.

79. "Woman Isolates an Organism as Cause of Multiple Sclerosis," *NYT*, June 8, 1957, 1:5, 20.

80. "Ray of Hope?," 99.

81. Charles C. Limburg, "The Geographic Distribution of Multiple Sclerosis and Its Estimated Prevalence in the United States," in Association for Research in Nervous and Mental Disease, *Multiple Sclerosis and the Demyelinating Diseases*, 15–24.

82. Leonard T. Kurland and Knut B. Westlund, "Epidemiologic Factors in the Prognosis of Multiple Sclerosis," in *The Status of Multiple Sclerosis*, ed. Roy Waldo Miner, *Annals of the New York Academy of Sciences* 58 (1954): 682–701. Leonard T. Kurland, "The Frequency and Geographic Distribution of Multiple Sclerosis as Indicated by Mortality Statistics and Morbidity Surveys in the United States and Canada" (Ph.D. diss., Johns Hopkins University School of Hygiene and Public Health, 1951), in U.S. Congress, House of Representatives, Subcommittee of the Committee on Veterans' Affairs, *Three-Year Presumption of Service Connection for Multiple Sclerosis*, 82nd Cong., 1st sess., Mar. 20, 1951.

83. See O. E. Buckley, introduction to Miner, *Status of Multiple Sclerosis*, i, and Leonard T. Kurland and Knut B. Westlund, "Epidemiologic Factors in the Etiology and Prognosis of Multiple Sclerosis," in ibid., 692. Leonard T. Kurland, "The Frequency and Geographic Distribution of Multiple Sclerosis as Indicated by Mortality Statistics in the United States and Canada," *American Journal Hygiene* 55 (1952): 457–76.

84. Richard Lechtenberg, M.D., *Multiple Sclerosis Fact Book*, 2nd ed. (Philadelphia: F. A. Davis Co., 1995), 7. Lawrence Steinman, "Autoimmune Disease," *Scientific American* (Sept. 1993): 108–9.

85. Kolb continued: "This computation was given to the Society by me as a guess some years back when the Association for Research in Nervous and Mental Diseases had their meeting on multiple sclerosis in New York City. The guess was made on the basis of the studies already conducted and Dr. Lindbergh's statistical investigations. At that time I was the chairman of the Statistical Committee for the Society. . . . I can assure you that the use of figures in such a manner is really detrimental to the Society. There has also already been an attack upon your figures in an article, which appeared in *Harper's Magazine* in January of this year, dealing with medical statistics. The statement not only discredits the Society but also discredits the National Institute. It is my hope that in future issues of the annual report, or statements to the public, such a statement will not appear again. . . . There is no need to plead the cause of multiple

sclerosis on the basis of numbers. The longer term disability produced by the illness makes it manifest to me, at least, that it is the most significant neurological problem now existing." Lawrence C. Kolb, M.D., Mayo Clinic, Rochester, Minn., to Harold R. Wainerdi, M.D., Medical Director, NMSS, Apr. 24, 1953, NARA/NIH, RG 443, ser. 47, box 2, folder "NMSS, vol. I."

86. See Patient Examination Record [hereafter PER], Aug. 26, 1952, folder P31, box 5, Tracy Jackson Putnam [TJP], M.D., Collection, 1938–75, Louise Darling Biomedical Library, University of California, Los Angeles [hereafter TJP Collection]. PER, May 29, 1958, folder P65, box 9, ibid. PER, Sept. 4, 1947, folder P75, box 10, ibid. PER, July 17, 1962, folder P80, box 10, ibid. PER, May 11, 1949, folder P82, box 10, ibid. TJP, "Hospital Report," Dec. 15, 1961, folder P107, box 12, ibid.#27, M.D., Beverly Hills, Calif., to Dr. #28, Los Angeles, May 21, 1948, folder P34, box 5, ibid. TJP to #19, M.D., San Francisco, Jan. 24, 1956, folder P158, box 19, ibid. TJP to #17, M.D., Salt Lake City, Jan. 8, 1954, folder P162, box 19, ibid. See also "Physician's Notes," June 18, 1963, folder U246, University of California, Los Angeles, Hospital Records, Los Angeles [hereafter UCLA Records]. "Intern's Admission Note," Mar. 18, 1956, folder U 235, ibid. "Physician's Notes," Apr. 5, 1956, folder U255, ibid. "Physician's Notes," Mar. 27, 1962, folder U213, ibid.

87. "Physician's Notes," May 13, 1959, folder U234, UCLA Records.

88. For the American myth of self-transformation, see Howard I. Kushner, *American Suicide: A Psychocultural Exploration* (New Brunswick, N.J.: Rutgers University Press, 1991), 54, 184–85, 193, 199.

89. De Kruif, "Patient Is the Hero," 71.

90. Ibid., 72.

91. Ibid.

92. Ibid., 74.

93. "Sentence Commuted," 17.

94. Benge, "I Escaped a Wheelchair," 62.

95. Ibid., 65.

96. Joan McCarthy as told to Alma Morris, "My Victory over MS," *Cosmopolitan* 148 (Apr. 1960): 69.

97. Ibid., 66.

98. Ibid., 68.

99. NMSS, *Self Help.*

100. Sterling, "Today Is What Counts," 64.

101. Ibid., 68.

102. NMSS, *Light on a Medical Mystery,* 7.

103. NMSS, *Self Help,* 3. It should be noted that this is the representation of patient clubs that the NMSS promulgated. A detailed ethnography of the patients' illness experience and these clubs would be desirable; however, I have not been able to locate sources to make this possible.

104. NMSS, *Light on a Medical Mystery,* 12–14. Cornelius H. Traeger, M.D., *Analysis for the Layman of Research Projects Supported by the National Multiple Sclerosis Society* (New York: NMSS, 1949), 12, NYAM. "Sclerosis Clinic Will Open Today," *NYT,* Mar. 29,

1948, 23. Lawry, "Fighting 'M.S.,'" 13. "Medical Research Aided," *NYT*, Aug. 16, 1948, 21. Minutes of the Meeting of the Medical Advisory Board, Sept. 25, 1953, Washington, D.C., Chapter, NMSS, NARA/NIH, RG 443, ser. 47, box 1, folder "NMSS, Washington D.C. Chapter."

105. Pearce Bailey, M.D., NINDB, Bethesda, Md., to Director, NIH, memorandum, "Weekly Report," Feb. 19, 1954, NARA/NIH, RG 443, ser. 47, box 19, folder "2 Reports and Statistics (Weekly Reports)."

106. Minutes of the Meeting of the Medical Advisory Board, Sept. 25, 1953, Washington, D.C., Chapter.

107. Sentence Commuted," 16.

108. Grant, "I've Got the Most Mysterious Disease," 27.

109. Moley, "Weapons against a Pitiless Enemy," 100.

110. Leonard Scheele, M.D., Surgeon General, USPHS, NIH, National Advisory Neurological Diseases and Blindness Council, Minutes of Meeting, Feb. 23–Feb. 24, 1952, NARA/NIH, RG 443, ser. 12, box 27, folder "Committees 1–7 National Advisory Neurological Diseases and Blindness Council Minutes of Meeting 2/23 and 2/24, 1952." For the new postwar consensus, see David J. Rothman, *Strangers at the Bedside: A History of How Law and Bioethics Transformed Medical Decision Making* (New York: Basic Books, 1991), 51–53. Victoria A. Harden, *Inventing the NIH: Federal Biomedical Research Policy, 1887–1937* (Baltimore: Johns Hopkins University Press, 1986), 181–83. Starr, *Social Transformation of American Medicine*, 333, 343.

111. R. Keith Cannan, M.D., Vice Chairman, Division of Medical Sciences, National Research Council, copy of speech delivered to American Rheumatism Association, May 28, 1953, in Harold R. Wainerdi, M.D., Consulting Medical Director, NMSS, NYC, to Members of the Medical Advisory Board, NMSS, memorandum, June 11, 1953, NARA/NIH, RG 443, ser. 47, box 2, folder "NMSS, vol. I."

112. National Multiple Sclerosis Act, 39.

113. Grant, "I've Got the Most Mysterious Disease," 26.

114. "Sclerosis Society Marks Tenth Year," *NYT*, Oct. 16, 1956, 27. For the postwar hopes for medical science and the resulting research explosion, see Patterson, *Dread Disease*, ix, 171–72.

115. Rothman, *Strangers at the Bedside*, 51.

116. Starr, *Social Transformation of American Medicine*, 340.

117. Patterson, *Dread Disease*, 118.

118. Ibid., 131.

119. Rothman, *Strangers at the Bedside*, 53. Harden, *Inventing the NIH*, 183. Patterson, *Dread Disease*, 171–72.

120. Federal Security Agency, USPHS, NIH, Ad Hoc Committee on Research Fellows, held Sept. 26, 1949, 3, 9, NARA/NIH, RG 443, ser. 2, box 143, folder not named.

121. L. A., Spokane, Wash., to Senator W. G. Magnuson, Washington, D.C., Oct. 16, 1950, NARA/NIH, RG 443, ser. 1, box 1, folder "Congressional Mail October."

122. Katherine Cannell, Lincoln, Nebr., to Senator Kenneth S. Wherry, Washington, D.C., July 29, 1948, NARA/NIH, RG 443, ser. 1, box 13, folder "Multiple Sclerosis."

123. Rep. Mike Mansfield, Washington, D.C., to Dr. Leonard A. Scheele, Surgeon General, USPHS, Washington, D.C., Aug. 24, 1948, RG 443, ser. 1, box 13, folder "Multiple Sclerosis."

124. National Multiple Sclerosis Act, 8–9.

125. Ibid., 19.

126. Ibid., 39.

127. Ibid., 29.

128. Ibid., 35.

129. W. K., "Research Begun on Multiple Sclerosis," 6.

130. National Multiple Sclerosis Act.

131. Pearce Bailey, "Incidence of Multiple Sclerosis in United States Troops," *Archives of Neurology and Psychiatry* 7 (1922): 582.

132. Surgeon General, unsigned, to Major General Malcolm C. Grow, Air Surgeon, Headquarters, U.S. Air Force, Washington, D.C., Nov. 10, 1947, NARA/NIH, RG 443, ser. 2, box 142, folder "Research Grants," 1.

133. David E. Price, M.D., Chief, Division of Research Grants and Fellowships, Federal Security Agency, Public Health Service, NIH, to All Advisory Council and Study Section Members, memorandum, Apr. 1, 1949, NARA/NIH, RG 443, NIH, ser. 2, box 143, folder not named. See also Supplement 1, National Advisory Neurological Diseases and Blindness Council, June 22–23, 1953, 9–17, NARA/NIH, RG 443, NIH, NINDB Council Minutes of Meetings, Nov. 1950 to Nov. 1954, ser. 12, box 27, folder "Public Health Service, National Institutes of Health, National Advisory Neurological Diseases and Blindness Council, Minutes of Meeting June 22–23, 1953."

134. David E. Price, M.D., Chief, Division of Research Grants and Fellowships, Bethesda, Md., to C. H. Traeger, M.D., NMSS, NYC, Nov. 3, 1948, NARA/NIH, RG 443, ser. 2, box 142, folder "0745 Research Grants."

135. U.S. Congress, House of Representatives, Subcommittee of the Committee on Veterans' Affairs, *Three-Year Presumption of Service Connection for Multiple Sclerosis,* 82nd Cong., 1st sess., Mar. 20, 1951, 130–33.

136. U.S. Congress, *National Health Plan,* hearings before a subcommittee of the Committee on Interstate and Foreign Commerce, House of Representatives, 81st Cong., 1st sess., on HR 4312 and HR 4313 (identical bills) and HR 4918 and other identical bills, 611–15. Lawry interview, Nov. 10, 1994. H. Houston Merritt, "Tracy Jackson Putnam, 1894–1975," *Transactions of the American Neurological Association* 100 (1975): 272.

137. *Congressional Quarterly Almanac,* 81st Cong., 2nd sess., 1950, vol. 6 (Washington, D.C.: Congressional Quarterly News Features, 1950), 182–83. Surgeon General to Senator Scott W. Lucas, Washington, D.C., Sept. 15, 1950, NARA/NIH, RG 443, ser. 1, box 1, folder "Congressional Mail—Sept. 1950." Surgeon General, Washington, D.C., to Julian B. Snow, Arlington, Va., Dec. 20, 1950, NARA/NIH, RG 443, ser. 1, box 1, folder "Cong. Correspondence—Dec 1950."

138. The data for the table come from Pearce Bailey, M.D., NINDB, Bethesda, Md., to Harold R. Wainerdi, M.D., Acting Medical Director, NMSS, NYC, Feb. 20, 1953,

NARA/NIH, RG 443, ser. 47, box 7, folder "9 Diseases and Conditions Detailed Statistical Information All Disease." Pearce Bailey, M.D., Director, NINDB, Bethesda, Md., to Cornelius H. Traeger, M.D., Consultant Medical Director, NMSS, NYC, Aug. 24, 1954, NARA/NIH, RG 443, ser. 47, box 2, folder "Associations 1, NMSS, Vol II Jan 1954 through."

139. Pearce Bailey, M.D., to Harold R. Wainerdi, M.D., Feb. 20, 1953. Pearce Bailey, M.D., to Cornelius H. Traeger, M.D., Aug. 24, 1954.

140. Pearce Bailey, M.D., to Harold R. Wainerdi, M.D., Feb. 20, 1953. Pearce Bailey, M.D., to Cornelius H. Traeger, M.D., Aug. 24, 1954.

141. Information Resource Center, NMSS, NYC, to Colin Talley, San Francisco, Dec. 31, 1997, in possession of author. "$102,000 Medical Gift," 5.

142. "Requests for Admission to the Clinical Center, June 10, 1952 to July 10, 1953," typescript, 19 pp., NARA/NIH, RG 443, ser. 47, box 14, folder "1-Patients Admissions, Inquiries (General-(unnamed patients) Filed Date Order."

143. Leonard Scheele, M.D., Surgeon General, USPHS, NIH, National Advisory Neurological Diseases and Blindness Council, Minutes of Meeting, Feb. 23–24, 1952, NARA/NIH, RG 443, ser. 12, box 27, folder "Committees 1–7 National Advisory Neurological Diseases and Blindness Council Minutes of Meeting 2/23 and 2/24, 1952," 4.

144. Leonard A. Scheele, M.D., Surgeon General, to Harold R. Wainerdi, M.D., Associate Medical Director, NMSS, NYC, Feb. 9, 1953, RG 443, ser. 47, box 2, folder "NMSS vol. I." USPHS, NIH, National Advisory Neurological Diseases and Blindness Council, Minutes of Meeting, Nov. 16, 1950, NARA/NIH, RG 443, ser. 12, box 27, folder "Committees 1–7 National Advisory Neurological Diseases and Blindness Council Minutes of Meeting 11/16/1950." Harold R. Wainerdi, M.D., Medical Director, NMSS, NYC, to Pearce Bailey, M.D., NINDB, Apr. 8, 1955, NARA/NIH, RG 443, ser. 47, box 2, folder "Assoc. I. NMSS, vol II Jan. 54–." USPHS, National Advisory Neurological Diseases and Blindness Council, Minutes of Meeting, June 4, 1951, NARA/NIH, RG 443, ser. 12, NIH, National Research Institutes, NINDB, National Advisory Neurological Diseases and Blindness Council Meetings, Nov. 50–Nov. 54, box 27, folder "Committee 1–7, National Advisory Neurological Diseases and Blindness Council Minutes of Meetings, June 4, 1951." See also Minutes of Meetings, June 24–25, 1955. Cornelius Traeger, NMSS, NYC, to Frederick L. Stone, Chief, Extramural Programs, NINDB, Washington, D.C., Nov. 28, 1951, NARA/NIH, RG 443, ser. 47, box 2, folder "National Multiple Sclerosis Society, volume I." Frederick L. Stone, Chief Extramural Programs, NINDB, Washington, D.C., to Cornelius Traeger, NMSS, NYC, to Dec. 7, 1951, RG 443, ser. 47, box 2, folder "National Multiple Sclerosis Society, volume I." Frederick L. Stone, NINDB, to Cornelius Traeger, NMSS, Apr. 10, 1952, NARA/NIH, RG 443, ser. 47, box 2, folder "NMSS, volume I." Frederick L. Stone, NINDB, to Cornelius Traeger, NMSS, Dec. 3, 1951, NARA/NIH, RG 443, ser. 47, box 2, folder "National Multiple Sclerosis Society, volume I."

145. Cornelius Traeger to Members of the Medical Advisory Board, NMSS, Aug. 29, 1952, NARA/NIH, RG 443, ser. 47, box 2, folder "NMSS, volume I." Harold R. Wainerdi to Members of the Medical Advisory Board, NMSS, Jan. 12, 1954, RG 443, ser. 47, box

2, folder "Associations 1, NMSS, Vol II Jan 1954 through." Sylvia Lawry to Oliver E. Buckley, NMSS, memorandum, Sept. 8, 1952, NARA/NIH, RG 443, ser. 47, box 2, folder "NMSS vol. I." Harold Wainerdi to Pearce Bailey, Feb. 16, 1953, NARA/NIH, RG 443, ser. 47, box 2, folder "NMSS volume I." Pearce Bailey to Roland P. Mackay, NARA/NIH, RG 443, ser. 47, box 2, folder "NMSS, volume I."

146. American Cancer Society, Annual Report, 1955, 27. In the 1957 Annual Report, the American Cancer Society noted that "ACS Directors also serve the NCI. The agencies collaborate closely at policy making levels which helps avoid undesirable duplication. . . . Dr. John Heller, Director of the National Cancer Institute, has said, 'The principal credit for the headway made against cancer is due to cooperative effort in research. Cancer research is financed by cooperative funds, coordinated by cooperative organizations and performed by cooperating skills. NCI and ACS cooperation is effective, it complements, it augments, and it supplements one with the other" (47). American Cancer Society, Annual Report, 1959, 5.

# CHAPTER 3

# Competing Medical Cultures, Patient Support Groups, and the Construction of Tourette Syndrome

HOWARD I. KUSHNER

Prior to the 1960s, Tourette syndrome, like more recent emerging illnesses, lacked an accepted etiology, a definitive diagnostic test, and even a concise case definition. Physicians frequently interpreted the involuntary behavioral tics associated with Tourette's as psychogenic in origin. For those living with these symptoms, psychogenic explanations implied that their illness was somehow "not real." The parents of Tourette's children also felt that a psychogenic diagnosis carried with it the stigma of bad parenting. A psychogenic diagnosis could also prevent a patient from obtaining disability and medical benefits. Howard Kushner's chapter highlights the role of Tourette's support groups in defining both scientific and lay understandings of emerging illnesses.

For sufferers of Tourette's and similar emerging illnesses, it is not enough simply to gain public recognition for their illnesses, to make them visible. It is also necessary to make them "real" by identifying an organic cause. The Tourette Syndrome Association (TSA), established in the early 1960s by the parents of children exhibiting Tourette's symptoms, is a prime example of an association of illness sufferers dedicated to establishing the physiological basis for their illness. Working in cooperation with biomedical researchers committed to an organic explanation for Tourette's, the TSA mobilized medical opinion, sponsored research into the organic causes of Tourette syndrome, and flooded the media with materials supportive of an organic explanation.

Kushner concludes that the dominance of organic explanations for Tourette's in North America, as opposed to the French preference for psychogenic explanations, was largely the product of the TSA's success in defining the terrain of research and discourse about Tourette syndrome in the United States and the relative absence of similar communities of suffering in France. As noted in Chapter 1, other patient advocacy groups, like the Fibromyalgia Association of America and the Chronic Fatigue Immunity Deficiency Syndrome Association of America, have lobbied energetically to have their illnesses defined as organic rather than psychogenic in origin.

---

Tourette syndrome (TS), widely diagnosed in North America and Britain, is almost never diagnosed in France, where it was originally identified and labeled in 1885. Why this is the case has as much to do with cultural and historical factors as it does with medical evidence. In the United States, unlike in France, people with this disorder and their families have exerted extraordinary influence on the transformation of Tourette's from a psychogenic to an organic disorder. Thus, medical research and clinical findings in themselves were necessary but insufficient to transform psychiatrists' views of the etiology of TS.

Gilles de la Tourette syndrome is a disorder in which afflicted persons display an array of sudden, rapid, recurrent, nonrhythmic, and stereotyped motor and vocal tics. The motor tics, which occur frequently throughout the day, generally involve head and neck jerking, eye blinking, tongue protrusions, shoulder shrugs, and various torso and limb movements. Vocalizations may include barks, grunts, yelps, coughs, repetition of one's own or others' words (echolalia), uttering obscenities (coprolalia), and blurting out inappropriate remarks. Other associated behaviors may include the inappropriate touching of oneself or others. Often these are coupled with compulsive behaviors, for instance, a repeated series of actions that must be performed before entering or leaving a room.[1]

Once thought to be a rare occurrence, TS represents one of the fastest-growing diagnoses in North America, with a prevalence of from 2.9 to 5.2 per 10,000.[2] A British team recently reported that 3 percent of thirteen- to fourteen-year-old secondary school children (299 per 10,000) fit the *Diagnostic and Statistical Manual of Mental Disorders-III-R* (DSM-III-R) classification of TS, which is four times higher than the previous highest rates, reported in 1990.[3] All studies find that males are four to five times more likely to become afflicted than females. One recent investigation reported that among boys the prevalence may be as high as 5 per 100.[4] Even at the low end, estimates would make the prevalence of TS more than double that of teen suicide in the United States.[5]

Part of the reason for contradictory findings about prevalence is attributable to physicians' and researchers' disagreements about which symptoms to include and which to exclude in their definitions of Tourette's. That is because TS is a *syndrome* rather than a *disease*. Although the term *disease* can mean many things, including simply distress, a disorder generally graduates from its status as a syndrome only when its underlying pathological causes, or etiologies, as complex as they may be, are uncovered. Measles, polio, smallpox, and sickle-cell anemia are labeled as diseases and not as syndromes because a tentative diagnosis based on signs and symptoms is confirmed or rejected through a laboratory test that indicates infection by a pathogen or the presence of deformed red blood cells. In contrast, the cause of a syndrome remains unknown. With syndromes, like Tourette's, schizophrenia, and chronic fatigue, diagnoses depend on the identification of a list of possible combinations of signs and symptoms that a person must display over an assigned period of time. But as Terra Ziporyn notes, this list may vary from physician to physician: "One researcher, for example, may have defined chronic fatigue syndrome as including anyone who feels tired most of the time, while her colleague may have restricted it to persistent fatigue coupled with flu symptoms and perhaps even associated with a virus. . . . Without an agreed-on definition, more sophisticated correlations and predictions become meaningless. If you don't know what chronic fatigue syndrome is, for example, how can you say what happens to people with chronic fatigue syndrome?"[6] The signs and symptoms that have been grouped together as Tourette's often seem to fit a spectrum, in which a single cause can result in a series of different symptoms. Alternatively, because its underlying pathology remains unknown, these signs and symptoms could result from a variety of disparate causes.[7]

Grouping certain signs and symptoms, but not others, under a single "name" or category inevitably influences practitioners' and patients' understanding of the possible causes and course of a disorder. The purpose of syndrome construction is to focus a practitioner's view of a patient's illness by privileging certain symptoms while downplaying others. What physicians believe constitute the legitimate symptoms of a disorder can have a profound influence on a sufferer's experience. Certainly it influences which symptoms become the focus for treatment and which are viewed as unrelated to the syndrome.

Since the late nineteenth century, when Tourette's was initially described, psychiatrists have disagreed, often vigorously, over which signs, symptoms, and behaviors should be included as part of TS. These controversies continue. Beginning in the 1980s a serious dispute arose among experts over whether obsessive-compulsive behaviors should be included within the Tourette's typology.[8] More recently a division has emerged among those who believe that conduct disorders are part of TS and those who resist too wide a spectrum of symptoms.[9] In addition, clinicians disagree over the length of time a symptom must be present to be considered for a diagnosis and whether onset must occur before the age of eighteen.

Although one of the purposes of the American Psychiatric Association's diagnostic handbook, the DSM, is to provide uniform definitions of psychiatric syndromes, a comparison of the definitions and classifications of TS in the revisions of DSM (III-R) reveals that there has been no absolute agreement over time about what symptoms are necessary or how long they must persist for a diagnosis of Tourette's.[10] An effort at a uniform definition was made in the construction of Tourette's categories in the DSM-IV (1994),[11] but dissension over the typology erupted among both members of the committee that had been commissioned to write this section and other experts as soon as the volume appeared.[12] By 1997 a team of experienced Tourette's researchers at Yale University Medical School's Child Study Center admitted that continuing disagreements over what constitutes the symptoms that are part of TS (its phenotype) continue to frustrate attempts to locate its underlying pathogenesis (causes).[13]

Nevertheless, the vast majority of researchers are persuaded that, no matter what its constituents, Tourette's is an organic disorder. The most persuasive evidence for a physiological basis of TM is that motor tics and involuntary vocalizations can often be controlled by drugs that act to suppress the transmission of the neurotransmitter dopamine in that part of the brain called the basal ganglia, which is responsible for certain motor movements. However, these drugs can only provide relief from the symptoms of Tourette's and do not eliminate the causes of the tics and vocalizations. Researchers continue to search for the underlying causes, which may reside in a genetic malfunction or an autoimmune reaction to a previous infection, or a combination of these and other cofactors.[14]

Established in the early 1970s in Bayside, Long Island, the U.S. Tourette Syndrome Association today has more than thirty thousand members, with chapters in almost every part of the United States. It boasts a medical board of directors made up of respected research neurologists, psychiatrists, and pediatricians from premier North American medical schools in the United States, Canada, and Britain, as well as prominent researchers from the National Institutes of Mental Health.[15] Parallel organizations have been set up in Canada, Britain, and, recently, Norway. The number of diagnoses of TS in the United States and Canada now exceeds 100,000. Millions of dollars a year are expended for research on the causes and treatment of Tourette's, about half of which is aimed at its possible genetic causes.

Yet in France, where the behavior was first identified more than a century ago, the psychiatric community is essentially unified in its belief that there is no such thing as Gilles de la Tourette *syndrome*. French psychiatrists and neurologists find syndromes in general to be North American inventions; they are extremely critical of the primary North American psychiatric diagnostic tool, the DSM. Instead they prefer the 9th (rather than the 10th) edition of the *International Classification of Disease* (ICD) or their own "triaxiel" classification system of childhood psychiatric disorders developed at the Centre de Binet in Paris, neither of which recognizes Tourette's as a distinct and separate entity.[16] French psychiatrists believe that what North Americans describe as TS does not exist, although allowance is made for an extremely rarely occurring Gilles de la Tourette's *disease* (*maladie de Gilles de la Tourette*), whose etiology is seen within a psychodynamic (psychoanalytic) framework.

Many French psychiatrists insist that the entire diagnosis of TS is fraudulent. One of the most well respected and widely read contemporary French psychoanalysts, Lucien Israël, wrote in 1993 that Gilles de la Tourette syndrome was an invention of the American Tourette Syndrome Association and American physicians, especially Oliver Sacks.[17] As if to underline this position, one of France's most famous psychiatrists, Cyrille Koupernik, wrote in a letter to me, "I will not hide from you that I was astonished to learn that there were 30,000 members of the Tourette Syndrome Association in the United States; I see this as analogous with the multiple personality phenomenon [in the United States] and I am not certain that it is the most hopeful evolution of a psychiatry which has so much influence in the rest of the world as that of your great and admirable nation."[18]

In contrast to the association in North America, the Association Française de Troubles Obsessionelles Compulsifs et du Syndrome Gilles de la Tourette (AFTOC), established in the late 1980s, was run, until its demise in 1997, by an elementary school teacher in the northern industrial city of Lille, from the bedroom of his parents' house. Currently, with help from the TSA and the Norwegian Tourette Association, an American woman living near Paris is attempting to resurrect a French support group.[19] The AFTOC's only serious medical support came from an allergist, Dr. Pierrick Hordé, who writes popularized tales of bizarre medical conditions and has no connection with any French medical school or research institute.[20]

All this is somewhat surprising because what today is called Tourette syndrome is linked to a disorder identified in 1885 by a twenty-eight-year-old Parisian neurologist, Georges Gilles de la Tourette, at the behest of his mentor, Jean-Martin Charcot, director of the Salpêtrière Hospital in Paris and the most influential neurologist in late-nineteenth-century France.[21] Charcot, who subsequently named the disorder *maladie des tics de Gilles de la Tourette*, remains an icon of French neuropsychiatry.[22] Yet at the very place where Charcot lectured on the disorder, the Salpêtrière Hospital, no neurologist or psychiatrist (both the neurology and psychiatry departments are literally housed adjacent to the Bibliothèque Charcot) would today make a diagnosis of Gilles de la Tourette syndrome.

How can two medical communities faced with patients displaying the same symptoms come to such opposite conclusions? Are the French simply insular? Are they wrong? Is it possible that the French are immune to Gilles de la Tourette syndrome? Or is it the Anglo–North Americans who are so totally misled?

The reason for this strange disparity rests as much on cultural and historical interpretations as it does on medical evidence. These cultural and historical disputes center on, as they generally do in Franco-American disagreements, issues of language and classification. Virtually every French practitioner with whom I have talked or corresponded has insisted on the designation Gilles de la Tourette's *disease* and objected strenuously to the Anglo–North American designation Tourette *syndrome*. However, they also insist that this disease is psychiatric, tied to obsessive-compulsive ideation rather than an idiopathic organic substrate.

Connected with issues of language and classification are differences in national medical culture and politics. These differences include national/ linguistic contrasts between Anglo–North American views and those of France and its francophone offsprings. Similar cleavages are found also *within* anglophone and francophone medical communities. Here the contests are between and among those practitioners and medical researchers who have seen the disorder from two distinct (generally mutually exclusive and often hostile) perspectives: psychological and organic. Approaches within these two categories of understanding have often themselves divided into hostile camps. Thus, psychological explanations generally have been seen as either behavioral (bad habits) or psychoanalytic (early childhood conflict). Organic explanations, though often more eclectic, have nevertheless sometimes been reduced to one of three approaches: genetic/inherited, biochemical/physiological, or (post)infectious. To complicate matters, there seem to be distinct national influences on the way in which professional rivalries have emerged in each society.

Having said this, I want to make it perfectly clear that in no way am I suggesting that those who display the symptoms of TS or other syndromes do not suffer from real distress. I have worked and continue to work in the clinic and the lab as well as with support groups, and I am actively involved in medical research. Thus, my point of view is not that of an unsympathetic outsider to medical culture and medical politics.

The medical culture and politics that have informed the construction of TS are most accessible through the contested histories that each medical community (both national and professional) relies on to give meaning to patients' presentations of symptoms. This is perhaps most clearly illustrated if put in the context of the controversies surrounding Gilles de la Tourette's initial construction of tic disease as a distinct disorder.

### The Contests over *Maladie des Tics*

In 1885 Gilles de la Tourette published a two-part article that identified a combination of multiple motor tics and "involuntary" vocalizations as constituting a distinct disorder that he called *maladie des tics convulsifs avec coprolalie* (convulsive tic disease with coprolalia).[23] Gilles de la Tourette's article laid out a classification of symptoms and prognosis based on his description of nine patients' case histories. In contrast to the more coordinated and bilateral movements of Sydenham's chorea, convulsive tic disease was progressive. It began with childhood motor and vocal tics that over time increased in number and variety, with the eventual appearance of coprolalia.[24] According to Charcot, unlike choreas and hysterias, *maladie des tics* might wax and wane, but it ultimately resisted all interventions.[25] There was no hope of "a complete cure," agreed Gilles de la Tourette, for clinical experience demonstrated that "once a ticcer, always a ticcer."[26]

Citing Théodule Ribot, Gilles de la Tourette explained that convulsive tic disease had a "degenerative" hereditary etiology.[27] Advocates of degeneration theory argued that diet and habits such as alcoholism and immoral behavior had a cumulative destructive effect on the nervous system that were inherited by succeeding generations.[28] Charcot concurred with Gilles de la Tourette's assessment, explaining that tic disease was "the direct product of [hereditary] insanity."[29] As an example of degenerative etiology, Ribot cited Jean Itard's 1825 case of the marquise de Dampierre, a French noblewoman notorious for publicly shouting out, in the middle of conversations, inappropriate or obscene words, especially "shit and fucking pig."[30] Gilles de la Tourette appropriated Itard's description of the marquise de Dampierre as his first and prototypical example of the progressive course of tic disease and its inevitable obscene outbursts.[31]

Because they had identified a set of symptoms, a course of illness, and a predisposing cause, Gilles de la Tourette and Charcot insisted that they had described a *disease,* which Charcot designated *maladie des tics de Gilles de la Tourette.* Although what today is called Tourette *syndrome* claims its pedigree from Gilles de la Tourette's 1885 article, the current designation draws only on Gilles de la Tourette's description of symptoms and ignores his and Charcot's attribution of the underlying causes and degenerative outcome. For Charcot and Gilles de la Tourette, however, a diagnosis of convulsive tic disease could not be made solely on the basis of symptom presentation. Only when symptoms persisted, resisting all interventions, and when a patient revealed a history of degeneration was the diagnosis *maladie des tics de Gilles de la Tourette.* Thus, Gilles de la Tourette's disease is not the same as what today is labeled Tourette syndrome.

Gilles de la Tourette's construction of convulsive tic disease was challenged by two of his Salpêtrière contemporaries, Georges Guinon and Edouard Brissaud, who claimed that case histories of their patients revealed that multiple motor tics, coprolalia, and echolalia were found in both hysterics and choreics.[32] As a result of these and other criticisms, Charcot and, to a lesser extent, Gilles de la Tourette refined their claims.[33] They conceded that motor and vocal tics and coprolalia were also symptoms of hysteria and that therefore these symptoms and signs were not unique to Gilles de la Tourette's disease.[34] What distinguished convulsive tic disease from a hysterical illness with similar symptoms was whether the *cause* was degenerative and whether the symptoms could be eradicated. If a patient had a degenerative family history and no intervention (such as hypnosis) was able to cure or ameliorate the course of the disease, a diagnosis of *maladie des tics* was justified. If, on the other hand, a patient improved and the family history was inconsistent with degeneration, a diagnosis of hysteria was appropriate.[35]

Rejected by critics within the Salpêtrière, *maladie des tics de Gilles de la Tourette* was increasingly marginalized. The publication of Henry Meige and E. Feindel's influential 1902 study, *Les tics et leur traitement,* furthered this process.[36] Translated into English by British neurologist S. A. Kinnier Wilson in 1907, *Tics and Their Treatment* became the standard for the diagnosis and treatment of motor and vocal tics for the next half century.[37] Expanding the phenotype to include all varieties of tics, Meige and Feindel insisted that only a minority of ticcers fit Gilles de la Tourette's subset.[38]

Tics and involuntary vocalizations, according to Meige and Feindel, resulted from uncorrected infantile habits in a population with hereditary weakness. Unless these habits were restrained, simple childhood tics would evolve to debilitating behaviors. Contrary to Gilles de la Tourette's assertions, Meige and Feindel insisted that convulsive tics were curable if early and vigorous interventions were adopted to reinforce the patient's will.[39] Meige and Feindel's hereditary view proved compatible with eugenics, and it also segued into Freudian explanations of early childhood sexual repressive conflict.

## Enter Psychoanalysis

Having no ticcing patients of his own, the Hungarian psychoanalyst Sandor Ferenczi appropriated those discussed by Meige and Feindel and used their descriptions as the basis for his 1921 landmark article claiming that tics were an outward expression of repressed masturbation. Ferenczi's explanation grew to become the bedrock of psychoanalytic interpretations.[40] By 1943 the influential American child psychiatrist Margaret Mahler wrote that tic behaviors, like those displayed by Dr. Arthur Shapiro's patient (see the section "Toward a New Master Narrative" later in this chapter), resulted from repressed sexual conflict that "renders the individual defenseless against overwhelming emotional and psychodynamic forces."[41] Although Mahler resurrected the designation *maladie de Gilles de la Tourette,* she insisted that these behaviors should be interpreted and treated as manifestations of a common underlying repressed autoerotism rather than as a specific and distinct disorder.

As a result, Gilles de la Tourette's disease, as a separate and legitimate disorder, had largely disappeared from British and North American psychiatry.[42] So complete was the triumph of psychoanalysis that the fourth series of the *Index Catalogue of the Surgeon General's Office* (1941), the official bibliographical medical publication of the U.S. government, instructed physicians looking under the heading "Gilles de la Tourette's Disease" to "see neurosis."[43]

Although psychoanalysts increasingly were forced to admit that psychoanalysis was ineffective in treating vocal and motor tics, they interpreted these failures in a psychoanalytic framework. Tic disorders, explained a prominent psychoanalyst in 1953, resisted psychoanalytic interventions because of "the role of the tic as the last 'desperate defense against psychosis.'"[44] Thus, the failure of psychoanalytic interventions did not interfere with the belief that the etiology of these behaviors was psychogenic. In 1958 psychoanalyst Z. Alexander Aarons, writing in the prestigious *Psychoanalytic Quarterly,* described a fifteen-year-old patient's "twitching and yelping" and "neuromuscular spasms" as "a masturbatory equivalent," enabling him "to ward off the danger of submission to his passive feminine wishes."[45]

By the mid-1950s antipsychotic phenothiazine drugs, especially chlorpromazine (thorazine), had been used with some success in the management of schizophrenia, anxiety, and hyperactive mania.[46] The effect of these psychotropic agents had two kinds of implications for the diagnosis and treatment of motor and vocal tics. First, psychoanalysts often viewed ticcing patients as having mild forms of schizophrenia, and therefore any intervention that ameliorated schizophrenic symptoms logically should be tried on motor and vocal tics. This made it and other similar-acting phenothiazines a reasonable pharmacological candidate to control tics at a gross level in some patients. Thus, in 1959 British psychiatrist S. Bockner reported that chlorpromazine treatment yielded "a rapid and remarkable response" in reducing tics and coprolalia in two patients.[47]

Second, for the most part, psychoanalytic psychiatrists reacted to the success of some pharmacology in controlling psychiatric symptoms in much the same way as they had reacted to psychosurgery or electroconvulsive shock. That is, they were persuaded that the control of symptoms by heroic means demonstrated nothing about the underlying causes of a disorder. Thus, they characterized psychoactive drugs as powerful and essentially nonspecific agents for the control and management of difficult cases.[48]

It was more difficult, but not impossible, to ignore reports of success of the dopamine antagonist haloperidol. Positive results of tic suppression with haloperidol were announced almost simultaneously in the early 1960s in Italy, France, and the United States, but these reports, particularly in the United States, emphasized the management of symptoms rather than the underlying causes of tic behaviors.[49] By the mid-1960s the effectiveness of haloperidol in treating tics was documented in a series of clinical reports published mainly in the *American Journal of Psychiatry,* the official organ of the American Psychiatric Association. However, psychiatrists reporting success in controlling their patients' symptoms with haloperidol continued to defer to psychoanalytic assumptions. In general, they presented haloperidol as a useful adjunct in making psychotherapy more effective for resisting patients.[50] Given their deference to psychoanalysis, none of these studies, either individually or in concert, helped to persuade the psychoanalytic establishment that motor tics and coprolalia should be rethought in an organic frame. This was the context in which psychiatrists, patients, and their families attempted to negotiate the meaning and treatment of involuntary tics and vocalizations by the mid-1960s.

## Toward a New Master Narrative

In April 1965 a twenty-four-year-old woman was referred for psychotherapeutic treatment to a New York psychiatrist, Dr. Arthur K. Shapiro. Shapiro, then associate clinical professor at the New York Hospital–Cornell Medical Center, reported that the woman's "symptoms were striking and bizarre: spasmodic jerking of the head, neck, shoulders, arms and torso; various facial grimaces; odd barking and grunting sounds; frequent throat clearing; and periodic and forceful protrusion of the tongue."[51] Although she displayed no coprolalia during the initial visit, the woman reported that at other times she could not restrain herself from occasionally shouting out the word *cocksucker*. The "twitching and jerking of the head and neck" began at age ten. These increased, and by high school her movements had led her classmates to refer to her as "Twitchey." Although they waxed and waned over the years, the twitching and jerking always returned, "insidiously worsened." At age twenty-two "short, loud screams began," followed six months later by "throat-clearing, rasping and barking noises, and coprolalia." Although the woman managed to confine her screaming and coprolalia to when she was home, when she was out in public "other symptoms increased, such as throat-clearing, barking noises, and tongue-protrusion." Social tensions, wrote the Shapiros, exacerbated her symptoms (p. 345).

Noticing that "the patient could control" her symptoms for brief periods of time, the mother assumed that her daughter's problem was definitely not organic, a view shared by the referring psychiatrist, whose diagnosis was "habit tic with hysterical personality associated with 'la belle indifference'" (345–350). The patient herself admitted that when she "occasionally saw someone twitching and jerking, her immediate reaction was that the person was crazy" (p. 345). Indeed, it was difficult for laypersons and physicians alike to explain the young woman's involuntary cursing, especially given its highly charged sexual content, in other than psychological terms.

Thus, the woman's initial diagnosis meshed with the dominant view of American psychiatry in 1965. For instance, a widely consulted and highly regarded standard pediatric text, Ford's *Diseases of the Nervous System in Infancy, Childhood, and Adolescence* (five editions from 1937 to 1966), instructed pediatricians to treat motor and vocal tics as "psychogenic disorders simulating organic disease."[52]

Contrary to psychoanalytic assumptions, Shapiro was convinced that his patient's behaviors were organic. Hospitalizing the woman for the next several months, Shapiro administered thirty-six neuroleptic and antidepressant drugs and combinations of drugs, settling finally on the major neuroleptic tranquilizer haloperidol. "From the first day of treatment," wrote Shapiro, the "symptoms disappeared."[53]

In their report announcing the apparently successful treatment of the young woman's symptoms with haloperidol, Shapiro and his spouse and coauthor, Elaine S. Shapiro, a clinical psychologist, concluded that, despite the claims of psychotherapists, the etiology of these symptoms was an idiopathic "organic pathology of the central nervous system." The Shapiros went further, presenting their findings as evidence of the therapeutic and intellectual paucity of psychoanalytic psychiatry. "Diagnosis for the patient described in this report," the Shapiros reminded their readers, "based on psychiatric impression and psychological testing was 'personality disorder with inadequate and passive dependent character traits'" (p. 349). "Physicians probably communicated to the patient and her family," wrote the Shapiros, "that emotional illnesses caused the symptoms. A harmful effect on her fantasies and character formation is likely" (p. 349). The Shapiros decried "the fashion in medicine to attribute symptoms and diseases without demonstrable organic pathology to a psychological wastebasket diagnosis." Psychoanalytic treatments of these symptoms, warned the Shapiros, "may result in iatrogenic [physician-induced] psychopathology. Physicians should be sensitive and cautious about the possible harm to patients of premature psychological diagnosis" (p. 349). For the Shapiros, then, the recognition of Gilles de la Tourette's description as a legitimate disorder was necessary so that those presenting with its symptoms would no longer be subject to inappropriate psychoanalytic diagnosis and treatment. In retrospect, it is not surprising, given the dominance of psychoanalysts on the editorial boards of American psychiatric journals in the 1960s, that the Shapiros' article was rejected for publication by every major American psychiatric journal, finding a home only in 1968 in the *British Journal of Psychiatry*.[54]

The Shapiros' article laid out a sort of "master narrative" for all the media stories and testimonials that would follow. This original narrative did two other things. First, it gave the disease a name. Second, it made certain that the name, Gilles de la Tourette syndrome, would be synonymous with an organic disorder. The Shapiros emphasized Gilles de la Tourette's 1885 description of symptoms, but they downplayed his view that tics had a degenerative etiology. They established the label "Gilles de la Tourette's syndrome" as the descriptor of a neurological disorder, which by definition stood in opposition to psychoanalytic claims. From the Shapiros' perspective, psychoanalytic treatment most likely "contributed to the patient's shyness, inhibited aggression, passivity, and fantasies of insanity."[55]

These views were underlined in a symposium panel on Gilles de la Tourette syndrome organized and chaired by Arthur Shapiro at the May 1968 meeting of the American Psychiatric Convention in Boston.[56] Shapiro launched into an attack on psychoanalytic therapeutics. "The tendency of physicians to attribute Tourette's syndrome prematurely to psychogenic factors," Shapiro argued, "contributes to an iatrogenic onus to patients already so incredibly burdened that it is a wonder to me and a credit to them that they can maintain their sanity."[57] Reviewing the discussion, Shapiro reported that "most participants agreed that haloperidol was the best and most predictable of all available treatments." Often it abolished all tics and had proved between "30 to 90 per cent" effective.[58]

The Shapiros' subsequent publications and papers elaborated the positions laid out in the 1968 paper and Boston symposium. Thus, before the first meeting of what would become the Tourette Syndrome Association, the central thesis of the association's official point of view already had been laid out. The Shapiros had attempted to reverse the rhetorical burden of proof. From then on, based on their results with haloperidol, they would claim that the organic factors of TS were established. Those arguing for a psychogenic etiology would have to demonstrate their conclusion through the scientific method, even though the Shapiros conceded that the action of haloperidol could not meet this test.[59] In a 1973 paper entitled "Organic Factors in Gilles de la Tourette's Syndrome," the Shapiros and their New York Hospital colleagues Henriette Wayne and John Clarkin concluded that "although considerable indirect evidence supporting an organic aetiology has been presented, definitive evidence is not yet available." But they stated emphatically, "Our data, and our interpretations of the literature, do not support a psychological aetiology, and the burden of proof is on those claiming it."[60]

Arthur and Elaine Shapiro's book *Gilles de la Tourette Syndrome,* coauthored with psychiatrist Ruth Bruun and neurologist Richard Sweet and published in 1978, is an impressive and comprehensive overview of what was known about TS. It concluded with a series of patient testimonials. "The Tourette Syndrome Association put us in touch with Drs. Elaine and Arthur Shapiro," wrote a mother who, along with her son, had been diagnosed with TS. "We visited them in New York in June 1974 and were both diagnosed as having Tourette's. I cannot begin to tell you what our visit to them meant to us. It has been such a relief to know what we have."[61] A father reported that "it was a relief to hear the phrase, 'Gilles de la Tourette syndrome' applied to our son's illness, after 27 long frustrating years . . . of counseling, psychiatric evaluations and testing, a month in a mental institution and, worst of all, four sad, frightening years in a home for emotionally disturbed children."

The father was relieved that he was able "to finally meet with an unusually competent psychiatrist who stated with assurance that the tic symptoms did not indicate that our son was psychotic, neurotic, or emotionally disturbed because of his family environment and parental inadequacy."[62]

A mother testified to the long, fruitless search for diagnosis and treatment for her daughter. "Finally, we located . . . Dr. Arthur Shapiro who put our daughter on Haldol immediately. Although she suffered from many of the drug's side effects, she had about an 80 percent reduction in her tics." Like the others, this mother was extremely grateful. "At last my child has been diagnosed. She has Tourette syndrome. Somehow a monster is less frightening when you call it by a name."[63] That the name was foreign and loosely based on a forgotten disease classification provided additional authenticity.

### The French View

Ironically, at the very moment when psychoanalysis was put on the defensive in North America and Britain,[64] in France it was gaining, rather than losing, a foothold. By the late 1980s the French psychiatric community was dominated by psychoanalysts who would see motor and vocal tics in much the same way as they had been viewed in the United States until the late 1960s.[65]

The extraordinary influence of psychoanalysis in France was the result of a confluence of a number of cultural and historical circumstances. Unlike its Anglo–North American relatives, French psychoanalysis, especially as it related to ticcing behaviors, drew more on the psychological writings of its countryman Pierre Janet (1859–1947) than on the more orthodox psychosexual interpretations put forth by Sandor Ferenczi and Margaret Mahler. In contrast to Ferenczi and Mahler, who generally characterized tics as repressed masturbation, Janet had stressed that tics were a "signal symptom" of an obsessional psychoneurosis.[66]

Tied to this particularly French psychoanalytic interpretation of the function of tics, and again in contrast to the situation in the United States, biological explanations for mental illness were identified in France with the Nazi and Vichy expulsions and persecutions of psychoanalysts (1940–1944), almost all of whom were of Jewish or émigré origins or both.[67] Although the majority of psychoanalysts who remained in occupied Paris suspended their work, a few continued to practice clandestinely. To this day psychoanalysis remains strong because it portrays itself as a bulwark of defense against detested racist medical practices. Criticisms of psychoanalysis are often labeled as protofascist and anti-Semitic.[68]

Among those who had continued to practice secretly during the occupation was the Parisian Serge Lebovici, whose Jewish émigré psychiatrist father was arrested and sent to die in Auschwitz.[69] In 1951 Lebovici published a psychoanalytic study, *Les tics chez l'enfant*, that, merging Janet's view of tics with more orthodox Freudian views, would become the most influential postwar French statement on motor and vocal tics.[70]

As claims continued to mount, particularly from North American researchers, that haloperidol was more effective than psychoanalytic interventions, Lebovici and his colleagues reacted. Reviewing the newest work, particularly the numerous studies by Arthur and Elaine Shapiro, Joseph Alliez and Serge Audon published a historical and clinical review in 1976 in *Annales Médico-Psychologiques*. Conceding that there was no longer any doubt that neurotransmission played a role in the presentation of these symptoms, Alliez and Audon nevertheless were convinced that the effect of haloperidol did not rule out psychogenic and social factors. They remained persuaded that convulsive tic disease should be understood in "a multidimensional framework where one considers neurological, psychiatric, psychological, sociological, psychoanalytic, and cultural approaches, in order to better exploit the dynamic aspect of all these factors in a therapeutic sense, without always juxtaposing them."[71]

By the 1980s, French psychoanalysis, split internally between orthodox and Lacanian factions and threatened externally by the psychopharmacological revolution in treatment, regrouped around the issue of convulsive tics.[72] Increasingly on the defensive, the official psychoanalytic position toward an integrated view of tics and coprolalia became less flexible and more orthodox, privileging psychogenic factors. By the 1990s, leading French psychoanalysts would take the offensive, decrying organic explanations of Gilles de la Tourette's syndrome as a construction, almost a conspiracy, of the North American medical establishment.

The first shot was fired, appropriately enough by Serge Lebovici, at the January 1982 meeting of the French Society of Psychiatry of Children and Adolescents in Paris, which was devoted to "les tics chez l'enfant." Central to the discussion was the role of psychoanalysis and psychotherapy in the face of the success of chemotherapy for the treatment of tics. According to Lebovici, part of the conflict over psychoanalysis versus haloperidol could be tied to classification of symptoms. "It is difficult," he believed, "to ascertain the limits between tic, properly speaking, and nervous habits that can become veritable obsessions." It was important to recognize that "the psychopathological structure that underlies tic is differently valued and it is variable.

Patients with obsessional symptoms tend to be emphasized. But one can observe tics in obsessional neurotics or in grave phobic syndromes, as in certain psychoses."[73]

Complicating the issue were trends coming from North America. "Today," asserted Lebovici, "we must measure the respective effects of psychoanalytic cures and psychotherapy, on the one hand, with chemotherapy with haloperidol on the other hand. In the United States, and we will see this in France, growing importance is given to the latter [haloperidol]," often "with excessive therapeutic risks," because Americans will not tolerate "anything that interferes with the school and social life of the child ticcer." Lebovici suggested that this was the result of "the direct participation of families at the therapeutic level." This interference with basic clinical research "has appeared as a new phenomenon, most represented in Anglo-Saxon countries." Clearly, this was not a trend that Lebovici favored. As he noted sarcastically, "in the United States, the *Tourette Syndrome Association* publishes a *newsletter* where summaries of research and publications of abstracts of social meetings are destined to be picked over by the 100,000 Americans who suffer from la maladie des tics."[74] Lebovici assumed that the interference of laypersons with diagnosis, treatment, and research was so outrageous that his colleagues needed only to learn of it in order to be appalled at this trend in American medicine.

Lebovici's paper signaled a new strategy for the defense of psychoanalytic theories of tic. The threat would no longer be portrayed as psychopharmacological; rather, fitting the more general French national attitude of the eighties, the issue was transformed into a wider defense of French cultural institutions from American medical and cultural imperialism. If the American Tourette Syndrome Association provided the appropriate symbolic target for the deficiencies of American popular culture, the American Psychiatric Association's DSM classification system represented the small-minded and literal view that American psychiatrists had of mental processes. Building on Lebovici's lead, psychiatrist P. Moran of Toulouse suggested that the Americans had inflated the incidence of TS by including many conditions that in France would not be classified as maladie des tics.[75] Martine Lefevre joined with Lebovici to attack the DSM-III's alleged descriptive precision. The DSM, they argued, had transformed disorders like Tourette's into a behavioral scale in which psychogenic perspectives, necessarily based on a more general view of the mechanisms of mental pathology, were "totally overwhelmed."[76]

Lebovici's psychoanalytic views carried the added authority of his antifascist credentials. Attempts to explain the etiology of tics in organic and non-psychoanalytic terms would be met with charges of anti-Semitism.[77] The connection between TS and the Nazis was made explicit in 1993 by Lebovici's colleague, psychoanalyst Lucien Israël, in his preface to Henry Meige's 1893 doctoral thesis describing the so-called Wandering Jew disease.[78] Meige's explicitly anti-Semitic analysis, including numerous racist photographs, was originally published in *La Nouvelle Iconographie de la Salpêtrière*, of which he was managing editor and whose editor in chief was Charcot.[79] Gilles de la Tourette also served on the editorial board. Meige would emerge in the early twentieth century as the foremost authority on convulsive tics. His and Feindel's book *Les tics et leur traitement,* which explained the genesis of tics in terms of hereditary degeneration, became the most influential twentieth-century statement about the behavior. Israël retraced these connections and, by implication, attached them all to the emergence of the diagnosis of TS in America in the 1970s:

> Let us recall the curious evolution of "maladie" des tics. Was not Gilles de la Tourette one of the editors of the *La Nouvelle Iconographie de la Salpêtrie* of which Meige was the managing editor? This maladie de Gilles de la Tourette, that hardly had any historical interest until recently, has known a veritable resurrection in the U.S., under the impetus of [Oliver] Sacks (author of *The Man that Mistook His Wife for a Hat*). A fine clinician and a man of a great generosity [Sacks] attracted attention to this syndrome by discovering three [sufferers] in only one day, just by strolling the streets of New York. The "American Tourette Association" is founded on the same basis as that of alcoholics anonymous, obesity anonymous, or the emotionally anonymous. This is the perspective into which we should place "Meige's Wandering Jews" or, more precisely, Meige and Charcot's. Making human misery into an illness reveals a great deal. Hitler and Mussolini, not to mention other lesser dictators, were paranoids. Everyone else is healthy. . . . It is the racists who are sick.[80]

Thus, Israël reaffirmed Lebovici's accusation that TS was a creation of the American Tourette Syndrome Association, a claim that continues to be repeated by many prominent French psychoanalysts today. From one point of view this accusation is essentially accurate. In France, lay patient and family support and advocacy groups remained almost nonexistent at least until the late 1990s, but in North America they were legion, drawing no doubt on a long tradition of voluntary associations reaching back to the nineteenth century. In the case of Tourette's, in which practitioners increasingly blamed parents for their children's illness, beginning in the 1970s the parents themselves, convinced of the organic nature of their children's affliction, led a revolt against psychoanalytic assertions.

## The Tourette Syndrome Association

The TSA's roots can be traced to a letter that appeared in the *New York Post* in December 1971, written by a distraught father, Martin Levey of Flushing, New York, and his wife. "We have a son who has a very rare neurological disorder called 'Gilles de la Tourette's syndrome,'" wrote Levey to the *Post*'s editor.[81] The Leveys, according to a *Wall Street Journal* article published the following June, had spent more than fifty thousand dollars in "an eleven-year search for a doctor who could cure their son" Bill, who "began making slight noises at the age of six." Soon, "the noises became obscenities, shouted at the top of his lungs, and were accompanied by body tics." Mrs. Levey reported, "The really tragic part of it was that the psychiatrists we went to would all blame me for Bill's problems. Right in front of me, they would ask him, 'What has your mother been doing to you?'"[82]

Although the letter appeared to be a desperate plea for medical advice, Bill had been under the Shapiros' care since 1967, when he was seventeen years old. By 1971, when the Leveys' letter was published, Bill had been taking haloperidol for four years, and his symptoms had almost totally abated.[83] The purpose of their letter was to organize a group to lobby physicians, the public, and the government to support acceptance of Tourette's as an organic disorder.

In early 1972, as a result of the letter in the *Post,* a few families met at the Leveys' home.[84] They agreed on the necessity of an activist organization if they were to find suitable treatments and ultimately a cure for this disorder. The Shapiros agreed to inform their other patients about the group's existence.[85] The members of the group decided to pursue five interrelated strategies: legal status, publicity, recruitment, medical advice and support, and research funding.[86]

They applied for and received a charter from the state of New York as a tax-exempt, nonprofit organization called the "Gilles de la Tourette Syndrome Association, Inc." They published personal notices in the *New York Times* inviting those who displayed ticcing symptoms, or whose family members displayed ticcing symptoms, to contact the association.[87]

Arthur Shapiro had agreed to act as an adviser. In these early years Arthur and Elaine Shapiro were the main connection the group had with the medical community. In fact, the Shapiros played such a central role in the formative years of the association that it is impossible to separate their views on the etiology and treatment of Tourette's from those of the association. After 1972, as each of the Shapiros' new studies appeared or as a paper was delivered at a professional meeting, it would be presented simultaneously in association meetings, and its conclusions would appear in the TSA newsletter. Often, the TSA would reprint and distribute these studies to its membership, to those inquiring about Tourette's, and to physicians requesting assistance and information.[88]

Most of the association's members had children under the Shapiros' care and agreed to supply the Shapiros with subjects for their clinical studies.[89] Few involved with the association believed that Tourette's was a rare disorder, but no one had any idea of its actual prevalence. However, the members all agreed that to obtain funding support, they had to alter the perception that Tourette's was a rare disorder. Certain from the overwhelming response to their advertisements and their personal experiences that the media and physicians had greatly underestimated the number of persons afflicted with TS, association leaders made a determined effort to correct this misimpression. Judy Wertheim, another of the association's founders and its president from 1980 to 1985, later confessed that she had "made up" the number of Americans with TS that Jane Brody published in her 1975 *New York Times* article—"conservatively, an estimated 10,000." Wertheim was absolutely convinced that even this number was an underestimation.[90]

The group continued to encourage any opportunities for research. Most important was its ability to supply patients for researchers interested in studying disorders that might have some connection with Tourette's.[91] By January 1975, Oliver Sacks were persuaded to "begin . . . intensive clinical studies of Tourette patients and their families."[92] By the mid-1970s the association had inextricably attached itself to and encouraged every new and promising route toward identifying and treating TS as an organic disorder. Notably absent were any projects that assumed that Tourette's had psychological causes or even studies that sought to treat the psychological problems that might result from the affliction.[93]

By April 1975, the association newsletter listed and supported six ongoing and new research projects.[94] Simultaneously, it prepared exhibits at national and regional medical meetings. The association launched what would emerge as one of its central functions, providing venues for researchers to meet and exchange views about the causes and treatment of TS.

Meanwhile, the TSA publicized Tourette's in an effort to recruit persons who suffered from it and to inform the public and the media that the affliction was an organic disorder and that those who had it should be viewed with compassion. The general information mass mailer was entitled "Tourette Syndrome: A Neurological Disorder Characterized by Multiple Tics."[95] A new one-page advertisement, "We're Looking for People Who Twitch, Yelp, and Grunt Uncontrollably," was created, and the association persuaded a number of national magazines, including the *Saturday Review, U.S. News and World Report,* and *Medical Economics,* to print it as a public service announcement. The advertisement described typical symptoms, noting that "most victims accept it as an emotional problem, *which it is not.* Many doctors, too, misdiagnose it as a mental disorder. *But it is physical!* There is a drug that can help control the symptoms in many people." Readers of the ad were told, "If you or your children or friends show any of these symptoms, please contact us immediately for more information."[96]

Not only was TSA instrumental in providing information for full-length articles in national and regional magazines and journals, but also it was successful in placing smaller informational articles in the *Reader's Digest* (1973), *Science Digest* (1973), and *Psychology Today* (1976).[97] In addition, the association inserted letters in the medical advice columns of *Newsday* and the *New York Daily News* whose answers publicized the group.[98] Eventually, this strategy resulted in letters published in nationally syndicated advice columns, including "Dr. Joyce Brothers," "Ann Landers," and "Dear Abby." The association also got its message across through television and film.[99] McNeill Laboratories (the manufacturer of Haldol) agreed to provide funds for a teaching film for physicians. Along with each showing of the film, McNeill also agreed to pay for and distribute reprints of the 1973 article by the Shapiros and Wayne, "Treatment of Tourette Syndrome," as well as the TSA brochure.[100]

By now, what the association considered favorable publicity began to appear on its own.[101] The TSA increased its efforts to "educate" physicians about the organic nature of Tourette's by attempting to place exhibits at every relevant national and regional medical meeting. In addition, those active in Tourette's research, especially the Shapiros and others, continued to make presentations about TS at national and regional medical meetings.[102] A medical advisory board was established, which, in addition to Bruun, Arnold Friedhoff, Sacks, A. Shapiro, and Sweet, included Thomas Chase, director of scientific research of the National Institute of Neurological and Communicative Diseases and Stroke, and Robert Good, president and director of Sloan-Kettering Institute. By 1976 a new generation of medical researchers had been stimulated to examine organic etiological factors in TS.

Already, chapters had been organized in northern and southern California, northern and southern Florida, Connecticut, Massachusetts, New Jersey, Illinois, Michigan, Maryland, and Washington, D.C. A separate Canadian affiliate, the Tourette Syndrome Foundation, was also established.[103] By the end of 1976, each of the goals outlined by the group that had met at New York Hospital in 1972 was firmly established. If Tourette syndrome was not yet a household word, an extraordinary transformation in attitude of patients, families, and physicians was under way.

By 1982 the TSA had printed and distributed twenty-six thousand copies of the Shapiros' *Tics, Tourette Syndrome and Other Movement Disorders: A Pediatricians' Guide*.[104] The pamphlet told pediatricians that "recent clinical studies indicate that Tourette syndrome is caused by primary organic disease of the central nervous system." But the authors admitted that this conclusion rested on circumstantial evidence: "While it is recognized that the evidence is inconclusive because a specific lesion has not been demonstrated, circumstantial evidence supports an organic etiology for tics." They also warned practitioners that because "the symptoms of Tourette Syndrome can be so bizarre, frightening and troublesome," and because afflicted children often are "disruptive in school, create interpersonal difficulties with peers, upheaval in the family, and often interfere with, or sharply limit, normal social intercourse," physicians' "usual response . . . is to recommend psychological treatment." But the Shapiros had "concluded that psychological treatment including psychotherapy is clearly ineffective as a primary treatment for tics and Tourette Syndrome." In fact, they were convinced that, because "psychological studies . . . failed to support the frequently advanced notion that psychological conflicts cause tics or Tourette Syndrome," "psychological treatment . . . for tic symptoms" was "both ineffective and inappropriate."[105]

The association actively lobbied the U.S. Congress and other legislative bodies for regulations that would help persons with Tourette's and their families gain access to drugs and receive legal protection from harassment. At a hearing in Washington, D.C., on June 26, 1980, people with TS and family members gave impassioned testimony about Tourette's and the need for federal support for orphan drugs like pimozide. The hearing received national press coverage, including major stories in both the *New York Times* and the *Los Angeles Times*. These accounts caught the attention of the producer of the popular weekly television series *Quincy,* starring Jack Klugman, who decided to devote an episode (March 4, 1981) to Tourette's. *Quincy* not only educated the American public about Tourette's as an organic disorder, but it also helped get the then stalled "Orphan Drug Bill" passed by the House of Representatives and eventually, in January 1983, signed into legislation that made it much easier for pharmaceutical firms to undertake the development of so-called orphan drugs.[106]

By 1982, the association had begun to attract sizable donations from private individuals, corporations, and foundations to support research on TS and to cover administrative costs. These donations, along with the new legitimacy accorded to research on Tourette's, resulted in a flood of newly funded studies, presentations at professional meetings, and published journal articles.[107] In May 1981 the TSA had sponsored the First International Gilles de la Tourette's Symposium in New York City. Almost three hundred physicians and researchers from China, Japan, India, Britain, Belgium, France, Denmark, Germany, Canada, and the United States attended. Participants included all the major North American researchers. Medical panels discussed topics including clinical diagnosis, animal models, genetics, epidemiology, and clinical pharmacology.[108]

"The Tourette Syndrome Association," proclaimed its president, Judy Wertheim, in the spring of 1982, "is directly or indirectly responsible for all the research currently being conducted on behalf of Tourette patients. This statement may appear to be overly presumptuous, but it is not! Ask yourself who was doing research on Tourette Syndrome more than ten years ago? Who understood or cared about the plight of victims with Tourette? Practically no one!" According to Wertheim, "the endeavors of the TSA, both medical and general publicity, have made everything happen."[109] Arthur Shapiro told the tenth anniversary membership meeting in May, "We cannot discuss Tourette Syndrome without discussing the TSA, which has advanced the recognition of this disorder, educated countless numbers of educators, psychologists, social workers, and communities, and aided patients and families, contributed to disseminating information and understanding of the illness, and is now supporting research."[110]

Although the remarkable success of the TSA owed much to the energy of its leaders, it also drew on a wider North American historical tradition that authorized narrowly focused grassroots movements to call professional expertise into question. In contrast, the resilience of a psychoanalytic frame for convulsive tics in France reflects, at least in part, a counter historical tradition, in which laypersons continued to defer to experts, especially in the arena of medicine and disease.

From this perspective, Lucien Israël's critique is exactly correct. That is, Tourette syndrome was the creation of the Tourette Syndrome Association. Although we do not have to endorse Israël's corollary that this creation represents a fictional disorder, we can remain persuaded that without such support groups Anglo–North American ticcing patients might well have continued to be viewed and treated as suffering from a psychogenic disorder, as they are in France today. In a way, the current power of psychoanalysis in France mirrors the situation that faced the Shapiros and their patients in the late 1960s.

## Conclusion

The difference between French and Anglo–North American medical views about Tourette syndrome reminds us that medical research and clinical findings in themselves often are insufficiently persuasive in the face of long-held assumptions about the etiology and treatment of syndromes. Rather, as with the case of Tourette's, successful challenge to a longstanding medical hypothesis requires not only new and robust clinical findings but also the cultural, political, and economic support from the afflicted and their families. Moreover—and no less important—as with all political movements, the construction of a new medical paradigm always rests on a particular reading of the past. Thus, the development of treatment of involuntary tics with haloperidol and other neuroleptics led to very different histories of TS. In one history, largely produced by North American and British clinicians, Georges Gilles de la Tourette emerges as focal point and hero of the story, whereas from a psychoanalytic point of view, Gilles de la Tourette disappears entirely. From this perspective this chapter has been less interested in deciding which, if in fact any, of these histories is "correct" than in examining how the different views that medical researchers hold about the onset and course of this condition have led them to read the past in a certain way.

NOTES

1. American Psychiatric Association, *Diagnostic and Statistical Manual of Mental Disorders,* 4th ed. rev. (Washington, D.C.: American Psychiatric Association, 1994), 101–103.

2. Caroline M. Tanner and S. M. Goldman, "Epidemiology of Tourette Syndrome," *Neurology Clinics* 15 (1997): 395–402; Caroline M. Tanner, "Epidemiology," in *Handbook of Tourette's Syndrome and Related Tic and Behavioral Disorders,* ed. Roger Kurlan (New York: Marcel Dekker, 1993), 337–344, see esp. 339–341; Theodore Fallon Jr. and Mary Schwab-Stone, "Methodology of Epidemiological Studies of Tic Disorders and Comorbid Psychopathology," in *Tourette Syndrome: Genetics, Neurobiology, and Treatment,* ed. Thomas J. Chase, Arnold J. Friedhoff, and Donald J. Cohen (New York: Raven Press, 1992).

3. Anne Mason, Sube Banerjee, Valsamma Eapen, Harry Zeitlin, and Mary Robertson, "The Prevalence of Gilles de la Tourette's Syndrome in a Mainstream School Population: A Pilot Study," *Developmental Medicine and Child Neurology* 40 (1998): 292–296. Mason et al. claim that "TS . . . is more common, milder and is associated with fewer other psychopathologies than has been suggested by previous studies" (296); Also see Valsamma Eapen, Mary M. Robertson, Harry Zeitlin, and Roger Kurlan, "Gilles de la Tourette's Syndrome in Special Education Schools: A United Kingdom Study," *Journal of Neurology* 244 (1997): 378–382.

4. David E. Comings, J. A. Himes, and Brenda J. Comings, "An Epidemiological Study of Tourette's Syndrome in a Single School District," *Journal of Clinical Psychiatry* 51 (1990): 463–469.

5. Howard I. Kushner, *Self-Destruction in the Promised Land: A Psychocultural Biology of American Suicide* (New Brunswick, N.J.: Rutgers University Press, 1989), 114–117.

6. Terra Ziporyn, *Nameless Diseases* (New Brunswick, N.J.: Rutgers University Press, 1992), 1–2.

7. Alfred S. Evans, *Causation and Disease: A Chronological Journey* (New York: Plenum, 1993), 2. Also see David E. Comings, "Tourette's Syndrome: A Behavioral Spectrum Disorder," in *Advances in Neurology: Behavioral Neurology of Movement Disorders*, ed. William J. Weiner and Anthony E. Land (New York: Raven Press, 1995), 293–303; Mary M. Robertson, "The Gilles de la Tourette Syndrome: The Current Status," *British Journal of Psychiatry* 154 (1989): 147–169.

8. Arthur K. Shapiro and Elaine S. Shapiro, "Evaluation of the Reported Association of Obsessive-Compulsive Symptoms or Disorder with Tourette's Disorder," *Comprehensive Psychiatry* 33 (1992): 152–165.

9. David E. Commings, "Tourette Syndrome: A Hereditary Neuropsychiatric Spectrum Disorder," *Annals of Clinical Psychiatry* 6 (1994): 235–247.

10. Not that there hasn't been a concerted effort. See The Tourette Syndrome Classification Study Group, "Definitions and Classifications of Tic Disorders," *Archives of Neurology* 50 (1993): 1013–1016.

11. DSM-IV, 101–103.

12. Donald Cohen to Sue Levi-Pearl (re: DSM-IV Criteria), June 2, 1994, Tourette Syndrome Association Archives, Bayside, Long Island, N.Y. [hereafter TSA Archives]. Roger Freeman, Diane Fast, and M. Kent, "DSM-IV Criteria for Tourette's," *Journal of the American Academy of Child and Adolescent Psychiatry* 34 (1995): 400; David E. Comings, "Letter to the Editor, re: DSM-IV," *Journal of the American Academy of Child and Adolescent Psychiatry* 34 (1995): 401; M. A. First, A. Frances, and H. Pincus, "DSM-IV Editors Reply," *Journal of the American Academy of Child and Adolescent Psychiatry* 34 (1995): 402. Drs. Arthur K. and Elaine S. Shapiro, interview with author, September 21, 1994, Scarsdale, N.Y.

13. James F. Leckman, Bradley S. Peterson, George M. Anderson, Amy F. T. Arnsten, David L. Pauls, and Donald J. Cohen, "Pathogenesis of Tourette's Syndrome," *Journal of Child Psychology and Psychiatry* 38 (1997): 119–142; see discussion on 122–123. For a similar view, see Roger Kurlan, "Future Direction of Research in Tourette Syndrome," *Neurologic Clinics* 15 (1997): 451–456.

14. For a comprehensive discussion of this history, see Howard I. Kushner, *A Cursing Brain? The Histories of Tourette Syndrome* (Cambridge: Harvard University Press, 1999).

15. Sue Levi-Pearl, "The Tourette Syndrome Association, Inc.," in Kurlan, *Handbook of Tourette's Syndrome*, 515–521.

16. The ICD-9, like its predecessors, is organized around etiology. In contrast, the ICD-10, like the DSM, is structured to reflect symptom presentation of psychiatric disorders. In both ICD-9 and -10, there is a category labeled "tic disorders," but following

the French psychiatric view, ICD-9 does not list Tourette's as a separate subcategory. Rather, ICD-9 has a subcategory, "convulsive tics" (307.20), that also includes both "habit" tics and "psychogenic" tics. In addition, in the ICD-9 there is a separate subcategory for "compulsive" tics (307.22). In contrast, the ICD-10 has a subcategory (F95.2) designated "combined vocal and multiple tic disorder [de la Tourette]" and has eliminated the subcategories for habit, psychogenic, or compulsive tics (p. 384). Finally, the ICD-9 is entitled *International Classification of Diseases,* but the ICD-10, again reflecting the DSM, added the word "Statistical" to its title. Largely owing to French insistence, the ICD-9 continues to be updated annually and exists as a diagnostic tool side by side with the ICD-10. *International Classification of Diseases,* 9th rev., Clinical Modification, 4th ed., 3 vols. (Geneva: World Health Organization, 1995); *International Statistical Classification of Diseases and Related Health Problems,* 10th ed., 3 vols. (Geneva: World Health Organization, 1992), 3: 384. *International Classification of Diseases,* 9th rev., Clinical Modification, 4th ed., 3 vols. (Geneva: World Health Organization, 1995).

For a discussion of French psychiatry's rejection of the DSM in favor of the ICD, see L. Singer, "La place de la psychiatrie française dans la psychiatrie européenne," *Annales Médico Psychologiques* 151 (1993): 256–259, esp. 256. For the Binet system, see Martine Lefevre and Serge Lebovici, "L'application de la nouvelle classification americaine dite DSM III, a la psychiatrie de l'enfant et de l'adolescent," *Psychiatrie de l'Enfant,* 26 (1983): 459–505; Serge Lebovici, "Note a propos de la classification des troubles mentaux en psychopathologie de l'enfant," *Psychiatrie de l'Enfant* 31 (1988): 135–149.

17. Lucien Israël, "Preface to Henry Meige," in *Le Juif-Errant à La Salpêtrière,* ed. Lucien Israël (Paris: Collection Grands Textes, Nouvelle Objet, 1993), 4–5. See also S. Lebovici, J.-F. Rabain, T. Nathan, R. Thomas, and M.-M. Duboz, "A propos de la maladie de Gilles de la Tourette," *Psychiatrie de L'Enfant* 29 (1986): 5–59, esp. 57–58; Tobie Nathan, *L'influence qui guérit* (Paris: Éditions Odile Jacob, 1994), 223–332.

18. Cyrille Koupernik to author, August 28, 1995, letter in author's possession.

19. The woman, Bridget Haardt, reports that, although her child had classic symptoms, French physicians refused to make a diagnosis of Tourette's. Haardt finally took her son back to the United States, where he was quickly diagnosed. Bridget Haardt to Chris Melbye, February 7, 1997; Sue Levi-Pearl to Haardt, March 26, 1997, both in TSA Archives.

20. Pierrick Hordé, *Novelles histories incroyable de la médecine* (Paris: Editions Filipacchi, 1994). See his discussion of TS on 55–73.

21. Georges Gilles de la Tourette, "Étude sur une affection nerveuse caractérisée par de l'incoordination motrice accompangnée d'écholalie et de coprolalie (jumping, latah, myriachit)" [Study of a nervous affliction characterized by motor incoordination, accompanied by echolalia and coprolalia (jumping, latah, myriachit)], *Archives de Neurologie* 9 (1885): 19–42, 158–200. For a complete discussion of Gilles de la Tourette's claims, see Kushner, *A Cursing Brain?,* 21–44.

22. *Revue Neurologique* 150, nos. 8–9 (1994).

23. Gilles de la Tourette, "Étude sur une affection nerveuse," 19–42, 158–200.

24. Ibid., 184

25. Jean Martin Charcot, *Leçons du Mardi à la Salpêtrière Policliniques, 1887–1888,* Notes de Cours de MM. Blin, Charcot, et Colin. Handwritten and printed (Paris: Bureaux du Progrès Médical, 1887–1888), 86–87.

26. Georges Gilles de la Tourette, "La maladie des tics convulsifs," *La Semaine Médicale* 19 (1899): 153–156; Gilles de la Tourette, "Étude sur une affection nerveuse," 171, 188, 194, 200.

27. Théodule Ribot, *Les maladies de la volonté* (Paris: Félix Alcan, 1883); Gilles de la Tourette, "Étude sur une affection nerveuse," 200.

28. Ian R. Dowbiggin, *Inheriting Madness: Professionalization and Psychiatric Knowledge in Nineteenth-Century France* (Berkeley: University of California Press, 1991), 1–10, 116–143.

29. Charcot, *Leçons du Mardi,* 61.

30. Ribot, *Les maladies de la volonté,* 66–67; Jean M. G. Itard. "Mémoire sur quelques fonctions involontaires des appareils de la locomotion, de la préhension et de la voix," *Archives Générales de Médecine* 8 (1825): 385–407; see 405.

31. Gilles de la Tourette, "Étude sur une affection nerveuse," 194. None of Gilles de la Tourette's actual clinical observations, however, sustained his claims about the course or outcome of the disorder. Although Dampierre lived to be eighty-five years old, contrary to a variety of assertions since 1885, neither Gilles de la Tourette nor Charcot ever examined or formally met her. See Howard I. Kushner, "Medical Fictions: The Case of the Cursing Marquise and the (Re)construction of Gilles de la Tourette's Syndrome," *Bulletin of the History of Medicine* 69 (1995): 224–254.

32. Georges Guinon, "Sur la maladie des tics convulsifs," *Revue de Médecine* 6 (1886): 50–80; Georges Guinon, "Tics convulsifs et hystérie," *Revue de Médecine* 7 (1887): 509–519; Edouard Brissaud, "La chorée variable des dégénérés," *Revue Neurologiqu,* 4 (1896): 417–431.

33. Howard I. Kushner, "Freud and the Diagnosis of Gilles de la Tourette's Illness," *History of Psychiatry* 8 (1998): 1–25; see 8–21.

34. Charcot, *Leçons du Mardi,* 129, 152–153, 209–213; Jean Martin Charcot, "Des tics et des tiqueurs," *La Tribune Médicale* 19 (1888): 571–573.

35. Jacques Catrou, *Étude sur la maladie des tics convulsifs (jumping-latah—myriachit),* thèse pour le doctorat en médecine (Paris: Faculté de Médecine de Paris, Henri Jouve, 1890), 40–45; Kushner, "Freud and the Diagnosis of Gilles de la Tourette's Illness," 17–18.

36. Henry Meige and E. Feindel, *Les tics et leur traitement,* with a preface by Professor Brissaud (Paris: Masson, 1902).

37. Henry Meige and E. Feindel, *Tics and Their Treatment,* with a preface by Professor Brissaud, trans. and ed. S. A. K. Wilson (New York: William Wood and Co., 1907).

38. Ibid., 225.

39. Ibid., vans.

40. Sandor Ferenczi, "Psycho-Analytical Observation on Tic," *International Journal of Psycho-Analysis* 2 (1921): 1–30. Ferenczi distinguished obsessive behaviors from tics.

"Obsessive actions are chiefly differentiated from the tics," wrote Ferenczi, "by their greater complexity; they are real actions that aim at alteration of the external world (chiefly in an ambivalent sense) and in which narcissism plays no part or else a subordinate one" (30 ).

41. Margaret Schoenberger Mahler and Leo Rangell, "A Psychosomatic Study of Maladie de Tics (Gilles de la Tourette's Syndrome)," *Psychiatric Quarterly* 17 (1943): 579–603, quotation from 593.

42. For a contemporary discussion of the rarity of the diagnosis, see W. P. Mazur, "Gilles de la Tourette's Syndrome," *Canadian Medical Association Journal* 69 (1953): 520–522, esp. 520.

43. *Index Catalogue of the Surgeon General's Office* 6 (1941): 291.

44. Josef E. Heuscher, "Intermediate State of Consciousness in Patients with Generalized Tics," *Journal of Nervous and Mental Disorders* 117 (1953): 29–38, quoted in R. P. Michael, "Treatment of a Case of Compulsive Swearing," *British Medical Journal* 1 (1957): 1506–1508, quotation from 1507; also quoted and endorsed in Leon Eisenberg, Eduard Ascher, and Leo Kanner, "A Clinical Study of Gilles de la Tourette's Disease (Maladie des Tics) in Children, *Journal of American Psychiatry* 115 (1959): 715–723; see 722.

45. Z. Alexander Aarons, "Notes on a Case of Maladie des Tics," *Psychoanalytic Quarterly* 27 (1958): 194–204, quotation from 203–204.

46. See Solomon H. Snyder, *Drugs and the Brain* (New York: W. H. Freeman & Co., 1986), 71–73; Nancy C. Andreasen, *The Broken Brain: The Biological Revolution in Psychiatry* (New York: Harper and Row, 1985), 192–193.

47. S. Bockner, "Gilles de la Tourette's Disease," *Journal of Mental Science* 105 (1959): 1078–1081, quotation from 1080.

48. This point of view persisted until the 1980s. For instance, see Stanton Peele, "Reductionism in the Psychology of the Eighties: Can Biochemistry Eliminate Addiction, Mental Illness, and Pain?," *American Psychologist* 36 (1981): 807–818.

49. M. J. N. Seignot, "Un cas de maladie de tics de Gilles de la Tourette guéri par le R-1625," *Annales Médico Psychologique* 119 (1961): 578–79; G. Caprini and V. Melott,. "Una grave sindrome ticcosa guarita con halopéridol," *Revista Sperimentale Freniatra* 86 (1962): 191–196; George Challas and William Brauer, "Tourette's Disease: Relief of Symptoms with R.1625," *American Journal of Psychiatry* 120 (1963): 283–284; James L. Chapel, Noel Brown, and Richard L. Jenkins, "Tourette's Disease: Symptomatic Relief with Haloperidol," *American Journal of Psychiatry* 121 (1964): 608–610; Diane H. Kelman. "Gilles de la Tourette's Disease in Children: A Review of the Literature," *Journal of Child Psychology and Psychiatry* 6 (1965): 219–226.

50. For instance, a group of Iowa State University Medical School psychiatrists reported that although their patients with multiple motor tics and coprolalia had improved dramatically when treated with haloperidol, "it should be obvious that the problem of Tourette's disease is not resolved by control of the tics through medication." James L. Chapel, Noel Brown, and Richard L. Jenkins, "Tourette's Disease: Symptomatic Relief with Haloperidol," *American Journal of Psychiatry* 121 (1964):

608–610, quotation from 610. For a more detailed discussion see Kushner, *A Cursing Brain?*, 133–143.

51. The original report of this case appeared in Arthur K. Shapiro and Elaine Shapiro, "Treatment of Gilles de la Tourette's Syndrome with Haloperidol," *British Journal of Psychiatry* 114 (1968): 345–350. (Parenthetical page numbers that appear subsequently in the text are from this article.) A more accessible version serves as the opening for the introductory chapter of Arthur K. Shapiro, Elaine S. Shapiro, Ruth D. Bruun, and Richard D. Sweet, *Gilles De La Tourette Syndrome* (New York: Raven Press, 1978), 1–9, quotation from 1.

52. Frank R. Ford, *Diseases of the Nervous System in Infancy, Childhood, and Adolescence,* 5th ed. (Springfield, Ill.: Charles S. Thomas, 1966), 1292–1293.

53. Shapiro and Shapiro, "Treatment of Gilles de la Tourette's Syndrome with Haloperidol," 347.

54. Shapiro and Shapiro interview, September 21, 1994.

55. Shapiro and Shapiro, "Treatment of Gilles de la Tourette's Syndrome with Haloperidol," 349.

56. Arthur K. Shapiro, "Symposium on Gilles de la Tourette's Syndrome," Panel Roundtable at American Psychiatric Association Meeting, May 16, 1968, published in the *New York State Journal of Medicine* (September 1, 1970): 2193–2214.

57. Arthur K. Shapiro, "Dangers of Premature Psychologic Diagnosis," *New York State Journal of Medicine* (September 1, 1970): 2210–2213, quotation from 2211. The Shapiros would return to this theme often. See Arthur K. Shapiro and Elaine Shapiro, "Clinical Dangers of Psychological Theorizing, Giles Tourette Syndrome," *Psychiatry Quarterly* 45 (1971): 159–171.

58. Arthur K. Shapiro, "Discussion," *New York State Journal of Medicine* (September 1, 1970): 2214.

59. Arthur K. Shapiro, Elaine Shapiro, and Henriette L. Wayne, "Treatment of Tourette's Syndrome with Haloperidol, Review of 34 Cases," *Archives of General Psychiatry* 28 (1973): 96.

60. Arthur K. Shapiro, Elaine Shapiro, Henriette L. Wayne, and John Clarkin, "Organic Factors in Gilles de la Tourette's Syndrome," *British Journal of Psychiatry* 122 (1973): 659–664, quotation from 663. Similar arguments are found in Arthur K. Shapiro, Elaine Shapiro, Henriette L. Wayne, John Clarkin, and Ruth D. Bruun, "Tourette's Syndrome: Summary of Data on 34 Patients," *Psychosomatic Medicine* 35 (1973): 419–435.

61. "Testimonial of a Mother and Son Diagnosed in 1975," reprinted in Shapiro et al., *Gilles de la Tourette Syndrome,* 399–402, quotation from 402.

62. "Experience of a Father with a Son with Tourette Syndrome," June 1975, reprinted in Shapiro et al., *Gilles de la Tourette Syndrome,* 402–405, quotations from 402, 405.

63. "Experience of a Mother and Two Children," August 1975, reprinted in Shapiro et al., *Gilles de la Tourette Syndrome,* 405–408, quotations from throughout.

64. As a result of Nazi takeovers in the 1930s, many central and western European

psychoanalysts immigrated to the United States, Britain, or Canada, where, by the end of the Second World War, they joined forces and strengthened an already growing native psychoanalytic movement. By the end of the Second World War psychoanalytic psychiatrists dominated almost every aspect of American psychiatric psychiatry, from medical schools to control of the editorial boards of psychiatric journals. Psychoanalytic psychotherapy of one sort or another flourished, even among that majority of practitioners who had not been trained in psychoanalytic institutes and thus were not "officially" designated as psychoanalysts. But, as the discussion in this chapter indicates, the discovery of the effect of neuroleptics in the 1960s led to a revolution by somatic psychiatrists in the 1970s, which by the late 1980s had all but marginalized psychoanalytic psychiatry in the United States and Canada and, to a lesser extent, in Britain as well.

65. For instance, see J. P. Guéguen, "Le syndrome de Gilles de la Tourette (maladie des tics)," *Soins Psychiatrie* 110/111 (1989): 25–27. Guéguen argues that despite current organic and genetic explanations, *maladie des tics* is a psychopathological disorder as described by Lebovici. Also see Lebovici et al., "A propos de la maladie de Gilles de la Tourette," 5–59.

66. Janet quoted in Serge Lebovici, *Les tics chez l'enfant* (Paris: Presses Universitaires de France, 1951), 131.

67. This story is told in great detail in Elisabeth Roudinesco, *Jacques Lacan & Co.: A History of Psychoanalysis in France, 1925–1985*, trans. Jeffrey Mehlman (Chicago: University of Chicago Press, 1990), esp. 183–184. See also Alain De Mijolla, "La psychanalyse et les psychanalystes en France entre 1939 et 1945," *Revue Internationale d'Histoire de la Psychanalyse* 1 (1988): 167–223. For general background on the history of French psychiatry, see L. Singer, "La place de la psychiatrie française dans la psychiatrie européenne," *Annales Médico Psychologiques* 151(1993): 256–259. Although I do not agree with all his conclusions, see also P. Pichot, "Die Geschichte der deutschen Psychiatrie aus der Sicht der französischen Psychiater," *Fortschritte der Neurologie, Psychiatrie* 60 (1992): 317–328.

68. This was so, even though French psychoanalysts themselves would split into a number of warring factions that would result in serious schisms in the 1950s.

69. Roudinesco, *Jacques Lacan & Co.,* 183–184.

70. Lebovici, *Les tics chez l'enfant.*

71. Joseph Alliez and Serge Audon, "La maladie des tics de Gilles de la Tourette," *Annales Médico-Psychologiques* 2 (1975): 489–522, see 509–510, quotation from 514.

72. For a history of the conflicts between Lacanian and orthodox psychoanalysts, see Roudinesco, *Jacques Lacan & Co.,* 223–477, and Elisabeth Roudinesco, *Jacques Lacan,* trans. Barbara Bray (New York: Columbia University Press, 1997), 399–427.

73. Lebovici, "Les tics chez l'enfant: Introduction à la discussion," in *Les tics chez l'enfant,*169–170, quotation from 169.

74. Ibid., 170

75. P. Moran, "Aspects semniologiques," *Neuropsychiatrie de L'Enfance et de l'Adolescence* 31 (1983): 170–177, see esp. 170.

76. Martine Lefevre and Serge Lebovici, "L'application de la nouvelle classification americaine dite DSM III, à la psychiatrie de l'enfant et de l'adolescent," *Psychiatrie de l'Enfant* 26 (1983): 459–505. Also see Serge Lebovici, "Note à propos de la classification des troubles mentaux en psychopathologie de l'enfant," *Psychiatrie de l'Enfant* 31 (1988): 135–149.

77. Which did not stop some psychoanalysts, especially Jacques Lacan, from criticizing Lebovici for his refusal to support Lacan's revisionist Freudianism. See Roudinesco, *Jacques Lacan & Co.*, 237–238.

78. Israël, *Le Juif-Errant à la Salpêtrière*, 1–19.

79. Henry Meige, "Le Juif-Errant à la Salpêtrière," *Nouvelle Iconographie de la Salpêtrière* (in three parts), 6 (1893): 191–204, 277–291, 333–358.

80. Israël, *Le Juif-Errant à la Salpêtrière*, 5–6.

81. Mr. and Mrs. Martin Levey to Mrs. Schiff (editor of the *New York Post*), December 13, 1970, TSA Archives.

82. Barry Kramer, "Rare Illness Reduces Its Victims to Shouts, Grunts—and Swearing," *Wall Street Journal*, June 20, 1972, 15. The son's name was not actually "Bill."

83. Ibid.; Elaine S. Shapiro, interview with author, May 23, 1997, Scarsdale, N.Y.

84. Judy Wertheim, Elaine Novick, Betti Teltscher, and Erika Feinholtz, interview with author, May 12, 1997, New York City.

85. Abby Avin Belson, "The Tourette Syndrome Association," published as an appendix to Shapiro et al., *Gilles de la Tourette Syndrome*, 409–411; TSA newsletter [hereafter TSA NL] 4 (April 1977).

86. "Gilles": Official Newsletter of the Gilles de la Tourette's Syndrome Association 1 (Spring 1974): 2; TSA NL 4 (April 1977): 1.

87. Sy Goldis, interview with author, October 10, 1997, Jericho, N.Y.

88. Among those articles were Shapiro, Shapiro, and Wayne, "Treatment of Tourette Syndrome with Haloperidol: Review of 34 Cases"; Elaine Shapiro, Arthur K. Shapiro, Richard D. Sweet, and Ruth D. Bruun, "The Diagnosis, Etiology and Treatment of Gilles de la Tourette's Syndrome," in *Mental Health in Children*, vol. 1: *Genetics, Family and Community Studies*, ed. D. V. Siva Sankar (Westbury, N.Y.: PJD Publication, 1975), 167–173; Arthur K. Shapiro and Elaine S. Shapiro, *Tic Syndrome and Other Movement Disorders: A Pediatricians Guide*, pamphlet produced for the TSA and funded by the Laura B. Vogler Foundation, Inc., Bayside, New York, 1980.

89. "Gilles," 2; TSA NL 4 (April 1977): 1.

90. Judy Wertheim, interview with author, May 12, 1997, New York City; Jane Brody, "Bizarre Outbursts of Tourette's Disease Victims Linked to Chemical Disorder in the Brain," *NYT*, May 29, 1975, 70.

91. TSA NL (Spring 1974); TSA NL (Fall 1974); letter to members, October 23, 1974, TSA Archives.

92. TSA NL (January 1975); Oliver W. Sacks to Sheldon Novick, July 25, 1974, London, England, Novick correspondence, in author's possession.

93. Peter B. Neubauer to Novick, July 16, 1976; Novick to Margaret Mahler, July 21, 1975; Novick to Z. Alexander Aarons, August 19, 1975, Novick correspondence.

94. Including Friedhoff's neurotransmitter study, Van Woert's serotonin study, Sacks's neurological and emotional family study; Jose Yaryura-Tobias's (director of research at North Nassau Mental Health Center) chlorimipramine trial for patients who did not respond to haloperidol; and Miller's "new medication trial."

95. TSA Archives.

96. *Saturday Review,* April 3, 1976, 21 (italics in original); *U.S. News & World Report,* July 19m 1976; *Medical Economics,* September 20m 1976. The ad also appeared in the *Madison Avenue Magazine* in June 1976 and *Southern Living Magazine* in November 1976.

97. "New Drug Controls Tourette's Syndrome," *Reader's Digest* (March 1973); "Cure for Foul Mouths," *Science Digest* (September 1973): 53; Daniel Goleman and Jerome Engel Jr., "A Feeling of Falling," *Psychology Today* (November 1976): 107–108.

98. Anita Ricterman, "Problem Line," *Newsday,* August 1, 1975; T. R. Van Dellen, M.D., "The Family Doctor," *New York Daily News,* December 10, 1976. Also see Dr. Lester Coleman, "Speaking of Your Health" (syndicated), September 29, 1976, cited in TSA NL 4 (October 1976): 4.

99. TSA NL 5 (October 1975).

100. Shapiro, Shapiro, and Wayne, "Treatment of Tourette Syndrome with Haloperidol: Review of 34 Cases." Also see TSA NL 5 (October 1975).

101. Cy Goldis to "Friends," February 18, 1975, TSA Archives.

102. TSA NL 6 (July 1976): 1, 4–5. Elaine Novick to TSA members and "Friends," June 24, 1975, TSA Archives.

103. TSA NL 6 (July 1976): 1–2; TSA NL 6 (October 1976): 3–4.

104. TSA NL 9 (Spring 1982): 1–3.

105. Shapiro and Shapiro, *Tic Syndrome and Other Movement Disorders,* 8, 14.

106. The bill provided subsidies for liability insurance for users of these drugs, support for researchers, and funds for studies to determine future needs for orphan drugs. In addition, if no private company could be persuaded to produce these drugs, a federal agency created by the legislation was authorized to do so. TSA NL 8 (April 1982): 2; TSA NL 9 (Spring 1982): 2; TSA NL 9 (Summer 1981): 6–7; TSA NL 10 (Spring 1983): 9.

107. TSA NL 9 (Spring 1982): 8–9; TSA NL 9 (Summer 1982): 11–12.

108. TSA, "Special Report on the First International Gilles de la Tourette Syndrome Symposium," New York, May 27–29, 1981," TSA Archives.

109. Judy Wertheim, "A 10th Anniversary Message. Who We Are, How We Help," TSA NL 9 (Spring 1982): 1–2, quotation from 2.

110. Arthur K. Shapiro, "Remarks at the Tenth Anniversary Membership Meeting of the Tourette Syndrome Association," May 22, 1982, Mount Sinai Hospital, New York City. Reprinted in TSA NL 9 (Summer 1982): 1–2.

CHAPTER 4

# Democracy, Expertise, and Activism for AIDS Treatment

STEVEN EPSTEIN

Several chapters note the divide between the lay concepts of disease held by those suffering from emerging illnesses and the biomedical understandings of medical clinicians and researchers. The success of communities of suffering in getting their illnesses on the public health agenda requires that this divide be bridged. In some cases the link between lay and biomedical perspectives is fostered by outside health professionals sympathetic to the concerns of illness sufferers. The university epidemiologists described by Ellen Spears and the rheumatologists discussed by Deborah Barrett are examples of this kind of outside mediators. In other cases, members of communities of suffering acquire biomedical knowledge relevant to their illness. This "expertization of the lay person" is the focus of Steven Epstein's contribution. Epstein shows how AIDS activists constructed themselves as lay experts to gain a seat at the table where decisions concerning research and treatment protocols were made. The social composition of AIDS activists included a large contingent of college-educated professionals (e.g., physicians, lawyers, and academics) and thus facilitated the acquisition of medical knowledge. In addition, the prior mobilization of the gay and lesbian community around gay rights greatly facilitated its subsequent organization around AIDS.

In looking at the consequences of the success that AIDS activists have had in influencing AIDS research, Epstein notes the danger that those who participate in scientific review panels and other research advisory boards may become distanced from their fellow activists and less attuned to patient interests. Michelle Murphy notes a similar distancing in her essay on the organization 9to5 and sick building syndrome. Here the leaders of the movement become increasingly removed from the experience and concerns of office workers as they attempt to gain scientific credibility for the movement's health concerns.

AIDS activism transformed the ways in which communities of suffering interacted with the medical establishment. The AIDS group ACT UP created a form of "street politics" aimed at influencing medical research—a heath advocacy strategy largely unknown before the AIDS epidemic. AIDS activists also succeeded in injecting themselves into the workings of medical science. Multiple sclerosis and Tourette's support groups formed close alliances with medical researchers, but they made few claims to having knowledge that could contribute to the biomedical project and did not engage in the politics of street theater to influence the course of medical research. On the other hand, more recent communities of suffering, formed around such illnesses as chronic fatigue syndrome, fibromyalgia, and various forms of environmental illness, have attempted to emulate the success of AIDS activists in gaining a seat at the table through a combination of expertization and street politics. (See, for example, Juanne Clarke, "The Search for Legitimacy and the 'Expertization' of the Lay Person: The Case of Chronic Fatigue Syndrome," *Social Work in Health Care* 30, no. 3 [2000]: 72–94.)

---

In 1987, thousands of AIDS activists around the United States began confronting doctors, biomedical researchers, and federal health officials in visually arresting, angry, and provocative demonstrations.[1] Although the targets varied, much of this new wave of AIDS activism focused on the organization and pace of research on AIDS treatments. The messages were not subtle. At one scientific forum, activists handed out cups of Kool-Aid as a prominent researcher came to the podium, likening the effects of his research methods on AIDS patients to the effect of cult leader Jim Jones on his followers in Jonestown, Guyana (Crowley 1991: 40). When the commissioner of the Food and Drug Administration came to speak at a public forum in Boston in 1987, activists in the audience held wristwatches aloft ("FDA allows" 1988): for people with AIDS, these activists implied, time was running out. In October of that year, more than a thousand demonstrators converged on FDA headquarters in Rockville, Maryland, to "seize control" of what some labeled the "Federal Death Administration" (Bull 1988).

Fast forward to 1992: A subset of these same activists now sat as regular voting members on the committees of the AIDS Clinical Trials Group (ACTG), the entity established by the National Institutes of Health to oversee all federally funded AIDS clinical research. Serving alongside the most prominent AIDS researchers in the country (including the one who had been compared to Jim Jones), activists now worked with scientists to determine the most profitable research directions, debate research methodologies, and allocate research funds. Activists also served on institutional review boards at research hospitals around the country, evaluating the methods and ethics of clinical trials of AIDS drugs. At conferences, where once they had shouted from the back of the room, activists now chaired sessions. And their publications, like San Francisco–based *AIDS Treatment News,* had become routine sources of information about AIDS therapies for many doctors around the world. Prominent experts widely acknowledged the aptitude of AIDS treatment activists for understanding such matters as the stages of viral replication, the immunopathogenesis of HIV, and the methodology of the randomized clinical trial. As Dr. John Phair (1994), a former chair of the Executive Committee of the AIDS Clinical Trials Group, commented in 1994: "I would put them up against—in this limited area—many, many physicians, including physicians working in AIDS [care]. They can be very sophisticated."

The unusual social movement trajectory that I have sketched—with the contrast between banging on the doors of biomedicine in 1987 and sitting at the table by 1991—is interesting for all sorts of reasons. Here I would like to focus on the *politics of knowledge and expertise:* How did a grassroots movement produce a cadre of activist-experts? More generally, in what circumstances can patients and laypeople challenge hierarchies of expertise and participate effectively in processes of biomedical knowledge production—or transform such processes? How do they gain entry to these privileged domains, and what are the consequences, intended or unintended, of these kinds of incursions?

What I argue here is that activist movements, through amassing different forms of credibility, can in certain circumstances bring about changes in the epistemological practices of science—our ways of knowing the natural world. Nothing guarantees that such changes will be useful in advancing knowledge or in curing disease, but in this case lay participation in science had some tangible benefits, though not without risks. This is a surprising finding, and one that is, of course, at variance with popular notions of science as a relatively autonomous arena with high barriers to entry. And it runs counter to the view that many might normally voice—that science must be safeguarded from external pressures in order to prevent the deformation of knowledge.[2]

By scientific "credibility," I refer to the capacity of claims makers to enroll supporters behind their claims and present themselves as the sort of people who can give voice to scientific truths.[3] I understand credibility as a form of authority that combines aspects of power, legitimation, trust, and persuasion (Weber 1978: 212–254). However, this case differs from other sociological studies of scientific credibility by suggesting the diversity of the cast of characters who may strive for credibility in scientific controversies and the variety of routes by which credibility may be made manifest. More typically in science the attestations of credibility are recognizable markers like academic degrees, research track records, institutional affiliations, and so on. In the case of AIDS research, when we examine the interventions of laypeople as well as conventional experts, we find a multiplication of the successful pathways to the establishment of credibility, a diversification of the personnel beyond the formally credentialed, and hence more convoluted routes to the resolution of controversy and the construction of belief.

This study also presumes a particular historical moment in which, perhaps especially in the United States, popular attitudes toward science and medicine are highly polarized: a deep faith on the part of the public proceeds hand in hand with skepticism and disillusionment. The emergence of a new epidemic disease, in a society inclined to consider itself as having advanced beyond such mundane risks, had the effect of amplifying this ambivalence. When experts appeared unable to solve the problem of AIDS, the resulting disappointment created space for unconventional voices. Therefore, the study of credibility in AIDS research must also be a study of the public negotiation of the "credibility crisis" surrounding biomedical science—and a study of how the boundaries around scientific institutions become more porous, more open to the intervention of outsiders, precisely in such moments. Particularly when scientific credibility is in crisis, science may become the site of wide-ranging credibility struggles. Again perhaps especially in the United States, interventions by outsiders may get organized in the form of full-fledged social movements.

## Origins of the AIDS Treatment Activist Movement

The AIDS treatment activist movement in the United States is best conceived as a subset of a much larger but considerably more diffuse "AIDS movement" that dates from the early years of the epidemic; that encompasses a wide range of grassroots activists, lobbying groups, service providers, and community-based organizations; and that now represents the diverse interests of people of various races, ethnicities, genders, sexual preferences, and HIV "risk behaviors." The AIDS movement has engaged in manifold projects directed at a variety of social institutions, including the state, the church, the mass media, and the health care sector (Altman 1994; Cohen 1993;

Corea 1992; Crimp and Rolston 1990; Elbaz 1992; Emke 1993; Geltmaker 1992; Indyk and Rier 1993; Patton 1990; Quimby and Friedman 1989; Treichler 1991; Wachter 1991)—though at times it has been less concerned with achieving institutional change than with posing general challenges to cultural norms (Gamson 1989).

In its emergence and mobilization, the AIDS movement was a beneficiary of "social movement spillover" (Meyer and Whittier 1994): it was built on the foundation of other movements and borrowed from their particular strengths and inclinations. Most consequential was the link to the lesbian and gay movement of the 1970s and early 1980s (Adam 1987; Altman 1982, 1986). In the wake of fierce debates in the 1970s over whether homosexuality should be classified as an illness (Bayer 1981), gay men and lesbians were often inclined toward critical or skeptical views of medical authorities (Bayer 1985). More generally, it mattered that gay communities had preexisting organizations that could mobilize to meet a new threat; these community organizations and institutions also provided the face-to-face "micro-mobilization contexts" that are particularly useful in drawing individuals into activism (Lo 1992). It mattered, too, that these communities contained (and were substantially dominated by) white, middle-class men with a degree of political clout and fund-raising capacity unusual for an oppressed group. And it was crucially important that gay communities possessed relatively high degrees of "cultural capital"—cultivated dispositions for appropriating knowledge and culture (Bourdieu 1990). Within these communities are many people who are themselves doctors, scientists, educators, nurses, professionals, or other varieties of intellectuals. On the one hand, this has provided the AIDS movement with an unusual capacity to contest the mainstream experts on their own ground. On the other hand, it affords important sources of intermediation and communication between "experts" and "the public." Treatment activists themselves have tended to be science novices, but ones who were unusually articulate, self-confident, and well educated—"displaced intellectuals from other fields," as Jim Eigo, a New York City treatment activist with a background in the arts, expressed it (Antiviral Drugs Advisory Committee 1991: 50).

The fact that many lesbians (and heterosexual women) who would become active in the AIDS movement were schooled in the tenets of the feminist health movement of the 1970s (Corea 1992; Winnow 1992)—with its skepticism toward medical claims making and insistence upon the patient's decision-making autonomy (Boston Women's Health Book Collective 1973; Fee 1982; Ruzek 1978)—also had important implications for the identity and strategies of the movement. Other activists, both men and women, had prior direct experience in social movements such as the peace movement (Elbaz 1992: 72).

Central to the early goals of the AIDS movement was the repudiation of helplessness or "victim" status and the insistence upon self-representation. "We condemn attempts to label us as 'victims,' which implies defeat, and we are only occasionally 'patients,' which implies passivity, helplessness, and dependence upon others. We are 'people with AIDS,'" read a widely reprinted manifesto of the New York–based People with AIDS (PWA) Coalition (PWA Coalition 1988: 148–149). This is "self-help with a vengeance," as Indyk and Rier (1993: 6) nicely characterize it—an outright rejection of medical paternalism and an insistence that neither the medical establishment, nor the government, nor any other suspect authority would speak on behalf of people with AIDS or HIV.

AIDS activism entered a new and more radical phase in the second half of the 1980s, in the face of increasing concern about the inadequacy of the federal response to the epidemic, the stigmatization of people with AIDS or HIV, and the lack of availability of effective therapies for AIDS or its associated opportunistic infections and cancers. The year 1987 marked the birth, in New York City and then elsewhere around the country, of a new organization, called the AIDS Coalition to Unleash Power but better known by its deliberately provocative acronym, ACT UP (*ACT UP / New York Capsule History* 1991). A magnet for radical young gay men and women in the late 1980s, ACT UP practiced an in-your-face politics of "no business as usual." Adopting styles of political and cultural practice deriving from sources as diverse as anarchism, the peace movement, the punk subculture, and gay liberation "zaps" of the 1970s, ACT UP became famous for its imaginative street theater, its skill at attracting the news cameras, and its well-communicated sense of urgency. ACT UP groups typically had no formal leaders, and meetings in many cities operated by the consensus process.

ACT UP was only the most visible of a diverse set of groups that became interested in issues of medical treatment and research for AIDS around the United States in the mid- to late 1980s—the constellation of organizations that can be called the AIDS treatment activist movement. "Buyers clubs," existing on the fringes of the law, supplied patients with unapproved or experimental treatments smuggled in from other countries or manufactured in basement laboratories. Project Inform, a San Francisco–based organization with a more conventional structure than ACT UP, emerged as an advocate for the use of such experimental therapies and evolved into a multifocal treatment advocacy organization with its own lobbying campaigns, publications, and educational projects. A range of grassroots treatment publications appeared, providing their readers with a rich mix of scientific information, political commentary, and anecdotes about treatments gleaned from patient reports (Arno and Feiden 1992; Kwitny 1992). *AIDS Treatment News,* the best known of these alternative publications, had been advocating for some time for greater attention to be paid to issues of drug research and regulation. "So far, community-based AIDS organizations have been uninvolved in treatment issues, and have seldom followed what is going on," wrote its editor, John James, a former computer programmer, in a call to arms in May 1986. "With independent information and analysis, we can bring specific pressure to bear to get experimental treatments handled properly. So far, there has been little pressure because we have relied on experts to interpret for us what is going on. They tell us what will not rock the boat. The companies who want their profits, the bureaucrats who want their turf, and the doctors who want to avoid making waves have all been at the table. The persons with AIDS who want their lives must be there, too." To "rely solely on official institutions for our information," James bluntly advised, "is a form of group suicide" (James 1986).

### Gaining Credibility

Although treatment activists began, as I described, by employing highly confrontational modes of direct action, they always assumed that effective solutions to AIDS would have to come, in large measure, from doctors and scientists. Therefore, they resisted the notion—found, for example, in the animal rights movement (Jasper and Nelkin 1992)—that the scientific establishment was "the enemy" in an absolute sense. "I wouldn't exaggerate how polite we were," Mark Harrington (1994), one of the leaders of ACT UP / New York's Treatment and Data Committee, reflected. "At the same time, I would just say that it was clear from the very beginning, as Maggie Thatcher said when she met Gorbachev, 'We can do business.' We wanted to make some moral points, but we didn't want to wallow in being victims, or powerless, or oppressed, or always right. We wanted to engage and find out if there was common ground."

How did this rapprochement proceed? In effect, activists (or some subset of them) accomplished an identity shift: they reconstituted themselves as a new species of expert—as laypeople who could speak credibly about science in dialogues with the scientific research community. I cannot here consider in detail the specific tactics that activists employed to construct their scientific credibility (see Epstein 1995), but I would argue that four tactics were most important. First, activists acquired cultural competence by learning the language and culture of medical science. Through a wide variety of methods—including attending scientific conferences, scrutinizing research protocols, and learning from sympathetic professionals both inside and outside the movement—the core treatment activists gained a working knowledge of the medical vocabulary. Second, activists presented themselves as the legitimate, organized voice of people with AIDS or HIV infection (or, more specifically, the current or potential clinical trial subject population). Once activists monopolized the capacity to say "what patients wanted," researchers could be forced to deal with them in order to ensure that research subjects would both enroll in their trials in sufficient numbers and comply with the study protocols.[4] Third, activists yoked together methodological (or epistemological) arguments and moral (or political) arguments, so as to multiply their "currencies" of credibility. For example, activists insisted that the inclusion of women and people of color in clinical trials was not only morally necessary (to ensure equal access to potentially promising therapies) but also scientifically advisable (to produce more fully generalizable data about drug safety and efficacy in different populations). Finally, activists took advantage of preexisting lines of cleavage within the scientific establishment to form strategic alliances. For example, activists struck alliances with biostatisticians in their debates with infectious disease researchers about appropriate clinical trial methodology.

A key victory for activists, at a time when many AIDS researchers remained deeply suspicious of the activist agenda, was the support of Dr. Anthony Fauci, prominent immunologist and AIDS researcher and director of the NIH's Office of AIDS Research. "Something happened along the way," Fauci told a reporter in 1989. "People started talking to each other. . . . I started to listen and read what [activists] were saying. It became clear to me that they made sense" (Garrison 1989: A-1). Of course, Fauci and others may have deemed it strategic to incorporate activists into the process: as Fauci (1994) later commented, the assumption was that "on a practical level, it would be helpful in some of our programs, because we needed to get a feel for what would play in Peoria, as it were." Prominent academic researchers also acknowledged the gradual acquisition of scientific competence on the part of key activists. Dr. Douglas Richman (1994), an important AIDS researcher from the University of California, San Diego, described how Harrington of ACT UP / New York, in an early meeting with researchers, "got up and gave a lecture on CMV [cytomegalovirus] . . . that I would have punished a medical student for—in terms of its accuracy and everything else—and he's now become a very sophisticated, important contributor to the whole process."

As this encounter between different social worlds unfolded, activists pressed for more substantial degrees of inclusion in the NIH decision-making apparatus. Crucial decisions about clinical trials—which ones to fund, but also quite specific details about how the trials should be conducted, how the data should be analyzed, and which patients should be eligible to participate—were being made by the academic researchers who made up the advisory committees of the ACTG. To the consternation of the researchers, activists demanded representation on these committees; when Fauci stalled, activists decided to "Storm the NIH." This demonstration, at the NIH campus in Bethesda, Maryland, on May 21, 1990, proved to be another graphic media spectacle, like the FDA protest two years earlier (Hilts 1990). Soon afterward, activists were informed that most ACTG meetings would be opened to the public and that there would be a representative of the patient community, with full voting rights, assigned to each ACTG committee. In addition, by the early 1990s, activists acted as informal representatives to FDA advisory committees charged with evaluating new drugs, as appointed members of "community advisory boards" established by pharmaceutical companies, and as regular members of "institutional review boards" supervising clinical studies at hospitals and academic centers around the country. Activists began to have an important say in how studies were conducted, which patients were allowed into studies, how results from studies were evaluated, and which lines of research should be funded (Arno and Feiden 1992; Epstein 1996; Jonsen and Stryker 1993; Kwitny 1992).

## Consequences

A defining moment was the publication in the *New England Journal of Medicine,* in November 1990, of an article by AIDS researcher Thomas Merigan (1990) of Stanford, "You *Can* Teach an Old Dog New Tricks: How AIDS Trials Are Pioneering New Strategies." Praising the new "partnership of patients, their advocates, and clinical investigators," Merigan proceeded to endorse precisely those methodological stances that activists had promoted. He argued, for example, that "all limbs [of a trial] should offer an equal potential advantage to patients, as good as the best available clinical care"; that no one in a trial should be denied treatment for his or her opportunistic infections; that trials should not be "relentlessly pursued as originally designed" when "data appeared outside the trial suggesting that patients would do better with a different type of management"; and that "the entry criteria for trials should be as broad as scientifically possible to make their results useful in clinical practice" (1341–1343).

By pressing researchers to develop clinically relevant trials with designs that research subjects would find acceptable, activists helped to ensure more rapid accrual of the required numbers of subjects and to reduce the likelihood of noncompliance. And by working toward methodological solutions that satisfy, simultaneously, the procedural concerns of researchers and the ethical demands of the patient community, AIDS activists have, at least in specific instances, improved a tool for the production of scientific facts in ways that even researchers acknowledge. In this sense, AIDS activists' efforts belie the commonplace notion that only the insulation of science from "external" pressures guarantees the production of secure and trustworthy knowledge.

Those who were critical of lay participation in medical research were quick to suggest, and with some reason, that AIDS activists had muddied the waters of knowledge in their haste to see drugs approved. Yet any such assessment has to consider the larger picture. Absent the activists, what sort of knowledge strategies would have been pursued? Pristine studies addressing less-than-crucial questions? Methodologically unimpeachable trials that failed to recruit or maintain patients? Inevitably, there are risks inherent in the interruption of the status quo. But these must be weighed against all the other attendant risks, including those that may follow from letting normal science take its course while an epidemic rages.

One reason this case is so important is because it is quite conceivable that these changes in the arena of AIDS research will have an enduring impact on biomedicine in the United States.[5] The past decade has seen a marked upsurge of health-related activism of a distinctive type: the formation of groups that construct identities around particular disease categories and assert political and scientific claims on the basis of these new identities. Just as the AIDS movement drew on the experiences of other movements that preceded it, now its own tactics and understandings have begun to serve as a model for a new series of challengers.

Most notably patients with breast cancer, but also those suffering from chronic fatigue, environmental illness, prostate cancer, mental illness, Lyme disease, Lou Gehrig's disease, and a host of other conditions, have displayed a new militancy and demanded a voice in how their conditions are conceptualized, treated, and researched (Barinaga 1992; Kingston 1991; Kroll-Smith and Floyd 1997). These groups have criticized not only the quality of their care but also the ethics of clinical research ("Are placebo controls acceptable?") and the control over research directions ("Who decides which presentations belong on a conference program?"). Although not every such group owes directly to AIDS activism, the tactics and political vocabulary of organizations like ACT UP would seem, at a minimum, to be "in the wind."

(Could one imagine, before the AIDS activist repudiation of "victimhood," people with muscular dystrophy denouncing the Jerry Lewis Telethon as an "Annual Ritual of Shame" and chanting "Power, not pity" before the news cameras ["MD Telethon" 1991]?) Few of these constituencies have engaged in epistemological interventions that approach, in their depth or extent, AIDS treatment activists' critiques of the methodology of clinical trials. But Bernadine Healy, then the director of the NIH, got it right in 1992 when she told a reporter: "The AIDS activists have led the way. . . . [They] have created a template for all activist groups looking for a cure" (Gladwell 1992).

Breast cancer activism is an intriguing instance of this new wave because the links to AIDS activism have been so explicit and so readily acknowledged. In 1991, more than 180 advocacy groups in the United States came together to form the National Breast Cancer Coalition. "They say they've had it with politicians and physicians and scientists who 'there, there' them with studies and statistics and treatments that suggest the disease is under control," read a prominent account in the *New York Times Sunday Magazine* (Ferraro 1993: 26). In its first year of operation, the coalition convinced Congress to step up funding for breast cancer research by $43 million, an increase of almost 50 percent. "The next year, armed with data from a seminar they financed, the women asked for, wheedled, negotiated and won a whopping $300 million more" (Ferraro 1993: 27). The debt to AIDS activism was widely noted by activists and commentators alike. "They showed us how to get through to the Government," said a Bay Area organizer with breast cancer. "They took on an archaic system and turned it around while we have been quietly dying." Another activist described how she met with the staff of *AIDS Treatment News* to learn the ropes of the drug development and regulatory systems (Gross 1991).

Of course, it would be rash to assume that AIDS activism has created an automatic receptiveness on the part of scientists or doctors to health movements of this sort and that the next round of activists can simply step up to the counter and claim their rewards. A more likely scenario is that AIDS activism will usher in a new wave of democratization struggles in the biomedical sciences and health care—struggles that may be just as hard fought as those of the past decade. It is worth remembering, too, how difficult this sort of activism is to sustain: organizing a social movement is arduous enough, without having to learn oncology in your spare time.

### Complications

Other qualifications to this story deserve notice. Certainly it should be clear that such activism, no matter how broad ranging it becomes, is unlikely to bring about the thorough transformation of the knowledge-based hierarchies that structure the society in which we live. In fact, my analysis suggests a profound tension built into AIDS treatment activists' own project of democratizing expertise. On the one hand, by pursuing an educational strategy to disseminate AIDS information widely, activists promoted the development of broad-based knowledge empowerment at the grass roots. On the other hand, as treatment activist leaders became quasi experts, they tended to replicate the expert/lay divide within the movement itself: a small core of activists became insiders who "knew their stuff"; others were left outside to protest in more traditional ways. Furthermore, as many of the treatment activists moved "inward," took their seat at the table, and became sensitized to the logic of biomedical research, their conceptions of scientific methods sometimes turned in more conventional directions.

"I've seen a lot of treatment activists get seduced by the power, get seduced by the knowledge, and end up making very conservative arguments," contends Michelle Roland (1993), formerly active with ACT UP in San Francisco. "They understand . . . the methodology, they can make intelligent arguments, and it's like, 'Wait a minute . . . okay, you're smart. We accept that. But what's your role?'" Ironically, insofar as activists start thinking like scientists and not like patients, the grounding for their unique contributions to the science of clinical trials may be in jeopardy of erosion. Researcher John Phair (1994) notes that activists "have given us tremendous insight into the feasibility of certain studies," but he adds that "some of the activists have gotten very sophisticated and then forget that the idea might not sell" to the community of patients.

Can one be both activist and scientist? Is the notion of a "lay expert" a contradiction in terms? I don't think there are any simple answers here. But arguably, it was not possible for the key treatment activists to become authorities on clinical trials and sit on the ACTG committees without, in some sense, growing closer to the worldview of the researchers—and without moving a bit away from their fellow activists engaged in other pursuits. Furthermore, the new hierarchies of expertise that emerged within the ranks of the activists did, to a certain, predictable degree, superimpose themselves upon the bedrock of other dimensions of social inequality, including racial, gender, and class differences among activists. And this led to sharp tensions and outright splits within several activist organizations (DeRanleau 1990; Epstein 1996: 290–294; Epstein 1997; Vollmer 1990).

These questions of identity and strategy among knowledge-empowered social movements deserve extended attention. Here I intend only to suggest two implications of my analysis. First, the AIDS activist project of reconfiguring the knowledge-making practices of biomedicine has been executed in ways that are tentative, partial, and shot through with some powerful contradictions. Second, it proves to be not enough to ask what impact AIDS activists have had on the conduct of biomedical research. In addition, we need to ask the reciprocal question: What impact does the encounter with science have upon the social movement? How does the "expertification" or "scientization" of activism affect the goals and tactics of a social movement, as well as its collective identity (Epstein 1997)? Without doubt, the reciprocal relation between AIDS activists and AIDS researchers was equally transformative in both directions.

Let me now raise a final worry about the democratization of science: the obvious risk that lay participation will interfere with the good conduct of science and indeed delay the goals that all want to see achieved. What is to prevent real harm from being done? In this regard, it is important to note that the process of activist intervention in biomedicine is not without some painful ironies. On the one hand, the enterprise appears significantly driven by the dictates of expediency and dire need—"I'm dying so give me the drug now!"—yet on the other hand, the core treatment activists have increasingly become believers in science (however understood) and desperately want clinical trials to generate usable knowledge that can guide medical practice. As David Barr of the Treatment Action Group (a spin-off of ACT UP / New York) put it: "My doctors and I make decisions in the dark with every pill I put in my mouth" (Cotton 1991: 1362)—and this is not an easy way to live.

Insofar as activists want clinical trials to succeed, they must wrestle with the consequences of their own interventions. Do such interventions enhance activists' capacity to push clinical research in the directions they choose? Or do activists and researchers alike become subject to the unintended effects of their actions, trapped within an evolving system whose trajectory no one really controls? Here is a sort of worst-case scenario of the spiraling consequences of community-based interventions in the construction of belief in antiviral drugs—a caricature sketch, to be sure, but one that combines elements from a number of cases in the late 1980s and early 1990s. Drug X performs well in preliminary studies, and an NIH official is quoted as saying that X is a promising drug. The grassroots treatment publications write that X is the up-and-coming thing; soon everyone in the community wants access to X, and activists are demanding large, rapid trials to study it. Everyone wants to be in the trial because they believe that X will help them; but researchers want to conduct the trial in order to determine whether X has any efficacy. Those who cannot get into the trial demand expanded access, while others begin importing X from other countries or manufacturing it in clandestine laboratories. As X becomes more prevalent and emerges as the de facto standard of care, physicians begin to suggest that patients get hold of it however they can. Meanwhile, participants in the clinical trial of X who fear they are receiving a placebo mix and match their pills with other participants. When the trial's investigators report potential treatment benefits, activists push for accelerated approval of X, leaving the final determination of X's efficacy to postmarketing studies. But who then wants to sign up for those studies, when everyone now believes that the drug works, since, after all, the FDA has licensed it and any doctor can prescribe it?

This is a scary scenario, but it must be pointed out that activists themselves have sought to control this troubling escalation and to extricate themselves from what they rightly call the "hype cycle." As Mark Harrington wrote in late 1993: "One disturbing but inevitable result of the urgency engendered by the AIDS crisis is that both researchers and community members tend to invest preliminary trials with more significance than they can possibly bear" (7). To the extent that activists can develop a critique of this phenomenon of expecting too much from research, and to the extent that they can communicate the *relative uncertainty* of clinical trials to the broader public of HIV-infected persons, it may be possible to imagine a clinical research process that more fully reflects the interests of those who are most in need of answers.[6]

#### NOTES

An earlier version of this material appeared in "Democracy, Expertise, and AIDS Treatment Activism," in *Science, Technology, and Democracy,* ed. Daniel Kleinman (Albany: State University of New York Press, 2000).

1. This analysis of AIDS treatment activists derives from a larger research project concerned with studying the conduct of science in the AIDS epidemic and the role of laypeople, particularly activists, in the transformation of biomedical knowledge-making practices (see Epstein 1996). This account is based on interviews conducted with AIDS activists, AIDS researchers, and government health officials at the National Institutes of Health and the Food and Drug Administration, as well as on analyses of the accounts, claims making, and framing of issues presented in scientific and medical journals, the mass media, the gay and lesbian press, activist publications, activist documents, and government documents.

2. On the politics of public participation in science and medicine, see, for example, Balogh 1991; Blume et al. 1987; Brown 1992; Cozzens and Woodhouse 1995; Di Chiro 1992; Indyk and Rier 1993; Irwin and Wynne 1996; Kleinman 1995; Martin 1980; Moore 1996; Nelkin 1975; Petersen 1984; Rycroft 1991; White 1993; and Wynne 1992.

3. My conception of credibility borrows from scholarship in science studies that includes Barnes 1985; Barnes and Edge 1982; Cozzens 1990; Latour and Woolgar 1986; Shapin 1994; Shapin and Schaffer 1985; Star 1989; and Williams and Law 1980.

4. To borrow Bruno Latour's (1987: 132) term, activists constructed themselves as an "obligatory passage point" standing between researchers and the trials they sought to conduct. Of course, activists also needed the researchers to conduct the trials, so the relationship is best seen as symbiotic. See also Crowley 1991.

5. Here I mean to go beyond the argument, now routinely heard, that AIDS has forever changed conceptions of the doctor-patient relationship. That may be true, although probably the old-fashioned model of the omnipotent physician and the dependent patient was already on the way out.

6. This perspective brings activists into alignment with sociologists of scientific knowledge who advocate that public understanding of science can be improved if the public acquires a greater appreciation of the high degree of uncertainty in science (see Collins and Pinch 1993).

REFERENCES

*ACT UP / New York capsule history.* 1991. New York: AIDS Coalition to Unleash Power.

Adam, Barry D. 1987. *The rise of a gay and lesbian movement.* Boston: Twayne Publishers.

Altman, Dennis. 1982. *The homosexualization of America.* Boston: Beacon Press.

———. 1986. *AIDS in the mind of America.* Garden City, N.Y.: Anchor Press/Doubleday.

———. 1994. *Power and community: Organizational and cultural responses to AIDS.* London: Taylor and Francis.

Antiviral Drugs Advisory Committee of the U.S. Food and Drug Administration. 1991. Meeting transcript. Food and Drug Administration, Bethesda, Md., February 13–14.

Arno, Peter S., and Karyn L. Feiden. 1992. *Against the odds: The story of AIDS drug development, politics and profits.* New York: Harper Collins.

Balogh, Brian. 1991. *Chain reaction: Expert debate and public participation in American commercial nuclear power, 1945–1975.* Cambridge: Cambridge University Press.

Barinaga, Marcia. 1992. "Furor at Lyme disease conference." *Science* 256:1384–1385.

Barnes, Barry. 1985. *About science.* Oxford: Basil Blackwell.

Barnes, Barry, and David Edge. 1982. "Science as expertise." In *Science in context: Readings in the sociology of science,* edited by Barry Barnes and David Edge, 233–249. Cambridge: MIT Press.

Bayer, Ronald. 1981. *Homosexuality and American psychiatry: The politics of diagnosis.* New York: Basic Books.

———. 1985. "AIDS and the gay movement: Between the specter and the promise of medicine." *Social Research* 52:581–606.

Blume, Stuart, Joske Bunders, Loet Leydesdorff, and Richard Whitley, eds. 1987. *The social direction of the public sciences.* Dordrecht, Holland: D. Reidel.

Boston Women's Health Book Collective. 1973. *Our bodies, ourselves: A book by and for women.* New York: Simon and Schuster.

Bourdieu, Pierre. 1990. *The logic of practice.* Stanford, Calif.: Stanford University Press.

Brown, Phil. 1992. "Popular epidemiology and toxic waste contamination: Lay and professional ways of knowing." *Journal of Health and Social Behavior* 33:267–281.

Bull, Chris. 1988. "Seizing control of the FDA." *Gay Community News,* October 16–22, 1, 3.

Cohen, Cathy Jean. 1993. "Power, resistance and the construction of crisis: Marginalized communities respond to AIDS." Ph.D. diss., University of Michigan.

Collins, Harry, and Trevor Pinch. 1993. *The golem: What everyone should know about science.* Cambridge: Cambridge University Press.

Corea, Gena. 1992. *The invisible epidemic: The story of women and AIDS.* New York: Harper Collins.

Cotton, Paul. 1991. "HIV surrogate markers weighed." *Journal of the American Medical Association* 265, no. 11: 1357, 1361, 1362.

Cozzens, Susan E. 1990. "Autonomy and power in science." In *Theories of science in society,* edited by Susan E. Cozzens and Thomas F. Gieryn, 164–184. Bloomington: Indiana University Press.

Cozzens, Susan E., and Edward J. Woodhouse. 1995. "Science, government, and the politics of knowledge." In *Handbook of science and technology studies,* edited by Sheila Jasanoff, Gerald Markle, James C. Petersen, and Trevor Pinch, 533–553. Thousand Oaks, Calif.: Sage.

Crimp, Douglas, and Adam Rolston. 1990. *AIDS demographics.* Seattle: Bay Press.

Crowley, William Francis Patrick, III. 1991. "Gaining access: The politics of AIDS clinical drug trials in Boston." Undergraduate thesis, Harvard College.

DeRanleau, Michele. 1990. "How the 'conscience of an epidemic' unraveled." *San Francisco Examiner,* October 1, A-15.

Di Chiro, Giovanna. 1992. "Defining environmental justice: Women's voices and grassroots politics." *Socialist Review* 22:93–130.

Elbaz, Gilbert. 1992. "The sociology of AIDS activism, the case of ACT UP / New York, 1987–1992." Ph.D. diss., City University of New York.

Emke, Ivan. 1993. "Medical authority and its discontents: The case of organized non-compliance." *Critical Sociology* 19:57–80

Epstein, Steven. 1995. "The construction of lay expertise: AIDS activism and the forging of credibility in the reform of clinical trials." *Science, Technology, and Human Values* 20:408–437.

———. 1996. *Impure science: AIDS, activism, and the politics of knowledge.* Berkeley: University of California Press.

———. 1997. "AIDS activism and the retreat from the genocide frame." *Social Identities* 3:415–438.

Fauci, Anthony. 1994. Interview by author. October 31. Bethesda, Md.

"FDA allows AIDS patients to import banned drugs." 1988. *Los Angeles Times,* July 24, 18.

Fee, Elizabeth, ed. 1982. *Women and health: The politics of sex in medicine.* Farmingdale, N.Y.: Baywood.

Ferraro, Susan. 1993. "The anguished politics of breast cancer." *New York Times Sunday Magazine*, August 15, 25–27, 58–62.

Gamson, Joshua. 1989. "Silence, death, and the invisible enemy: AIDS activism and social movement 'newness.'" *Social Problems* 36:351–365.

Garrison, Jayne. 1989. "AIDS activists being heard." *San Francisco Examiner*, September 5, A-1, A-8.

Geltmaker, Ty. 1992. "The queer nation acts up: Health care, politics, and sexual diversity in the County of Angels." *Society and Space* 10:609–650.

Gladwell, Malcolm. 1992. "Beyond HIV: The legacies of health activism." *Washington Post*, October 15, A-29.

Gross, Jane. 1991. "Turning disease into political cause: First AIDS, and now breast cancer." *New York Times*, January 7, A-12.

Harrington, Mark. 1993. *The crisis in clinical AIDS research*. New York: Treatment Action Group.

———. 1994. Interview by author. April 29. New York City.

Hilts, Philip J. 1990. "82 held in protest on pace of AIDS research." *New York Times*, May 22, C-2.

Indyk, Debbie, and David Rier. 1993. "Grassroots AIDS knowledge: Implications for the boundaries of science and collective action." *Knowledge: Creation, Diffusion, Utilization* 15:3–43.

Irwin, Alan, and Brian Wynne. 1996. *Misunderstanding science? The public reconstruction of science and technology*. Cambridge: Cambridge University Press.

James, John S. 1986. "What's wrong with AIDS treatment research?" *AIDS Treatment News*, May 9.

Jasper, James M., and Dorothy Nelkin. 1992. *The animal rights crusade: The growth of a moral protest*. New York: Free Press.

Jonsen, Albert R., and Jeff Stryker, eds. 1993. *The social impact of AIDS in the United States*. Washington, D.C.: National Academy Press.

Kingston, Tim. 1991. "The 'white rats' rebel: Chronic fatigue patients sue drug manufacturer for breaking contract to supply promising CFIDS drug." *San Francisco Bay Times*, November 7, 8, 44.

Kleinman, Daniel Lee. 1995. "Politics on the endless frontier: Postwar research policy in the United States." Durham, N.C.: Duke University Press.

Kroll-Smith, Steve, and H. Hugh Floyd. 1997. *Bodies in protest: Environmental illness and the struggle over medical knowledge*. New York: New York University Press.

Kwitny, Jonathan. 1992. *Acceptable risks*. New York: Poseidon Press.

Latour, Bruno. 1987. *Science in action: How to follow scientists and engineers through society*. Cambridge: Harvard University Press.

Latour, Bruno, and Steven Woolgar. 1986. *Laboratory life: The construction of scientific facts*. Princeton, N.J.: Princeton University Press.

Lo, Clarence Y. H. 1992. "Communities of challengers in social movement theory." In *Frontiers in social movement theory*, edited by Aldon D. Morris and Carol McClurg Mueller, 224–247. New Haven: Yale University Press.

Martin, Brian. 1980. "The goal of self-managed science: Implications for action." *Radical Science Journal* 10:3–16.

"MD telethon boycott urged." 1991. *San Francisco Examiner,* September 2, B-1.

Merigan, Thomas C. 1990. "Sounding board: You *can* teach an old dog new tricks: How AIDS trials are pioneering new strategies." *New England Journal of Medicine* 323:1341–1343.

Meyer, David S., and Nancy Whittier. 1994. "Social movement spillover." *Social Problems* 41:277–298.

Moore, Kelly. 1996. "Organizing integrity: American science and the creation of public interest organizations, 1955–1975." *American Journal of Sociology* 101:1592–1627.

Nelkin, Dorothy. 1975. "The political impact of technical expertise." *Social Studies of Science* 5:35–54.

Patton, Cindy. 1990. *Inventing AIDS.* New York: Routledge.

Petersen, James C., ed. 1984. *Citizen participation in science policy.* Amherst: University of Massachusetts Press.

Phair, John. 1994. Interview by author. November 15. Chicago.

PWA Coalition. 1988. "Founding statement of people with AIDS/ARC." In *AIDS: Cultural analysis, cultural activism,* edited by Douglas Crimp, 148–149. Cambridge: MIT Press.

Quimby, Ernest, and Samuel R. Friedman. 1989. "Dynamics of black mobilization against AIDS in New York City." *Social Problems* 36:403–415.

Richman, Douglas. 1994. Interview by author. June 1. San Diego.

Roland, Michelle. 1993. Interview by author. December 18. Davis, Calif.

Ruzek, Sheryl Burt. 1978. *Feminist alternatives to medical control.* New York: Praeger.

Rycroft, Robert W. 1991. "Environmentalism and science: Politics and the pursuit of knowledge." *Knowledge: Creation, diffusion, utilization* 13:150–169.

Shapin, Steven. 1994. *A social history of truth: Civility and science in seventeenth-century England.* Chicago: University of Chicago Press.

Shapin, Steven, and Simon Schaffer. 1985. *Leviathan and the air-pump: Hobbes, Boyle, and the experimental life.* Princeton, N.J.: Princeton University Press.

Star, Susan Leigh. 1989. *Regions of the mind: Brain research and the quest for scientific certainty.* Stanford, Calif.: Stanford University Press.

Treichler, Paula A. 1991. "How to have theory in an epidemic: The evolution of AIDS treatment activism." In *Technoculture,* edited by Constance Penley and Andrew Ross, 57–106. Minneapolis: University of Minnesota Press.

Vollmer, Tim. 1990. "ACT-UP/SF splits in two over consensus, focus." *San Francisco Sentinel,* September 20, 1.

Wachter, Robert M. 1991. *The fragile coalition: Scientists, activists, and AIDS.* New York: St. Martins.

Weber, Max. 1978. *Economy and society.* Vol. 1. Edited by G. Roth and C. Wittich. Berkeley: University of California Press.

White, Stuart. 1993. "Scientists and the environmental movement." *Chain Reaction* 68:31–33.

Williams, Rob, and John Law. 1980. "Beyond the bounds of credibility." *Fundamenta Scientiae* 1:295–315.

Winnow, Jackie. 1992. "Lesbians evolving health care: Cancer and AIDS." *Feminist Review* 41:68–77.

Wynne, Brian. 1992. "Misunderstood misunderstanding: Social identities and public uptake of science." *Public Understanding of Science* 1:281–304.

CHAPTER 5

# Communities of Suffering and the Internet

DIANE E. GOLDSTEIN

The Internet has played a key role in the process by which emerging illnesses have been made visible over the past ten years. As folklorist
Diane Goldstein argues in this chapter, the Internet provides a means
by which individuals suffering from unexplained symptoms, or recognized conditions, can reach out and become part of a virtual community of suffering, exchanging information on symptoms, diagnoses, illnesses experiences, treatments, and research with other
individuals scattered across the globe. These communities can significantly contribute to the process by which emerging illnesses move
along the pathway to visibility described in Chapter 1. Internet communities are particularly important at the beginning of the pathway
to visibility, where it is often difficult to distinguish between normal
biological change and pathology. Goldstein's chapter, while focusing
on a group of women suffering from the severe symptoms of menopause, a condition that is seldom thought of as an emerging illness,
provides important insights into the work of these virtual communities of sufferers. The menopause support group Goldstein describes is
composed of women who are unsatisfied with the ways in which medical professionals and others have understood their condition. They
are suffering not only from symptoms but also from the stigma of not
being taken seriously. In this sense, their experience with menopause
resembles the experience of individuals who suffer from symptoms
associated with chronic fatigue syndrome, multiple chemical sensitivity, or fibromyalgia.

The Internet creates intimacy through anonymity. It allow a place
where individuals feel free to speak openly and subjectively about
their illness experiences because they are anonymous. The Internet, in
fact, creates a culture that gives primacy to subjective experience and

in which the authoritative voice of biomedicine is somewhat muted. The tension between this lived experience, what Ellen Spears refers to in her chapter as "situated knowledge," and the clinical understandings of medical practitioners is a recurrent theme in the discussions that occur in these virtual communities of suffering.

Internet support groups are also privileged communities. Goldstein's group is made up of middle- and upper-class women, most with a college education, who have access to and knowledge of computers. This raises questions about the role of the Internet in shaping the demographic profile of populations associated with specific emerging illnesses. The early epidemiology of chronic fatigue syndrome, fibromyalgia, and multiple chemical sensitivity seemed to indicate that these illnesses were associated with a privileged subset of the population. Later research, particularly on chronic fatigue syndrome, has indicated that these illnesses may affect a much broader socioeconomic spectrum of people. If the Internet contributes to the visibility of emerging illnesses, does it also contribute to the invisibility of illnesses experienced primarily by marginalized and impoverished populations who lack access to the Internet?

The role of the Internet in coalescing groups of illness sufferers raises other questions. In comparison with pre-Internet communities of suffering that emerged around multiple sclerosis and Tourette's described in this volume by Colin Talley and Howard Kushner, respectively, Internet groups emerged more rapidly but also more anonymously. These earlier communities of suffering shared their experience by mail, through advertisements in popular media, and through face-to-face interactions. The community of suffering that formed among African Americans in the neighborhood of Newtown in Gainesville, Georgia, described by Spears, also communicated through face-to-face interactions. What qualitative difference has the Internet made in defining how communities of suffering form and function? The Internet, despite being restricted in terms of access, may create a kind of cyberspace democracy in which it is difficult to evaluate, much less control, the flow of information within the community of suffering. Any effort to exercise control, moreover, can lead to divisions within the community and the creation of new alternative Internet support groups. In contrast, the pre-Internet networks created by the National Multiple Sclerosis Society and the Tourette Syndrome Association were more hierarchical and controlled. What are the consequences of this difference? Does cyberspace democracy inhibit the development of consensus concerning case definitions, treatment, and research goals among communities of suffering? On the other hand, does the intimacy permitted by anonymity on the Internet provide an outlet for the expression of subjective experiences and understandings that is more difficult through other forms of communication? Does the Internet therefore provide a therapeutic function unavailable to communities formed through other kinds of communication? At the same time, by creating a culture that gives primacy to subjective experience, does the Internet accentuate the tensions that often occur between the sufferers of emerging illnesses and the medical communities to which they turn for therapeutic assistance? The Internet is a powerful tool for communities of suffering, but its impact on the process of illness emergence is not straightforward.

Few definitions of health and illness are as hotly contested and politicized as menopausal syndrome. Tied up in our beliefs and attitudes about aging, life cycle, fertility, nature, and social status as well as issues of biological control, gender roles, and the power of pharmaceutical companies, menopause has become a manipulated syndrome, rife with agendas that do not necessarily include the cessation of symptoms or improvement of quality of life. Battles over the physiological versus psychological versus sociological nature of the syndrome have resulted in incredible frustration among women who are currently trying to deal with their personal experience of menopause. Biomedical models in the treatment of menopause tend to take a disease-oriented approach, treating decline in ovarian function and decreased hormone levels as pathological (Lock 1982, 1993; McCrea 1983). Feminist models call for the treatment of menopause with a normal, life-change-oriented approach, arguing that most of the symptoms of menopausal syndrome are a response to psychosocial factors related to gender role conditioning, decreased status due to aging, and a lack of role alternatives (MacPherson 1985; Worcester and Whatley 1992; Defey et al. 1996). Many women involved in the menopausal experience resent the various agendas that shape the definitions of their condition and its treatment, feeling that both perspectives diminish the actual somatic nature of the syndrome and minimize the reality of their physical, psychological, and social situation. This chapter is about one such group of women.

On the surface it would appear easy to define menopause; "the universal (for women), biological event, which, in theory, is marked clearly by the last menstrual cycle" (Lock 1982: 263). This part is rarely contested. The scope of the syndrome, however—its beginning and end, the transitional period (called the *climacterium*)—is more problematic. The climacterium is the period, often taking several years, when ovarian functions and hormone levels decline. Laboratory data are sparse on the nature of the climacterium, but it is known that whereas the decline in some women is steady, in others hormone levels drop suddenly and without warning, and in some women the decline occurs in irregular and unpredictable patterns (Lock 1982; Gittleman and Wright 1999). Owing to the direct relationship between hormone production and the menstrual cycle, some women experience an abrupt end to menstruation, others experience intervals of unpredictable, heavy bleeding, and still others experience times of scanty bleeding. The unpredictability of hormone production during this period produces a lack of consensus among doctors, social scientists, and those interested in women's health about the signs, biological and behavioral, that are associated with menopause. One thing is generally agreed upon, that the "hot flash" is a "true" though not inevitable symptom of hormone decline (Lock 1982). The incidence, significance, and causes of other reported symptoms such as depression, fatigue, headaches, and dizziness are repeatedly called into question as sociocultural rather than physiological (and yes, we may wish to read this here as therefore "not real"). Gynecological textbooks have frequently carried such statements as "the emotionally mature, busy and happy woman usually has few if any difficulties in adjustment" (Pettit 1962: 353).

Pat, Gail, and Cindy, three women currently struggling with menopausal and premenopausal syndrome, have no doubts about the reality of their symptoms. Pat wrote: "I'm almost too depressed to type. I cry for no reason, feel too sick to eat, my periods are all over the place (spotting, gushing, skipping, you name it). Been to the GYN, had the D&C, the mammogram. Seems I'm fine! Do I have stress in my life? Some, nothing new. I have a 7 year old child, a sweet husband, and I'm turning into a screaming maniac with them. I am 46 years old and slowly losing my mind" (September 5, 1996).

Gail says:

I'm 47 now. Around 42 I suddenly started having heavy periods that never ended (well, almost never, they would cease for 2 days or so, then start all over again). My doctor told me I was too young (how many times have you heard that?) for menopause and basically that shit just happens when you turn 40. Went to two other doctors who pretty much said the same thing. In the meantime I was experiencing that debilitating crashing fatigue that is inexplainable to someone who hasn't experienced it, the mood swings, the RAGE, the depression, and (all the time) just gushing away! At the beginning of last year I hit one day when it was 90 degrees and I was laying on the couch in thermals, flannels, with a blanket on and freezing, too tired and cold to do anything and thinking Jeez, I'm just going to lay here and bleed to death. (January 18, 1997)

For Cindy, the ovarian pain was so bad that she contemplated suicide. She said:

There have been days that I've prayed to be released from the pain and emotional hell I've been living in and the doctors made me feel like I was in a self-imposed prison that I could walk out of any time I wanted to, which in turn made me feel much worse. Then I felt guilty of destroying my family and making them miserable too. I felt if I was out of the picture, they'd all be sad for a couple of months, but then be able to get on with their lives. I wanted out of this painful life so bad, I did not consider the pain I'd inflict on others because mine was so much more devastating that I couldn't see beyond it. When you have to struggle for each minute of your day, and you fear each approaching minute, life is worse than any hell [and] death becomes a panacea of hope and freedom. (January 20, 1997)

Pat, Gail, and Cindy are all participants in a menopause support group run over the Internet. Alt.support.menopause is an unmoderated discussion list created for the exchange of personal experiences, problems, ideas, and information related to menopausal syndrome.[1] Participants are primarily lay women, although a few physicians and partners of menopausal women occasionally enter into the discussion.

In the fall of 1996 I began to monitor a series of Internet health support groups, ultimately requesting permission from participants in alt.support.menopause to write about the group. Alt.support.menopause operates as an Internet community of suffering, one that bridges distance and diversity to share the understandings, concerns, and needs of sufferers from menopausal syndrome. Although many of the observations discussed here focus specifically on this group and this syndrome, the issues are much wider reaching. Computer technologies are rapidly changing the shape of such communities by connecting distant individuals who share common health issues and by changing lay access to health information. Internet groups like the one discussed in this chapter connect patients to other patients in ways never before available, and the result is a lay community of sufferers empowered by shared experience and knowledge exchange.

Although computer networks have been around for nearly thirty years, they have proliferated incredibly in recent years, resulting in thousands of specialized networks that cater to a huge variety of interests including those of sports fans, cooks, music enthusiasts, dancers, poets—practically all topics of general and sometimes quite specific interest. Health support groups exist in large numbers on the Internet,[2] combining mass electronic communication with what is generally a small-group phenomenon. Most of us think of health support groups as small intimate groups, limited in numbers and oriented toward allowing individuals to explore their questions, concerns, and experiences concerning illness together with others who share those experiences. The Internet, however, would seem to be anything but intimate and personal. After all, people communicating with one another via the Internet are frequently total strangers at computer terminals scattered all over the world—people who have never seen one another, who are not likely to meet, and who may not even know one another's names. Nevertheless, the list of titles beginning alt.support is vast—alt.support.epilepsy, alt.support.post-polio, alt.support.eating disorders. One might posit that these groups function more as electronic newsletters than as actual support groups, posting news of new treatments, announcing clinical trials, advertising products that improve quality of life, and offering general information. Surprisingly, the postings are frequently intensely personal, and the subscribers are supportive, knowledgeable, and oriented toward both observation and research.

The Internet support group epitomizes a phenomenon social scientists and communications scholars are just beginning to grapple with—the amazing ability of Net users to turn what is perhaps our most global form of communication into a seemingly local and vernacular form, a cozy little club. Folklorist John Dorst characterizes this by saying: "To some, Mc-Luhan's global village seems on the verge of realization. The computer networks appear to make possible communities that, though physically dispersed, display attributes of the direct, unconstrained, unofficial exchanges folklorists typically concern themselves with" (1990: 180).

Most scholars seem to conceive of this spatially. Bruce Mason notes, for example, "The internet is not the wide open expanse of the Internet Super Highway but a series of localizations" (1998: 8). Adapting Yi Fu Tuan's studies of space and place, Mason writes, "Tuan claims that 'place is security, space is freedom: we are attracted to one and long for the other.'" He continues, "It is my contention that cyberspace has been envisioned as a space for which we long but that our experiences of the internet are more place-like. . . . Rather than trying to exist in this huge undifferentiated wilderness we call cyberspace, we are busy creating cyberplaces and the two exist in relation to each other."

The spatial issues involved in this localization of the global are interesting, but they do not seem to really address how it is that an individual can communicate his or her most painful moments as a bulimic or an alcoholic in the most intimate and personal ways while a few hundred or potentially a few thousand others "listen in" to the conversation. This, one would think, is not an ideal condition for a health support group.

In part, what appears to be an ability to block out the public nature of Internet communication seems to be a result of how a sense of presence is achieved on the Net. Barbara Kirshenblatt-Gimblett notes that "presence in the medium is a function of interactivity—the more interaction, the more people are present to each other" (Kirshenblatt-Gimblett 1996: 74). If we reverse Kirshenblatt-Gimblett's assertion, we can posit that absence on the Net is perhaps achieved by noninteraction. In this sense, one's audience can be conceived of as only those directly involved in the interaction at that moment. Although most Internet support groups have hundreds of members, generally only a dozen or so are actually involved in direct interaction on any subject at any given time.

It would be a mistake here, though, to satisfy ourselves simply with the notion that Net users are able to make one another disappear through "out of sight, out of mind" reasoning and for us to pretend that that solves the support group puzzle. Primary among the characteristics of Internet communication is its anonymity, epitomized by a *New York Times* cartoon showing two dogs sitting in front of a computer with one saying to the other, "Isn't it great, on the Internet no one knows you're a dog" (cited in Mason 1998). In the Internet support group your comments are not linked to you in any way but your virtual presence. Physicians and therapists who counsel patients to enroll in support groups frequently encounter the comment "but I can't talk about this to other people—I feel too visible, like everyone is looking at me." On the Internet, however, no one can see you; no one knows who you are. A person using the Internet can construct an identity that he or she may not be as willing to express in daily life (Turkle 1995). The very thing that one would think distances Net users may, in fact, be what makes Internet health support work.

But beyond the question of how could such a thing work on the Internet is the question of content. What do these groups actually do? I believe that what they do is create their own separate and distinct medical culture, a culture that gives primary importance to the role of subjective experience. Although I do not buy into the notion of the Internet as a wide-open egalitarian social space, the health support groups seem to be, for the most part, communities of lay members, self-governed and self-maintained, with any hierarchical prominence given to the list maintainers, older members or so-called regulars (Mason 1998). The lists I have participated in do not give special status to either health professionals or survivors who participate. But the lists do have boundaries. List maintainers can exclude people from the list who break topic or engage in abusive behavior; they may enforce topic control and censor postings that are irrelevant or commercialized (such as drug advertisements). The lists may be tightly controlled, but list hierarchies fall along lines established within the group and *not* generally along lines that maintain normal medical status structures.

As settings for the lay discussion of health issues, the lists become sites not only of support but also of resistance to medical authority. Of central interest to the subscribers is the exertion of medical autonomy and the definition of their condition and its amelioration from within the boundaries of the group. Group definitional frameworks are experience-centered, beginning with the shared observations of sufferers and moving out to include relevant research. As is indicative of most lay health belief systems, members place high value on subjective experience and focus on ability to function and social well-being as evidenced by physical fitness, energy, vitality, absence of pain, feeling healthy, and the ability to maintain social relationships (Calnan 1987). This contrasts with the biomedical valuing of objective experience over the subjective and the emphasis placed in Western biomedical paradigms on health as the absence of disease and the appropriate functioning of biological and psychophysiological processes. What occurs within the group is the gradual construction of "vernacular health theory."

The term *vernacular* has been used for some time to refer to community-based forms of cultural expression, such as vernacular architecture, vernacular design, and vernacular language. Although I am using the term *vernacular theory* to refer to community-based theory, my meaning goes further, following on the work of Thomas McLaughlin (1996), who contrasts "vernacular theory" with "critical theory." McLaughlin defines vernacular theory by saying, "It refers to the practices of those who lack cultural power and who speak a critical language grounded in local concerns, not the language spoken by academic knowledge-elites. They do not make use of the language of analytic strategies of academic theory; they devise a language and strategy appropriate to their own concerns. And they arise out of intensely local issues that lead to fundamental theoretical questions" (1996: 5-6). Vernacular theory can be seen as parallel to Michel Foucault's notion of "subjugated knowledges." Foucault notes, "A subjugated knowledge is an autonomous, non-centralized kind of theoretical production, one ... whose validity is not dependent on the approval of the established regimes of thought" (Foucault 1980: 81). These knowledges, according to Foucault, "have been disqualified as inadequate to their task or insufficiently elaborated: naive knowledges, located low down on the hierarchy, beneath the required level of cognition or scientificity." Foucault contrasts these knowledges with what he calls "the tyranny of globalizing discourses" (83).

McLaughlin discusses the work of John Fiske, author of *Power Plays: Power Works*, in relation to vernacular theory. Although Fiske denies the notion of community-based *theory*, feeling it suggests too much external impact, he offers some interesting insights into the development of the local. Fiske explores oppositions between "the imperializing dominant culture which creates stations (such as institutions, texts) for imposing its power on subjects, and localizing cultures of resistance which try to change those 'stations' into 'locales' where people themselves feel some control and where their cultural needs are met" (Fiske 193: 19). Fiske argues, "Imperializing knowledges are rationalistic, constructing social categories that extend power throughout the culture" (19). In contrast, localizing knowledges "function not to extend a great vision over the world but to produce a localized social, ethnic, communal sense of identity. They create 'cultures of practice,' ones that develop ways of living in the world and which seek to control only those ways of living rather than the world in which they live."

In reflecting on our earlier questions related to the incredible ability of Net users to take the global and make it local—Foucault's comments on globalization and Fiske's comments on localization gain particular resonance. Support group users—those who share a particular form of suffering—are in some sense a local group spread out globally. By appropriating the global institution of the Internet and turning it into a locale, they gain control and create an opportunity for vernacular theory.

For those who are interested in public health, the Internet support group can become a way to access public understandings of health and illness. We must be aware, however, that the circumstances of Internet support groups create a different situation than would be found with a typical lone sufferer of a disease or syndrome. By couching what is created in the group in terms of vernacular theory, I point to several issues: (1) The absence of medical authority allows for the creation of authority based on experience rather than on "objective" information. (2) Shared experience creates patterns observed by the members as definitive of the syndrome and creates the basis for health theory. (3) The power of local control (that is, joint control over the internal discourse of the group) can provide the basis for proactive behavior extending beyond the boundaries of the group (despite Fiske's assertion that localizing knowledges seek only to control ways of living in the world and not to control the world in which they live). (4) Vernacular theory provides an experientially based, alternative construction of illness that, although subjugated in terms of medical authority, is likely to address the actual daily concerns, experiences, and worldview of those coping with illness. Vernacular theory raises questions about dominant cultural assumptions and, like all theory, begins in specific interpretive complexities, proceeds by local rules, uses local forms of discourse, and makes its fullest sense in the cultural context out of which it arises.

At this point, I will turn back to my case study on menopause, hoping that my discussion of this particular Internet health support group will ground my comments in the lives of people who, no matter how virtual they may seem, share real health issues.

New readers of alt.support.menopause would find it hard not to be struck immediately by the women's experiential construction of their syndrome as contrasted with biomedical, social scientific, and feminist constructions of menopause. Most of the women in the group are well educated, have done extensive research on their condition, and are convinced that they are more knowledgeable about menopause then are their doctors.[3] One of the members characterized the group by saying: "Most of us are abnormal because we are here—we have a computer and know how to use it, we have an internet account and we are literate and better educated then average, etc. . . . This is not a random group. We are a self-selected biased group skewered towards the top end of educational and socioeconomic groups and the bottom end of menopausal comfort" (August 18, 1997).

Pat, Gail, and Cindy, like the other participants, write in their symptoms, looking for affirmation that other women have had the same experiences and discussing their observations related to links between the symptom and hormonal changes, treatment strategies, or lifestyle issues. Within the group, no area of life or perceived change is discounted as irrelevant; confirmation comes from others with the same experience. In a post called "Is there art during menopause" Donna asked: "Are there any artist/writer types out there experiencing EXTREME loss of creativity during menopause, as am I? The physical symptoms are nothing to me compared with this! I feel intolerable loss of identity without my creative spirit. . . . Am I doomed to live out the remainder of my years in this fallow trench?" (September 6, 1996). Confirming posts poured in. One said: "I just logged onto this site and I can't believe it. Am here trying to find out about taking progesterone to help deal with my menopausal reactions. I am 50, had breast cancer at 44 and have been struggling with symptoms of headaches, hot flashes, crying jags, irregular periods, loss of memory, loss of libido to name the more prevalent, and I never thought that my difficulties getting it going in the studio could be related" (September 6, 1996).

There is no mind/body dualism here (a criticism the women level at their physicians);distinctions are not made between the psychological and physiological. Symptoms unconfirmed by others are not discredited; they simply remain unconfirmed, possibly to be brought up again in the future.

Sue wrote in asking about experiences with hair loss and received several responses including one that said: "Glad to see your post. I posted about hair loss a week ago and only had 1 response. I am 42 not even in perimenopause yet (according to the MD), but I am experiencing thinning and loss" (March 8, 1996). Confirmation, on the other hand, is its own reward—opening up discussion on cause and treatment.

The women in the group are proactive about the list of symptoms generated through discussion of experience. Together with the women from menopause, another Internet group, they have generated a reference list of the signs of menopause, and they recommend that the women write this list on a 3 × 5 card and give it to their physicians to keep in the top drawer of their desk. They suggest that the list be scanned whenever a woman over thirty consults an internist, gynecologist, neurologist, oncologist, surgeon, or any other specialist. The list is posted often, usually with the following introduction: "The purpose of this list of signs is to inform women and their physicians about menopause so that appropriate action can be taken, to evaluate the effectiveness of herbs, hormones or drugs, or wait it out. Varying subsets of signs occur in each individual. A common experience of the women who developed this list is misdiagnosis by a physician or psychiatrist and often years of anxiety about the symptoms and inappropriate treatment" (September 5, 1996). The list follows:

1. Hot flashes, flushes, night sweats and/or cold flashes, clammy feeling

2. Bouts of rapid heart beat

3. Irritability

4. Mood swings, sudden tears

5. Trouble sleeping through the night (with or without night sweats)

6. Irregular periods; shorter, lighter periods; heavier periods, flooding, phantom periods

7. Loss of libido

8. Dry vagina

9. Crashing fatigue

10. Anxiety, feeling ill at ease

11. Feelings of dread

12. Difficulty concentrating, disorientation, mental confusion

13. Disturbing memory lapses

14. Incontinence, especially upon sneezing, laughing, urge incontinence

15. Itchy crawly skin

16. Aching, sore joints, muscles, tendons

17. Increased tension in muscles

18. Breast tenderness

19. Headache change, increase or decrease

20. Gastrointestinal distress, indigestion, nausea

21. Sudden bouts of bloat

22. Depression

23. Exacerbation of existing conditions

24. Increase in allergies

25. Weight gain

26. Hair loss or thinning; head or whole body; increase in facial hair

27. Dizziness, light headedness, episodes of loss of balance

28. Changes in body odor

29. Electric shock sensation under the skin and in the head

30. Tingling in the extremities

31. Gum problems, increased bleeding

32. Burning tongue

33. Osteoporosis. (September 5, 1996)

Members recognize that many of the symptoms on the list are also indicative of other illnesses such as hypothyroidism, diabetes, and clinical depression. Where possible they stress the particular constellation that is indicative of menopausal rather then other etiologies.

The women in alt.support.menopause frequently discuss the difference in the nature of many of the symptoms when they are hormonally caused rather than caused by other etiologies. Members insist on a qualitative difference between endogenous and hormonal depression, indicating that the difference is experientially clear but difficult to articulate. One member commented about medical researchers: "As far as I can see they are relating verbal accounts or subject reports through questionnaires to equate what are subjectively very different events. Haven't many of us experienced a new kind of headache, a different kind of insomnia, as well as a different kind of depression since hitting perimenopause?" (September 9, 1996)

Women in the group point to a sort of catch-22 in the medical literature—symptoms are discounted as unrelated to menopause because they occur in the general female population, not simply in the population defined as menopausal—but members of the group then argue that the general female population is menopausal if the climacterium is included in the term's definition. One woman wrote:

> One of the problems with [these studies] is that they say these problems can't be related to menopause because the same problems were reported by women not defined as menopausal. Look who they exclude from this category— "approximately 30–50% of women did report these complaints, but they were reported equally by women who were pre, peri and postmenopausal." So what do they mean by menopause or menopausal? If peri and post don't count— then where does the menopause come in? I don't understand. . . . Unless they were referring to people in their 20's I'm not sure how they could be really sure themselves. (September 9, 1996)

Since members believe that it is the fluctuation of hormones that causes symptoms, they also believe that symptoms are constantly changing. Members continually send messages such as: "Remember as you read these posts that we are all different and our needs change at different times in this process" (September 7, 1996).

Although members include an almost limitless number of symptom cluster patterns in their understanding of the menopausal profile, many rely on hormonal tests to provide final confirmation and recommend blood and saliva testing to newcomers to the list. FSH, a blood test that measures hormonal levels at the time the blood was drawn, is recommended with the caution that fluctuation in hormonal levels may render a single reading useless. Members continually comment on the assumption by doctors that women are not menopausal based on their age, without ordering confirmational hormonal tests. Not uncommonly, the women order their own tests and discuss interpretation of results on the Net. One woman wrote:

> Got my saliva test back from Aeron Labs today. It's a real easy read. These are my results. I have 2.4 pgml Estradiol level. Normal level for premenopausal women is 0.5–5.0, so that's OK. Now for the big news: my progesterone level is .06. Normal level for a menstruating woman should be 0.1–0.5, which means I'm low, low, low. Just to make sure I called Aeron Labs to have this interpreted. Yes, it was confirmed, I am very low. The doctor there said if my estrogen was lower, I might be alright. However it is the ratio that concerns him. If you have normal estrogen levels, but if your progesterone is too low, in essence you have too much estrogen. That's why I've been bleeding every month like a geyser. . . . I switched doctors again recently because as soon as they hear blood they get their knives sharpened. He wanted me to have a sonogram and yet another

D&C, even with a normal pap smear result and a previously normal D&C. He said we might have to consider hysterectomy. I'd rather bleed. Not one doctor, would you believe, ordered a hormone test! If I hadn't taken this test, I would have feared cancer and would have submitted to yet another unnecessary surgery. (September 10, 1996)

Many of the women also self-medicate, playing with quantities of their own prescription drugs, using mail-order and Mexican pharmacies, and testing natural hormone supplements.

One woman wrote: "I've finally started just taking some of every estrogen supplement I have—which means I'm risking running out, but oh, how wonderful to feel good again!!! So I think I'm doubling or tripling the miserly doses those docs hand out. I'm going to be seeing a doc several hundred miles away from me who apparently is actually interested in the quality of life of her patients. Yipee! . . . with the right amount of hormones I feel fabulous!!!! Great change eh!! Tell Betty C. that noisy broad ain't whining no more" (November 16, 1996).

Although many women are concerned with the risks (such as endometrial or breast cancer) of hormone replacement therapies, the women on alt.support.menopause are more concerned with the risks (immediate and painful) of hormonal imbalance. The women share information on FDA-approved synthetic hormones (such as Premarin, made from mare's urine) and non-FDA-approved natural hormones or "phytoestrogens" (largely derived from soybeans or yams), which they purchase by mail order from places such as the Women's International Pharmacy. Whereas many of the women believe that the synthetic hormone replacements have become a dangerous cash cow for pharmaceutical companies (see Lee 1996; Love 1997), others feel that synthetic hormones are stronger and work faster than natural supplements. Studies cited by the group indicate that synthetic hormones are two thousand times stronger then their natural counterparts (September 9, 1997).

But the group firmly stands by a completely experiential understanding of treatment—go with anything that reduces the pain. As one woman said: "The moral is, you have to work out what works for you, regardless of the moralizing of others that if you take HRT [hormone replacement therapy] you are a tool of the capitalists" (March 17, 1997). Another said: "If we do succumb to a hype, and try the product, we soon find out if it works for us. Many here have written that HRT has not been the magic bullet. Some have had success with replacement. Sometimes there's no other way to tell but try it and see what happens" (September 8, 1996).

The women in the group object intensely to what they call the "don't worry, be happy, this is a natural process, approach" (September 5, 1996) associated with feminist menopausal models. One member wrote:

I agree that the drug companies are only concerned with money and they do probably have most doctors in their back pockets but when women are truly suffering and some do during menopause more than others, they need to know that there is help not just more of this grin and bare it philosophy. I also agree too many women are taking hormones that probably don't need them. Especially when they claim to get magical results in just days. That usually turns me off. But I've been living and dealing with menopause for 12 years now and time does not make your symptoms go away. (September 7, 1996)

In the 1960s and 1970s, North American feminists began to challenge medical authority by questioning the legitimacy of the disease model of menopause, arguing that menopause is not an illness but rather a natural process of aging, which most women pass through without difficulty (Worcester and Whatley 1992; Daly 1995). Feminist researchers argued that the biomedical disease model depicts women's physical and mental capabilities as dependent on the healthy functioning of their reproductive organs. Women in the group feel that by emphasizing the natural and unproblematic nature of menopause, feminist researchers have silenced the voices of women who are in need of medical attention.

One woman said: "Don't tell me to celebrate the passage into aging. If I do this crone stuff, go out and dance naked around a campfire, well the bleeding stop? The hair loss? The fatigue and flashes? Will I stop feeling like I'm dying?" (September 14, 1996).

The women in the group have a similar negative reaction to behavioral and cross-cultural research suggesting that menopausal distress is linked to cultural attitudes toward aging. Members cite numerous studies that explain symptom variation as a social, psychological, or cultural event, such as McKinley and Jeffreys's (1974) statistical study linking symptoms and class position or Bart's (1970) suggestion that severity of symptoms is related to overidentification with the mother role. By far the most cited study is Lock's (1993) work indicating that few Japanese women experience more than transitory distress during menopause. To simply accept the notion that menopausal symptoms are culture-specific dismisses the somatic nature of the menopausal experience shared by group members. Rather than accepting such studies as a comment on North American social roles and attitudes toward aging that shape menopausal experience, the women point to issues of diet and lifestyle that might play a preventative role in Japan.

One wrote: "Studies show that soy markedly reduces the adverse symptoms of menopause, particularly hot flashes, which are virtually eliminated. In Japan, where daily soy consumption is high, hot flashes among menopausal women are so rare that the Japanese don't even have a word for it" (December 6, 1996).

Other members of the group focus on *symptoms* that were reported in the Japanese case. One person wrote: "The most frequently reported symptom was stiff shoulder, given by 52% of Japanese women. There is no equivalent for the 'stiff shoulder' syndrome as described in Western Societies. So if we switch to [soy] we may be trading hot flashes for stiff shoulders" (December 8, 1996).

Stiff shoulder syndrome is a highly debated issue in the group, reported by some members as present in their particular symptom cluster and associated by others with parallel symptoms reported in North America, such as muscle stiffness as a result of tossing and turning from insomnia.

Like chronic pain, menopausal syndrome resists objectification. As Good notes (1994: 132), "Our medical practices are designed to localize suffering in a discrete site in the body, a site which can be made visible and subjected to therapeutic procedures. That which resists objectification is proclaimed as subjective, a functional disorder of the functional self, held responsible for producing its own suffering." Menopause appears to be ambiguous and amorphous; its onset and advancement, the nature and fluctuation of hormonal change, variation in symptom formation, lack of replicability in testing, variation in response to treatment, and ineffability all deobjectify. Through its resistance to objectification menopause lends itself to metaphorical construction in terms and concepts like deficiency, natural process, and culture-bound syndrome. Each metaphor, in turn, is literalized in definition. The work of alt.support.menopause is to reverse the deobjectification of menopause by gathering information compiled from experience to localize the syndrome. At the same time the group strives to subjectify medicine and academia, to create an epistemology that values experiential and phenomenological constructions of health and illness, and to give a voice to those who suffer. From Pat, Gail, and Cindy's point of view, lying on the couch freezing and bleeding to death while physicians play "create a drug—create a deficiency games" and while politics silence pain is just not good enough.

## NOTES

An earlier version of this chapter appeared in *Health: An Interdisciplinary Journal for the Social Study of Health, Illness and Medicine*, 4, 3 (2000), 309–323. It appears here with the permission of Sage Publications.

1. Alt.support.menopause can be found at www.tile.net/news/altsupportmeno-pause.html.

2. Surprisingly little has been published thus far on the Internet as a source of health information. The few published pieces that do exist are in general agreement that the Internet is profoundly and irreversibly revolutionizing patient-physician relationships by providing an interactive environment that transcends barriers of expertise and professionalization. (See, for example, Swartz 1998; Edwards 1999; Hardey 1999; and Reece 1999).

3. It is important that the analysis of lay understandings of health not be assessed strictly in terms of compliance with or adherence to biomedical models. The majority of lay health traditions vary from biomedical perspectives not only in treatment strategies but also in overall goals, values, and desired outcomes. Experientially based health care decisions often vary considerably from medical advice, but such conflicts do not in and of themselves necessarily constitute risk unless there are direct negative health implications. My observations suggest that most support groups are very open to discussions that weigh risks and benefits and that continually revisit the literature on causes, symptoms, and treatment. Health care choices that might be considered dangerous, questionable, or ineffectual are not made lightly or without considerable research and discussion.

## REFERENCES

Bart, P. B. 1970. "Mother Portnoy's Complaints." *Transaction* 8:69–74.

Calnan, M. 1987. *Health and Illness: The Lay Perspective*. London: Tavistock Publications.

Daly, Jeanne. 1995. "Caught in the Web: The Social Construction of Menopause as Disease." *Journal of Reproductive and Infant Psychology* 13:115–26.

Defey, D., E. Storch, S. Cardozo, O. Diaz, and G. Fernandez. 1996. "The Menopause: Women's Psychology and Health Care." *Social Science and Medicine* 42(10): 1447–56.

Dorst, John. 1990. "Tags and Burners, Cycles and Networks: Folklore in the Telectronic Age." *Journal of Folklore Research* 27:179–80.

Edwards, C. S. 1999."Internet Services Provide Peer Support." *Oncology Nursing Forum* 26(10): 1597.

Fiske, John. 1993. *Power Plays: Power Works*. London: Verso.

Foucault, Michel. 1980. "Two Lectures." In *Power Knowledge: Selected Interviews and Other Writings, 1972–1977*, translated by Colin Gordon, 78–108. New York: Pantheon.

Gittleman, Ann Louise, and Jonathan V. Wright. 1999. *Before the Change: Taking Charge of Your Perimenopause*. San Francisco: Harper.

Good, Byron. 1994. *Medicine, Rationality, and Experience: An Anthropological Perspective.* Cambridge: Cambridge University Press.

Hardey, Michael. 1999. "Doctor in the House: The Internet as a Source of Lay Health Knowledge and the Challenge to Expertise." *Sociology of Health and Illness* 21(6): 820-35.

Kirshenblatt-Gimblett, B. 1996. "The Electric Vernacular." In *Connected: Engagements with Media,* edited by G. Marcus, 21-65. Chicago: University of Chicago Press.

Lee, John R. 1996. *What Your Doctor May Not Tell You about Menopause: The Breakthrough Book on Natural Progesterone.* New York: Warner Books.

Lock, Margaret. 1982. "Models and Practice in Medicine: Menopause as Syndrome or Life Transition." *Culture, Medicine and Psychiatry* 6(3): 361-80.

———. 1993. *Encounters with Aging Mythologies of Menopause in Japan and North America.* Berkeley: University of California Press.

Love, Susan. 1997. *Dr. Susan Love's Hormone Book: Making Informed Choices about Menopause.* New York: Random House.

MacPherson, Kathleen I. 1985. "Osteoporosis and Menopause: A Feminist Analysis of the Social Construction of a Syndrome." *Advances in Nursing Science* 7(4): 11-12.

Mason, Bruce. 1998. "Guigné Award Presentation 1998." Manuscript. The Guigné International Ltd. Graduate Research Award in Folklore and Technology, St. John's, Memorial University, St. John's, Newfoundland.

McCrea, Frances B. 1983. "The Politics of Menopause: The Discovery of a Deficiency Disease." *Social Problems* 31(1): 111-23.

McKinley, Sonja, and Margot Jeffreys. 1974. "The Menopausal Syndrome." *British Journal of Preventative and Social Medicine* 28:108-15.

McLaughlin, Thomas. 1996. *Street Smarts and Critical Theory: Listening to the Vernacular.* Madison: University of Wisconsin Press.

Pettit, M. D. 1962. *Gynecologic Diagnosis and Treatment.* New York: McGraw-Hill.

Reece, R. L. 1999. "The Internet as an Equalizing, Energizing and Transforming Force in Patient and Physician Relationships." *Connecticut Medicine* 63(11): 683-88.

Swartz, M. K. 1998. "Evaluating Information on the Internet." *Journal of Pediatric Health Care* 12(6, 11): 335-37.

Turkle, Sherry. 1995. *Life on the Screen: Identity in the Age of the Internet.* Toronto: Simon and Schuster.

Worcester, Nancy, and Mariamne H. Whatley. 1992. "The Selling of HRT: Playing on the Fear Factor." *Feminist Review* 41:1-26.

# CHAPTER 6

# Illness Movements and the Medical Classification of Pain and Fatigue

DEBORAH BARRETT

Communities of suffering share a common interest in moving their illnesses onto the public health agenda; however, they may employ different methods and strategies to achieve this goal. Sociologist Deborah Barrett compares the process of emergence of two illnesses, chronic fatigue syndrome (CFS) and fibromyalgia. These two illnesses have much in common, including overlapping symptoms, the absence of an accepted etiology, and the fact that both conditions have produced skepticism among both lay and medical observers. Over the past two decades both have gained recognition as significant medical problems. Yet the histories of the emergence of CFS and fibromyalgia differ in significant ways—specifically, the makeup of patient advocacy groups and the methods they employ, the relationship between these groups and professional medical researchers, and the extent to which activist groups positioned themselves as insiders or outsiders in relation to the medical establishment.

Why did these two movements emerge in different ways? The early involvement of rheumatologists in establishing fibromyalgia as a legitimate disease entity and the absence of a similar medical link to the emergence of CFS were clearly critical. Yet this begs the question: Why did rheumatologists take an interest in fibromyalgia? There is no clear answer to this question. It may be that rheumatology in the 1960s was like neurology in the 1950s. Both were underfunded medical research areas in need of a disease. For neurologists, as Colin Tally suggests, it was multiple sclerosis, for rheumatologists it was fibromyalgia. In contrast, there was no clear biomedical constituency for CFS. The early engagement of virologists in CFS research led to inconclusive results.

This failure, combined with tensions between medical researchers and lay activists over control of the direction of CFS research, discouraged many researchers from becoming actively involved in the CFS cause. Moreover, at the time that CFS emerged in the 1980s, virology had its hands full with emerging infections including HIV/AIDS. Virologists did not need new research opportunities. Therefore, the insider/outsider strategies of the communities of suffering for these two illnesses may have been in part a reaction to the initial receptiveness of the biomedical community.

Both approaches, whatever their origins, have had their advantages and disadvantages. Fibromyalgia's insider approach brought quicker medical recognition. It also brought relatively extensive research support. In fact, fibromyalgia, although arguably less significant than CFS from an epidemiological perspective, has received an equal amount of state funding. On the other hand, the early association with rheumatology may have led to a premature narrowing of both research and treatment protocols. In contrast, CFS advocates' outsider approach has left the door open to a variety of research and treatment paths, even though CFS advocates, like their fibromyalgia colleagues, adamantly oppose avenues of inquiry that imply the existence of behavioral or psychogenic components. The downside of an outsider strategy and the consequent lack of definitional narrowing is that CFS serves as a kind of magnet attracting a broad range of sufferers with an array of unrelated health problems. This diversity complicates efforts to establish a clear case definition, let alone a disease etiology, for CFS. The multiple revisions of the CFS case definition, as well as the creation of multiple CFS advocacy groups, are in part a product of this diversity.

Pain and fatigue can indicate any of a number of medical conditions or simply that we are not taking care of ourselves. At times we all feel sick and tired. But if our aches and pains surpass an expected level and become chronic and debilitating, we enter into another category. We believe something is wrong. At that point, what happens if existing categories of illness cannot make sense of our experience? What if our doctors do not find anything "wrong"? What if we appear "normal" to the outside world, despite our suffering? This is what faces sufferers of unexplained symptoms.

Clusters of symptoms become meaningful only when they are interpreted through medical diagnoses. Throughout history, new diagnoses have been introduced when current categories become unsatisfactory. New illnesses become official through "illness movements."

During the last quarter of the twentieth century, new illness movements arose to promote new categories for poorly understood suffering. This chapter examines the social movements that have introduced two such categories: fibromyalgia syndrome (FMS) and chronic fatigue syndrome. Each has arisen in the United States as a separate diagnosis. The chief symptom of FMS is a history of widespread pain. Wolfe et al. (1990: 171) list the specific types of pain that are present: (1) pain in both sides of the body; (2) pain above and below the waist; (3) axial skeletal pain (cervical spine, anterior chest, thoracic spine or low back). Also, (4) low back pain is considered lower segment pain. The second criterion for FMS is "pain in 11 of 18 [specific] tender point sites on digital palpation." People with this syndrome may also experience fatigue, sleep disturbances, headache, temperature sensitivity, atypical patterns of numbness and tingling, exercise intolerance, morning stiffness, irritable bowel syndrome, anxiety, and other symptoms (Wallace and Wallace 1999: 16).

The symptoms of CFS are

[1] clinically evaluated, unexplained, persistent or relapsing chronic fatigue that is of new or definite onset (has not been lifelong); is not the result of ongoing exertion; is not substantially alleviated by rest; and results in substantial reduction in previous levels of occupational, educational, social, or personal activities; and 2) the concurrent occurrence of four or more of the following symptoms, all which must have persisted or recurred during 6 or more consecutive months of illness and must not have predated the fatigue: self-reported impairment in short-term memory or concentration . . . ; sore throat; tender cervical or axillary lymph nodes; muscle pain, multijoint pain without joint swelling or redness; headaches of a new type, pattern, or severity; unrefreshing sleep; and postexertional malaise lasting more than 24 hours. (Fukada et al. 1994: 956)

Fibromyalgia, then, is characterized mainly by widespread pain in muscles and connective tissues. CFS involves disabling fatigue. Both possess a combination of pain and fatigue, as well as other symptoms such as sleep disturbance, cognitive dysfunction, headache, and exercise intolerance.

These illnesses share more than a body of symptoms. They emerged in the last decades of the twentieth century as not yet fully recognized and poorly understood categories of suffering. Those who suffer from them have been dismissed at times as hypochondriacs and malingerers. Each illness has been at the center of debates in which experts argue whether symptoms are organic, psychological, or some complex combination of both. Modern medicine has a marked bias toward empirical proof of illness causes or mechanisms, such as a parasite, virus, lesion, or other observable abnormality in cell, tissue, or bone. Both illness categories suffer from their lack of clear etiology, and their diagnoses are based primarily on subjective reports of patients themselves. At present, no cure or effective treatments exist for either syndrome. The term *syndrome* itself signifies that they contain a constellation of associated symptoms, signs, and laboratory findings that can occur together but may result from different causes and that no disease entity has been found. For the most part, suffering takes place invisibly in that little or nothing is perceptible to onlookers. Suffering has become visible through the efforts of illness movements to move the conditions onto the public agenda.

### History of Pain and Fatigue Syndromes

Medical categories intended to make sense of disabling pain and fatigue are not new. As early as 1750, Sir Richard Manningham described a condition called febricula, or "little fever," marked by great weariness all over the body, "little flying pains," and forgetfulness (Wallace and Wallace 1999: 4). Over the next two centuries, a variety of conditions and labels were proposed for this cluster of mysterious symptoms, including neurasthenia (Beard 1869), railway spine (Caplan 1995), rheumatic muscle callous (1898), muscular rheumatism (1900), fibrositis (1904), rheumatic muscle sickness (1912), muscle gelling (1919), nodular rheumatism (1920), and muscle hardenings (1925) (Simons 1990).

These diagnoses competed with one another and slipped back and forth between psychological and physiological explanations. It has been suggested, for example, that neurasthenia remained popular as long as it was viewed as a nonpsychiatric, neurological illness caused by environmental factors that affected successful people and for which the cure was rest (Wesseley 1990). The diagnosis became psychological and dropped dramatically in popularity with its reinterpretation by Sigmund Freud (May 1999). More masculine concepts of battle fatigue emerged after major warfare: shell shock after World War I (Barke et al. 2000) and posttraumatic stress disorder after World War II (Kinzie and Goetz 1996).

In the early twentieth century, these diagnoses were hotly debated. Some complained that fibrositis, for example, constituted a useless wastebasket category, whereas others thought it was being unfairly neglected (Telling 1911; Llewellyn and Jones 1915). "There is perhaps no disorder concerning the nature of which speculation has been so rife," one textbook noted (Llewellyn and Jones 1915: vii).

Around midcentury, epidemics of mysterious disabling fatigue occurred episodically, each time taking on a new name. Following an atypical polio epidemic in the 1930s, a series of new diagnoses surfaced for debilitating aches and fatigue, including chronic nervous exhaustion, postinfectious chronic fatigue, and myasthenic syndrome (Friedberg and Jason 1998: 7). Names reflected geographic areas, such as Icelandic disease after a 1949 outbreak in Iceland (Jenkins 1991: 32). Two outbreaks involved medical staff rather than patients and attracted much attention: 1934 at the Los Angeles County Hospital, and 1955 at the Royal Free Hospital in London. In both, some cases resolved, but others became chronic. The latter generated the names "epidemic neuromyasthenia" and "myalgic encephalomyelitis" (ME), meaning painful muscles and inflammation of the central nervous system, a name still in use throughout Europe. Clinicians sought other names that reflected what they considered to be the condition's essential nature, such as "neuritis vegetativa" on the belief that it affects the autonomic nervous system or "benign ME" to emphasize its nonterminal nature (Jenkins 1991: 32). During the 1960s and 1970s, chronic brucellosis was often cited as the cause of chronic fatigue, a diagnosis that lost popularity when the condition began to be viewed as psychiatric (Shorter 1992: 306).

The rise of the psychological paradigm, beginning mainly in the 1950s, recast illnesses from organic to psychogenic, changing views of cause and treatment. The concept of psychosomatic rheumatism had arisen in the early 1940s and explained pain as a response to intense emotional stress (Halliday 1941). Halliday instructed clinicians to examine for fibrositis nodules in the muscles and, when absent, consider the diagnoses of hysteria, psychotic depression, or chronic anxiety. Through the 1970s, unexplained widespread pain was most often considered a manifestation of psychogenic rheumatism, and sufferers were often considered hysterical. Fibrositis was taught as a psychiatric problem in American medical schools (Copeman 1969: 504). Most sufferers, if diagnosed, were given psychogenic diagnoses that the illness movements later rejected as demeaning.

## Explanatory Categories of Emergent Illness

By the end of the twentieth century, illness movements had carved out a variety of separate illnesses that account for widespread pain and fatigue. Some object to these new categories, trying to discredit them by likening them to hysteria and neurasthenia (e.g., Bohr 1995; Showalter 1997; Hadler 1999). Nonetheless, new illness categories generate hundreds of peer-reviewed research articles and millions of dollars in medical research (figs. 6.1 and 6.2). To varying degrees, they are recognized by the governing bodies of the medical establishment and the state. This chapter examines the processes through which two illness movements have sought official recognition for new paradigms of otherwise unexplained symptoms of pain and fatigue and the consequences of the movements' efforts. It addresses two research questions:

1. How did illness movements generate new diagnostic categories of pain and fatigue?

2. Why have these illness movements resulted in medical legitimacy, research, and funding?

The dominant medical narrative of diagnosis formation describes the naming of new illnesses as an automatic reaction to advances in medical knowledge. This "discovery" paradigm contends that diagnostic categories are introduced when new diseases are identified or when signs and symp-

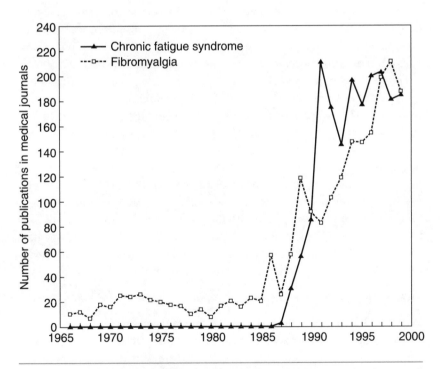

Fig. 6.1. Annual count of publications on chronic fatigue syndrome and fibromyalgia in medical journals, 1966–99. Source: Medline.

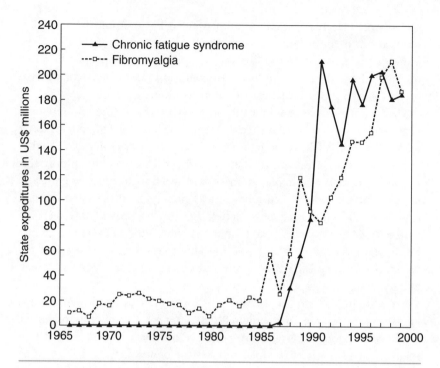

Fig. 6.2. State expenditures on chronic fatigue syndrome and fibromyalgia, in US$ millions, 1966–99. Sources: NIH, Office of the Director, and NIH, Office of the Budget, National Institute of Allergy and Infectious Diseases.

toms of an existing illness or its subcategories are understood and classified. However, even when pathologies are identified, as in a diseased tick in Lyme disease, a retrovirus in HIV/AIDS, or pathological cells in cancer, political and cultural factors determine if and how they are categorized. That cancer refers to metastasizing cells regardless of whether they inhabit the reproductive system, lungs, or brain represents a decision. Such decisions affect the organization of medicine and the mobilization of resources. The American Cancer Society, for example, provides funding toward the amelioration of all conditions considered cancer.

In contrast to the discovery approach, this chapter holds diagnoses as social inventions intended to distinguish somatic anomalies. Michel Foucault suggests: "Composed as they are of letters, diseases have no other reality than the order of their composition. In the final analysis, their varieties refer to those few simple individuals, and whatever may be built up with them and above them is merely Name" (1973: 118). This represents the social-constructionist paradigm: ailments can be interpreted in limitless ways, and it is the meaning attributed to them by individuals that determines their perceived reality. The ways symptoms are grouped, named, and interpreted are all social products. All conditions are socially constructed because all symptoms are subject to interpretation.

This chapter does not doubt the existence of severe pain and fatigue. I begin with the premise that such suffering has long existed, at least as far back as Job in the Old Testament, who suffered from persistent body aches and nights without rest. Rather than exploring the reality of symptoms, this chapter considers the explanation of and reaction to chronic pain and fatigue. Sufferers look to current cultural notions to make sense of their experience. Beliefs of the day—whether they involve the will of God (as Job believed), demonic forces, vital humors and bile, childhood trauma, genetics, or neurotransmitters—provide the language for making sense of the world. It is unsurprising that a multitude of explanatory systems, names, and reactions have appeared over the past two centuries for such suffering.

I seek to understand the ways in which interested actors have promoted explanatory categories of emergent illness. For a set of symptoms to become a medical category, it must receive a distinct diagnostic label by mainstream medicine. This chapter examines the processes of nosology, or the clustering of symptoms into stable diagnoses. Names put forward by illness movements hold the potential to transform an otherwise seemingly unrelated group of ailments into the coherent narrative of diagnosis. This chapter emphasizes the struggles of illness movements to bring illness categories recognition in the larger medical community.

In our society diagnoses are the basis for medical care. The first step in treatment is the struggle to view symptoms through an established category. Much of our medical apparatus is devoted to diagnosis, often involving invasive tests to search for lesions. Various scanning technologies, including magnetic resonance imaging (MRI), have become a standard practice in the high-tech quest to attribute symptoms to existing categories.

Diagnoses not only lie at the heart of treatment regimens but also are the linchpin of meaningful social organization. Medical research and funding revolve around diagnoses. Once recognized as a diagnostic category, symptom clusters gain worldwide recognition. The International Classification of Disease (ICD) embodies a complex codification scheme for current illness groupings. When a new diagnosis is added to the ICD, it becomes an available and legitimate way of viewing symptoms. In our highly bureaucratized world, ICD codes provide the basis for national data collection on mortality and morbidity as well as individual access to insurance coverage for health care. Diagnoses also provide avenues for identity construction and self-help movements, as evidenced by the flourishing of support groups and research and educational organizations created for every known diagnosis.

Because of the marked bias in modern medicine toward conditions with clear empirical markers, proponents for apparently vague conditions face significant struggles. The success of illness movements, I suggest, depends on (a) the social basis of its participants as well as (b) their strategies to frame their cause.

## Fibromyalgia Syndrome

In the mid-1970s fibromyalgia had yet to emerge as a diagnosis in the United States or elsewhere. The closest thing to it was the rheumatic condition fibrositis, meaning inflammation of the muscles. For the most part, physicians viewed fibrositis as psychogenic and its sufferers as malingerers and hypochondriacs. It did not help that the preponderance of its sufferers were women, given the medical establishment's long history of devaluing women's complaints (Todd 1989; De Ras and Grace 1997). It was common for patients who reported all-over aches, pains, and debilitating fatigue to be dismissed with the damning phrase "it's all in your head," so common and so detested by sufferers that fibromyalgia support groups on the Internet three decades later introduced the acronym "IAIYH" to refer to this experience.

The name "fibromyalgia" gained currency in 1981. That year, two separate studies concluded that "fibrositis" was a misnomer because patients did not typically have inflammation, as implied by the "-itis" suffix (Bennett 1981; Yunus et al. 1981). This "negative finding" led Yunus et al. (1981) to coin a new term that removed any etiological reference that might later be disproved: the name "fibromyalgia" simply described the subjective symptoms of sufferers—pain in the muscles—and therefore could not be easily invalidated. That they chose a scientific-sounding name, rather than common wording such as "muscle pain syndrome" or "achy muscle syndrome," lent a patina of authenticity.

Although critics of a fibromyalgia syndrome have persisted into the twenty-first century, it had gained a considerable measure of status by the 1990s. This occurred without any medical breakthrough concerning the basis of this condition. The success of fibromyalgia in becoming part of the mainstream medical lexicon, I argue, derives from the social basis of its main proponents and their strategy to frame fibromyalgia-cum-science.

The prime movers and shakers in the fibromyalgia illness movement were medical researchers and clinicians in rheumatology, the specialty that treats conditions such as arthritis. Approximately 20 to 30 percent of patients who enter rheumatologists' offices suffer from widespread pain of unknown origin (interview with Robert Bennett, January 3, 2000). Until recently, rheumatologists referred these patients to psychiatrists. According to an influential fibromyalgia proponent, Dr. Robert Bennett, rheumatologists like himself were becoming frustrated with the mental health approach: "I would send them to psychiatrists who would send them back. Among these patients were lawyers, doctors, and businessmen . . . they did not seem crazy" (Bennett interview). Yet no other paradigm existed to make sense of their experience. This began to change in the late 1970s with the publication of an article entitled "Two Contributions to Understanding of the 'Fibrositis' Syndrome" in the *Bulletin of Rheumatological Diseases,* a widely circulated professional newsletter. Fibromyalgia proponents consider this article pathbreaking and say it laid the founding ideas for the new paradigm (Russell 1996b: 1). In it, the authors report that fibrositis sufferers have both a sleep disturbance and a particular pattern of tender points on their body (Smythe and Moldofsky 1978). Dr. Smythe, a Canadian rheumatologist, examined a group of his patients with musculoskeletal pain and found tender areas in well-defined and reproducible locations (locations that are the same from patient to patient). A Canadian psychiatrist and sleep researcher, Dr. Moldofsky, studied Dr. Smythe's population for sleep problems. Moldofsky had become interested in pain patients while in England working with patients with "writer's cramp." Together Smythe and Moldofsky declared fibrositis "alive and well."

This was not the first time that a pattern of tender points was noted. As early as 1838, a French physician described patients with local pain points in a common distribution (Shorter 1992: 312). Nor were the sleep findings limited to this population—various pains have been found to interfere with deep sleep. Nor did the article report any discovery of the causes or mechanisms of these symptoms. Nonetheless, Smythe and Moldofsky's report provided something that could be measured by the physician-examiner beyond patient self-reports. This offered the possibility that unexplained symptoms of pain and fatigue might be physiological, rather than psychological, after all. The credibility of this article was sufficient to arouse interest in a small group of North American rheumatologists.

Although this and later articles described the maladies of fibrositis/fibromyalgia sufferers in some detail, the medical profession in general did not regard this condition seriously. There was a definite stigma associated with studying fibromyalgia, which was considered not a true scientific enterprise. Even though new "discoveries" are the stuff that make careers, university physicians who wanted to research fibrositis were warned against "academic suicide." One internationally renowned fibromyalgia researcher confided that his department chair instructed him to drop his research, calling it "a dead end that would end his career" (personal correspondence, May 30, 2000, subject who wishes to remain anonymous).

The story of how fibromyalgia gained acceptance revolves around the credentials of its main proponents, doctors of rheumatology, and their strategy, a decided adherence to the scientific and bureaucratic methods sanctioned by their profession. In the late 1980s, a group of interested rheumatologists, convinced that fibromyalgia suffering deserved medical attention, set out to make it legitimate: "It is increasingly evident that fibromyalgia is an important cause of persistent 'unwellness' and functional impairment which has a major impact on North American society. Its study does not yet have the same intellectual rigor and exactness demanded of the 'new biology, but it won't 'just go away' because some may choose to ignore it" (Bennett 1989: 1).

One strategy employed by these maverick rheumatologists was to organize scientific conferences to increase respectability, disseminate research findings, and generate new ideas. The first took place in 1986, sponsored by the pharmaceutical company Merck, Sharpe and Dome (MS&D), which was interested in promoting its product Flexoril, a muscle relaxant, as a fibromyalgia treatment. MS&D had funded clinical trials, and the results looked somewhat promising (according to researchers, the company later lost interest, when its patent expired). This meeting increased rheumatologists' attention to fibromyalgia—the "soft tissue club," which had very sparse attendance in the 1970s, attracted hundreds of rheumatologists in the 1980s (Bennett interview). In 1988, rheumatologists held a major conference in London, Ontario, in which no nonmedical participants were allowed. Interested laypersons were barred entry as a strategy to ward off stigma. The only fibromyalgia sufferer who managed to attend was Beth Ediger, on account of her nursing degree (interview with Mary Anne Saathoff, president, Fibromyalgia Association of America, November 29, 1999). Ediger went on to write one of the earliest educational booklets on fibromyalgia, which some activists point to as a landmark in the field (Ediger 1991). Later that year an Arthritis Foundation (AF) chapter sponsored a fibromyalgia symposium in Palm Springs, California. Participants from various medical specialties, many of whom had never before heard of fibromyalgia, were invited to "foster new ideas for research into the fibromyalgia/fibrositis syndrome" (Bennett 1989:1). In 1989, the first in what became a series of fibromyalgia conferences convened in Minneapolis. The International Symposium on

Myofascial Pain and Fibromyalgia, later referred to as MYOPAIN, was initiated by a dentist, Dr. James Friction, and a general physician, Dr. Essam Awad, who had studied fibrous tissue biopsies in the 1950s (Russell 1996a: xi). Rheumatologists spearheaded subsequent MYOPAIN conferences and renamed the series with more forceful words, "World Congress," meeting every three years: 1992 in Copenhagen, Denmark; 1995 in San Antonio, Texas; 1998 in Silvi Marina, Italy; 2001 in Portland, home of Dr. Bennett's Oregon organization; and 2004 planned for Munich, Germany. Existing international conferences also began to incorporate fibromyalgia into their agenda: the Eighth International Symposium of Physical Medicine Research Foundation met in 1995 to discuss "fibromyalgia, chronic fatigue syndrome, and repetitive strain injury" (Chalmers et al. 1995), and the Eighth World Congress on Pain held a satellite symposium in Vancouver, British Columbia, in 1996 called "Muscle Pain Syndromes and Fibromyalgia: Pressure Algometry for Quantification of Diagnosis and Treatment Outcome."

At the 1986 MS&D meeting, rheumatologists raised the idea that a "gold standard" was needed to make fibromyalgia a legitimate area of study. They were concerned that fibromyalgia relied heavily on patient self-reports and that no laboratory or imaging methods provided the sort of measure that the medical community and its biomedical model dictate. They decided on a strategy to elevate the status of fibromyalgia. Instead of redoubling efforts to hunt down the fibromyalgia pathology—a long-term strategy that involved risk and uncertainty—they formed a committee to design a "definitive" study that would provide findings to serve the function of diagnostic criteria. The Multicenter Fibromyalgia Criteria invited interested researchers to evaluate more than three hundred different factors in a group of approximately five hundred patients—roughly half were selected because they were presumed to have fibromyalgia; the others suffered from other causes of musculoskeletal pain.

The study's principal findings were that fibromyalgia patients can be identified by two criteria: (1) widespread pain and (2) tenderness in eleven of eighteen tender points. Although they found other symptoms, researchers argued that these two provided the most efficient markers. The pronouncement of these symptoms as reproducible, empirical regularities rationalized the condition, despite criticism from opponents (see Bohr 1995; Hadler 1999). The researchers were savvy in presenting their methods as consistent with those used for more established illness categories. They reported impressive statistical significance: their findings were "83 percent sensitive" and "81 percent specific" (Wolfe et al. 1989). The diagnostic criteria for fibromyalgia, they wrote, "have been developed in a fashion analogous to these for [rheumatoid arthritis and systemic lupus erythematosus] and as such represent the current state of the art with respect to epidemiological techniques for criteria development" (McCain 1989: 192). Neither of these findings, however, constituted a scientific "discovery." The criterion of widespread pain simply reiterated the subjective report of patients. Tender points involve physician observations: during an exam, the physician is to press lightly (just until the fingernail whitens) on specific tender points as well as control spots and evaluate the patient's response. To standardize pressure, an instrument called a dolorimeter was introduced. If a patient responds to a pressure of 4 kg with a visible or vocal response of discomfort, such as by flinching or jumping, the location is considered a tender point. In the absence of blood work or other lab tests, the appearance of science and objectivity in this test is critical. That the researchers recorded their findings systematically provided the appearance of objectivity required for considering the symptom as real and significant. But the tender point threshold of eleven (out of eighteen points) represents a judgment call based on the need for an agreed-upon quantitative measurement. A variety of earlier studies had proposed a wide variety of numbers and locations of pressure points and threshold levels (see Smythe and Moldofsky 1978; Yunus et al. 1981; Campbell et al. 1983; Wolfe and Cathey 1985; Simms et al. 1988). Furthermore, some physicians are critical of the entire enterprise: "The accuracy and reliability of the dolorimetry leaves much to be desired . . . the probing thumb is worse. Interestingly, the 'expert' tender-point assessors were less accurate, reliable and trainable than the naïve examiner. Many patients . . . with all sorts of chronic illnesses . . . and some subset of nonpatients, all find the pressure of the probing thumb particularly uncomfortable. However the clinical community that promulgates the fibrositis concept is not ready to give up on trigger points" (Hadler 1999: 31). Hadler criticizes the process as tautological because it relies on the existence of tender points to distinguish "fibromyalgia" patients from the control group, made up of other pain sufferers, and contends that tender points correlate with distress, not disease (28–39).

Such criticism and discrepancies over tender points could have brought skepticism to the findings of the multicenter study. Instead, in 1990, as planned, the American College of Rheumatology (ACR) adopted these findings as official diagnostic criteria. In 1992, medical experts from around the globe signed an international declaration that decreed fibromyalgia a "real and significant health problem" ("Consensus Document" 1993). The World Health Organization (WHO 1992–94) then endorsed this declaration and revised its fibromyalgia entry for the Tenth International Statistical Classification of Diseases (ICD-10) from "an unspecified disorder" to "a painful, non-articular condition," calling it the most common cause of "chronic, widespread, musculoskeletal pain" (Csillag 1992: 663). These monumental international events for this "emerging" condition were intended to place it on par with other recognized illness categories.

The publication of the ACR criteria gave a great boost to fibromyalgia research. As shown in figure 6.1, publications increased from an average of ten articles per year in the 1970s to twenty in the early 1980s and then quickly escalated around the time of the publication of the ACR criteria (1990) to more than one hundred annually. In addition to articles published in rheumatology journals, a home for fibromyalgia research was created in 1993. The *Journal of Musculoskeletal Pain* (*JMP*) was founded, according to an editorial in the first issue, in response to a perceived need "for more academic communication" on fibromyalgia and related conditions. The vision behind it involved Kristen Thorson, of the Fibromyalgia Network, a patient advocate group. Its goals were a "high quality, peer reviewed . . . journal . . . aimed at a professional readership" (Russell 1993: 1–2). The introduction to the first edition quotes Dr. Yunus, who "makes a strong appeal for careful adherence to the scientific method. . . . Anything less . . . will fail to meaningfully advance our meager knowledge about these painful syndromes" (6). The editor reassures readers that the papers have been subjected to "blinded peer review."

Participants in this movement presented their research and findings in the language of progress, with particular regard to the search for objective diagnostic markers. For example, the 1989 International Symposium on Myofascial Pain and Fibromyalgia in Minneapolis was called "remarkable progress toward finding useful objective tests" (Simons 1990: 2). Fibromyalgia research has produced numerous findings about neurotransmitters, hormone secretion, and central nervous system processes that appear to go awry in fibromyalgia patients. Because a multiplicity of theories exist, no single negative finding can discredit the diagnosis as a whole. At a recent international fibromyalgia conference, Dr. Bennett asserted that the admonition in medical articles that "the cause of fibromyalgia is not known" is no longer justified because of "impressive advances . . . made in . . . the neurobiology of chronic pain" (Bennett 2000).

But it is not the findings that give fibromyalgia legitimacy so much as the belief and confidence that fibromyalgia represents an organic disease and that answers will surface with additional research. This belief underlies the medical machinery that supports and generates additional studies, conferences, and research dollars. As a rheumatological condition, fibromyalgia was endorsed early on by the AF and became the project of the National Institute of Arthritis and Musculoskeletal and Skin Diseases (NIAMS), part of the National Institutes of Health. In a recent congressional report, the House Appropriations Committee urged NIAMS to consider additional steps to strengthen the NIH research effort for fibromyalgia (and numerous other conditions) (NIAMS 1999). NIAMS responded that it "continues to place a high priority on encouraging investigators to conduct studies on fibromyalgia" and has included members of a patient advocacy group (NIAMS 1999). Research money from NIH devoted to fibromyalgia, mainly from NIAMS, increased from under $1 million in 1993 to nearly $7 million a year at the close of the century (see fig. 6.2).

The standing of this condition rests on the scientific reputations of its researchers. Sufferers also became active in the fibromyalgia "illness movement," but the focus of their contribution was on the promotion of the scientific research effort.

In the mid-1980s fibrositis was virtually unknown by the public—it was absent in popular media, and patients generally lacked a diagnosis and therefore an identity for patient mobilization or the language of science to back them. While rheumatologists championed fibromyalgia on the part of their sufferers, a patient-oriented movement gained momentum through the medical specialty of physiatry, or physical medicine. Its leaders occupied the dual role of physician and sufferer. This activity developed separately from the rheumatology movement, although it gained information from the rheumatology publications.

In 1986, Dr. George Waylonis, head of Medicine and Physical Rehabilitation in Columbus, Ohio, and Ruth Buzard, a registered nurse (RN) in the hospital, both fibromyalgia sufferers themselves, sifted through files of former patients to identify those with widespread, unexplained pain. They compiled a mailing list and sent out information packets with an article describing fibromyalgia, written by Dr. Waylonis in 1984, and an invitation to an upcoming meeting. More than one hundred people turned out for the May 1986 meeting (interview with George Waylonis, May 10, 2000). With the financial backing of Bert Missikman, whose daughter had fibromyalgia, they created a nonprofit group under the wing of the hospital. At the first meeting was Mary Anne Saathoff, an RN who had suffered for years with pain and fatigue without any understanding of her problems. Saathoff became director of education and helped organize five educational meetings a year. Participants traveled from outside the Columbus area to attend. By 1995, the Ohio organization went national—the Fibromyalgia Association of America (FMAA). From the start the FMAA sought to build collegial alliances with medical institutions and established medical organizations, such as the Arthritis Foundation. The AF accepted fibromyalgia without a struggle, as a way to expand its scope and jurisdiction. At the third meeting of the Columbus fibromyalgia group, the local AF chapter sent a speaker, and the National Arthritis Foundation later introduced a self-help course and a book on fibromyalgia, modeled after the AF's existing lupus self-help activities.

Mary Anne Saathoff served as FMAA president from 1992 until December 31, 2000, when she and the organization both retired. According to Saathoff (interview, November 14, 2003), the FMAA formed when people knew little about the condition. As information became widely available and other good organizations flourished, the need for the FMAA diminished. When the FMAA closed its doors, it passed the torch (and its membership) to the Fibromyalgia Association of Greater Washington, Inc. (FMAGW), which then changed its name to the National Fibromyalgia Partnership, Inc. The FMAGW had emerged in the same way that the FMAA did in the early 1990s—it began as a small, local group (in Virginia) and grew steadily, all the time remaining dedicated to being medically credible.

Of the many fibromyalgia sufferer associations that emerged in the 1990s, almost all have followed the model of working in conjunction with the medical community to achieve and maintain the illness's credibility. Only one went in a different, more combative direction: the USA Fibrositis Organization, which eschewed the medical profession and maintained its anachronistic name into the early 1990s, despite estrangement from other fibromyalgia groups (Waylonis interview).

In sum, although debates persist, fibromyalgia won legitimacy in the ACR, WHO, Social Security, research, and funding. These outcomes are quite striking because findings about this condition are not of the type valued by research community, that is, definitive organic markers. This success lies in the social basis of the movement and its strategies. First, the main advocates have been rheumatologists, rather than laypersons. Second, the strategy of the illness movement has been to continue systematically to present research in the form that is acceptable to the scientific community. It presents findings as a disinterested group of experts; it is experts—or "rationalized others," as Meyer (1994) calls this group—who define cultural meanings for the wider society, for larger consumption. Fibromyalgia became a modern diagnosis because the small group of rheumatologists became suspicious of the psychological paradigm and decided to scientize the category. Patient groups began to organize when this professional group created the identity for them. The lay movements that have followed from medical definitions have been tethered to the medical community, but it did not welcome patient groups into fuller participation until after the adoption of medical criteria. Nor did the media's response to this syndrome precede the defining statements by the medical profession. This was an illness movement from within, highly regulated by the medical profession, which shouldered the burden of providing a convincing narrative of the "realness" of this condition.

## Chronic Fatigue Syndrome

The symptoms of CFS overlap significantly with those of fibromyalgia. The main distinction between the two is the balance between fatigue and pain—people with fibromyalgia tend to hurt more, whereas those with CFS feel more tired. But the disparity is not always significant; sufferers often receive both diagnoses. Many doctors and researchers now contend that the two are different manifestations of a shared underlying disorder. Moreover, research reveals that many patients diagnosed with one disorder fit the criteria for the other (Aaron et al. 2000). Which diagnosis patients receive depends in part on the type of doctor they see. Rheumatologists and physiatrists commonly diagnose fibromyalgia; CFS has been the province of internists and infectious disease specialists.

The nosology narratives of these two illnesses diverge more than their similarities may suggest. Unlike fibromyalgia, which seems to strike a geographically random population, CFS emerged as an infectious mini-epidemic. At various points in the twentieth century, outbreaks of mysterious flu-like illnesses with protracted symptoms including cognitive dysfunction and fatigue have been noted. Some were named after their cite of origin, such as Icelandic disease, others after already existing illnesses they resembled, like epidemic neuromyasthenia.

The syndrome that became understood as CFS emerged from two main events in the mid-1980s: (1) the publication of research that linked severe fatigue to infection by the Epstein-Barr virus, which causes mononucleosis, and (2) observations by two internists of a debilitating fatigue epidemic.

In 1984, Dr. Richard DuBois and his colleagues published data on fourteen patients with disabling fatigue who had evidence of an active Epstein-Barr (E-B) virus infection in their blood (DuBois et al. 1984). They likened this to a new "chronic mononucleosis." The following year similar descriptions appeared in the widely read *Annals of Internal Medicine,* a prestigious journal that reports on medical breakthroughs (Jones et al. 1985; Straus et al. 1985). Jones and his colleagues reported that thirty-nine of forty-four fatigued patients tested positive for the E-B virus (EBV) antibody. Steve Straus, the head virologist at the National Institute of Allergy and Infectious

Diseases, and his colleagues had similar results. The condition they described became labeled chronic E-B virus (CEBV) or chronic mononucleosis-like syndrome. Findings of CEBV rekindled interest among a small group of medical researchers. Doctors around the country began to check patients' Epstein-Barr titers.

Simultaneously, internists Dan Peterson and Paul Cheney had begun to observe an unexplained flu-like illness among their patients in Incline Village, Nevada, a resort town near Lake Tahoe. After reading about CEBV in the *Annals,* they sent patients' blood samples to a lab, which reported high levels of EBV antibody, evidence of the chronic mono-like condition described in the literature. The number of patients who presented with these symptoms startled these physicians. In 1985, Dr. Peterson contacted the U.S. Centers for Disease Control and reported what appeared to be the start of an epidemic. The CDC sent epidemiologists Gary Holmes and John Kaplan of the Division of Viral Diseases, who, following standard procedure, created a working case definition and evaluated patients (CDC 1986). As with earlier outbreaks of this type, a search ensued for its organic cause. The NIAID convened a consensus conference on CEBV that included several sufferers (Tobi and Straus 1985).

The media significantly influenced the framing of this condition. In the case of fibromyalgia, the earliest articles contained authentication from the Arthritis Foundation and American College of Rheumatology and described fibromyalgia in medical language, with sympathetic treatment of patient suffering (see, e.g., Voorhees 1989). In contrast, CFS headlined stories before it even had a name, in part because of the sensational aspect of contagion stories. Between 1984 and 1988, Lake Tahoe physicians reported more than four hundred cases of protracted flu-like symptoms (Daugherty et al. 1991). Before data were evaluated to scientific community standards, attention to this mysterious condition mounted. Television stations picked up the story, with special pieces appearing on national programs such as "Nightline" and "20/20." That this outbreak occurred in an affluent resort town excited a media frenzy. A skeptical press maligned sufferers of CFS, branded "yuppie flu" and "affluenza," as spoiled, bored, and the creators of their own suffering, likening them to the neurasthenics of Victorian times. The derisive moniker "yuppie flu" may have been a response to sufferers' efforts to portray themselves as high achievers to avoid being characterized as shirkers (Feiden 1990: 13). The pejorative name contributed to the defensiveness of sufferers, many of whom were not in the upper socioeconomic strata.

Soon the constellation of symptoms became public knowledge. Sufferers flocked to Peterson and Cheney's practice in Nevada, which kept the CDC abreast of new findings and increasing numbers. However, instead of rallying the larger medical community, their practice became increasingly marginalized in its efforts to convince the government that an infectious epidemic was afoot. Although like-minded internists and family physicians became active in the treatment of CEBV patients, most of these doctors did not conduct research. The burden of discovering the viral culprit fell on the shoulders of infectious disease specialists. At the CDC, a large governmental organization concerned with a huge array of diseases, CFS competed for attention with conditions already recognized and shown to be fatal, like HIV/AIDS.

Sufferers play a significant role in the illness movement. Even before the condition was named, support groups began to form. As the national media covered this illness, sufferers rallied. The CEBV Association–later renamed the CFIDS Society–opened its doors in Portland, Oregon, in 1985, following an article in the *Oregonian* newspaper profiling a sufferer. She received so many phone calls from local residents with similar symptoms that she organized a meeting, and within a few months, the national group emerged (Feiden 1990: 18). Another influential group formed in Charlotte, North Carolina, where Dr. Cheney moved from Lake Tahoe in 1987 and met long-time sufferer Marc Iverson. Under Iverson's leadership, the Charlotte group raised a great deal of interest and money to research CEBV. Support group leaders from the various organizations circulated surveys to learn about needs of doctors and patients nationally (interview with Orvaline Pruitt, president, National Chronic Fatigue Syndrome and Fibromyalgia Association, June 20, 2000).

Rheumatologists pioneered fibromyalgia activism on behalf of patient populations, who backed the mainstream medical approach when they ultimately mobilized. CFS, on the other hand, lacked a strong support network or a disciplinary base within the medical profession, and sufferers assumed more dominant roles in the movement, trying to generate medical attention and public legitimacy. According to the chronicler of the CFS/CFIDS Network, Karyn Feiden, sufferers were long frustrated with the slow progress of the medical community. "Thanks in good measure to the dedicated work of patients themselves, CFS has clearly emerged from that bleak period of neglect" (Feiden 1990: 5).

CFS patient groups both initiated medical conferences and constituted the largest percentage of attendants. Among the earliest was a meeting in Austin, Texas, that gathered all the big names in CEBV research, including Drs. Straus and Jones. Despite this star-studded lineup, the meeting failed to overcome skepticism and attract a medical audience. Originally scheduled for the University of Texas at Austin campus, it took place instead in a hotel. In 1987, patient groups organized a conference in Portland, Oregon, at a Holiday Inn. Shortly after, a patient group in Kansas City initiated a conference that led to the formation of the patient-run National CFS Advisory Counsel (Pruitt interview). In 1989, the First Chronic Fatigue Syndrome Conference, a state-sponsored meeting attended by approximately five hundred health care professionals, was organized by a patient group, the San Francisco CFIDS Foundation. The foundation influenced the meeting's agenda and guest list, excluding one of the most prominent CEBV researchers because his new research suggested that CFS may have a psychological component (Libman 1989).

Meanwhile, research on CEBV began to produce inconclusive data. Studies found healthy subjects with elevated levels of EBV antibodies and CFS sufferers who had never been exposed to the virus. Doubt grew that EBV could be the sole agent causing fatigue symptoms. In 1987, Holmes and other CDC researchers published a study of 134 patients in the *Journal of the American Medical Association* (*JAMA*), reporting that Epstein-Barr virus could not be pinned down as the cause of this syndrome (Holmes et al. 1987). Although nails were being hammered into the coffin of the CEBV paradigm, an infrastructure had been constructed to understand this virus, in both medical and lay circles. Researchers and clinicians familiar with this patient body did not lose interest. Popular journals continued to report on CEBV, private labs promoted blood testing, and fatigued patients requested tests from their physicians (Henig 1987). Government agencies, mobilized to react to the epidemic, received continued pressure to proceed with its work. The NIAID was deluged with letters from sufferers, including physicians with CFS. In 1987, the House Appropriations Committee encouraged the NIH and CDC to investigate this further (Johnson 1996).

That same year, the CDC convened a group of international experts to discuss terminology for the condition. Both CEBV and the British concept of postviral fatigue syndrome were rejected as misleading. The agency chose to keep the unwieldy phrase "chronic mononucleosis-like syndrome," which it had been using since 1985 because it lacked conclusive data to suggest an alternative. Under pressure from patient groups and Congress, the CDC convened another group in 1988 to devise an official working case definition for the condition. An international committee, chaired by Holmes of the CDC, decided to select a new name with no association to the cause or mechanism for the malady and agreed on "chronic fatigue syndrome" after its primary symptom. The definition appeared in the *Annals of Internal Medicine* (Holmes et al. 1988). The official adoption of a name and definition provided a certain amount of legitimacy for use by physicians and researchers. The U.S. Public Health Service, Congress, and patient groups adopted the name. But unlike the fibromyalgia diagnostic with its tender points as marker, chronic fatigue was described as a diagnosis of exclusion, made by ruling out dozens of other diseases able to produce similar symptoms. The modified ICD-9 included CFS under fatigue, along with psychogenic, nervous, combat, and other sorts of fatigue (though in Britain the term *myalgic encephalomyelitis* remains in use). Research on CFS shot up from sixty publications in 1988 to more than two hundred in 1990.

Research on CFS through the early 1990s continued to focus on viral causes. During the 1980s, human retroviruses were very much on the public's mind in the United States because of HIV/AIDS and activism. As the CEBV paradigm lost credibility, a more general virus paradigm gained momentum. The second suspected culprit was a newly discovered member of the common herpes virus family called HHV-6, also implicated in AIDS. Researchers found evidence that CFS sufferers had elevated rates more often than normal people (Yalcin et al. 1994; Patnaik et al. 1995).

More provocative news about the involvement of a specific retrovirus spread from a team of researchers at the Wistar Institute in Philadelphia in 1991. Because of the high level of patient involvement and interest, including direct funding of medical research, this lead went public rather than adhering to the peer-reviewed sequence. The researchers held a major press conference to promote their findings and created quite a stir—including rumors that CFS was related to AIDS—until it was squelched by subsequent research findings. As with the discovery of "cold fusion," scientists failed to reproduce the findings. Various researchers, including the Retro Virus Branch of the CDC, examined a slew of human and animal viruses—human herpesvirus 6, human spumavirus, simian retrovirus, simian T cell leukemia virus, and bovine leukaemia—and concluded that no relationship exists between any of the viruses studied and the symptomed population (Gow et al. 1992; CDC 1993; Folks et al. 1993). These negative findings delivered another blow to CFS advocates, who remained eager for documentation that CFS is a physiological disease entity.

In the absence of conclusive results, researchers also turned to psychiatric and psychosocial explanations. As psychiatrists entered the fray, the personality and mental status of CFS patients became a topic of analysis. Debates ensued over whether the constellation of symptoms was due to a new epidemic or an older, psychogenic condition like neurasthenia. Articles on CFS, with titles such as "Is It Real?" and "Fact or Fiction?" appeared in medical journals (Kroenke 1991; Rand 1991). Scientific conferences on CFS included papers on the psychological nature of the syndrome. Sufferers and advocates, who sought the legitimacy that came with organic disease, considered this turn of events to be highly disparaging. As Dr. Cheney explained: "I think the knee-jerk reaction is to call anything that we don't understand, or that we can't prove is organic by the definition of the day or the technology of the day, 'psychiatric.' And I don't think that's right. . . . Between the years 1980 and 1984, the only thing we could find in AIDS patients was disturbed immune function and a menagerie of other infections that didn't make any sense" (Dr. Paul Cheney, quoted by Jacobson 1989). Allusions such as this to AIDS proliferated within the CFS community. To gain credibility, activists tended to emphasize the debilitating nature of the disorder and framed CFS as "AIDS junior" and even a "non-HIV positive AIDS" (see Johnson 1996). Because CFS had been framed primarily as an infectious disease, it shared a scientific niche with AIDS and therefore competed with it for resources and attention. Comparisons were used to fix expectations. For example, one dedicated researcher explained the difficulty of determining the cause of CFS by comparing it to finding the cause of AIDS: "Even AIDS was simpler than this, because it was caused by one virus, and that virus is very new in the last 20 years. It is more likely (with chronic fatigue) that a whole bunch of other old viruses and other infectious agents are all able to cause some people's bodies to act in a particular way to cause this disease" (Dr. Anthony Komaroff, quoted by Libman 1989). CFS advocates and support groups condemned the government's response to CFS as inferior to its efforts against AIDS:

> The government's response to the illness has been slow, inconsistent, and we believe, inadequate. By contrast, another contemporary serious illness, AIDS, has captured considerably more attention from public health officials and scientists. (Friedberg and Jason 1998: xiii)
>
> We're still trying to recover from the AIDS epidemic, and nobody wants to admit there may be another epidemic around the corner. (Marya Grambs, director, San Francisco CFIDS Foundation, quoted in Painter 1989)

CFS activists also used early AIDS activism as a framework for their own illness movement, adopting ACT-UP tactics to prompt a groundswell of activity. Within the CFS community, a vocal portion has assumed an outsider advocacy strategy. Not all CFS activists fit this characterization. Some tried instead the approach favored by fibromyalgia groups: working within the medical establishment. The National Chronic Fatigue Syndrome Association in Kansas City, for example, sought to work with medical professionals to provide educational matter to the public. In 1992 this organization added

fibromyalgia to its name and agenda when its leaders realized that a significant proportion of its members were diagnosed with the ACR criteria. Local
support groups also began to advocate for both diagnoses, adding one or the
other to their original name. Yet even with the overlapping symptoms and
populations, rheumatologists hesitated in associating fibromyalgia with its
more controversial sister diagnosis. Only 10 percent of journal publications
on either condition discuss both.

The CFS movement's outsider approach is evident in its contestation of
the syndrome's name. Although the CDC and WHO had settled on "chronic fatigue syndrome," some patient groups felt this name trivialized their
suffering. Other names were discussed, including chronic fatigue and
immune dysfunction syndrome (CFIDS), a name that invoked the immune
system and an association with AIDS. The Charlotte group was first to adopt
CFIDS, beginning a trend among patient groups to reject the CFS moniker.
Not all sufferers supported this name change, arguing that it was premature
to adopt a name involving immune dysfunction without scientific proof
that this was the cause of their symptoms. However, massive interest on the
part of organized sufferers diffused this patient-initiated name.

CFS activists succeeded in bringing their concerns to national and international authorities. In 1991, the NIAID held a meeting in which conferees
proposed a new case definition that excluded fewer patients with psychiatric
conditions. Their recommendations were published in the *Annals of Internal
Medicine* (Schluederberg et al. 1992). In 1993, the CDC invited an international group of experts to review the working case definition because the
"current criteria for the chronic fatigue syndrome do not appear to define a
distinct group of cases" (Fukada et al. 1994: 954). It also invited public input
on the name and case definition. Insiders report that this motivation came
more from politics than scientific grounds. The expert group, which included several patient group representatives, published a revised definition in
the *Annals of Internal Medicine* (Fukada et al. 1994). The new definition
sought to "address some of the criticisms of [the 1988] case definition and to
facilitate a more systematic collection of data internationally" (Fukada et al.
1994: 957). It reduced the number to four out of a list of eight possible symptoms, consequently expanding the number of people who fit the diagnosis.
Like the 1988 definition, it included a list of conditions whose presence
excludes the diagnosis of CFS. This time, it also specified exceptions to these
exclusions. People who suffer from illnesses "defined primarily by symptoms that cannot be confirmed by diagnostic laboratory tests, including
fibromyalgia, anxiety disorders, somatoform disorders, nonpsychotic or
melancholic depression, neurasthenia, and multiple chemical sensitivity

disorder," were not to be excluded (Fukada et al. 1994: 956). The new definition concludes with an apologetic and defensive note that, although published in a professional medical journal, seems to engage in critiques within the lay community:

> The name "chronic fatigue syndrome" is the final issue that we wish to address. We sympathize with those who are concerned that this name may trivialize this illness. The impairments associated with chronic fatigue syndrome are not trivial. However, we believe that changing the name without adequate scientific justification will lead to confusion and will substantially undermine the progress that has been made in focusing public, clinical, and research attention on the illness. We support changing the name when more is known about the underlying patho-physiologic process. (Fukada et al. 1994: 957–958)

Despite its lack of a scientized name, the new criteria ushered in a new level of acceptance. Chronic fatigue syndrome, which was not represented in the ICD-9, except as "fatigue syndrome" or "neurasthenia," appears on the ICD-10 with postviral fatigue syndrome and benign myalgic encephalomyelitis, classified under disorders of the central nervous system (CDC 2001). Yet activist groups continued to protest. A survey conducted in June 1997 by the CFIDS Association of America showed that "an overwhelming majority of patients wants the name changed immediately" (CFIDS Association of America 1997).

By certain measures, CFS has secured a respectable spot as an institutionalized category of illness. Figure 6.1 shows that research publications on chronic fatigue syndrome outnumber those on fibromyalgia. Figure 6.2 illustrates that funding for CFS began earlier than that for fibromyalgia and was more plentiful. Although the first funding for research came from patient organizations—among others, the Chronic Fatigue Syndrome Research Foundation in Denver, Colorado—state funding has come from various sources, mainly Congress, the CDC, and the NIAID. Despite controversy over whether CFS is viral, the NIAID—the NIH's viral research arm—has increased its sponsorship over the years, from less than $1 million in 1987 to more than $7 million in a peak in 1995 (see fig. 6.2).

Throughout the 1990s, the CFS movement began to resemble the fibromyalgia movement. People with chronic fatigue remained quite active, still constituting the majority of participants at most CFS conferences. However, conferences increasingly included medical boards and other trappings of the medical approach. Theories of CFS are veering away from infectious viruses toward the central nervous system, much like fibromyalgia theories (Folks et al. 1993; Heneine et al. 1994). Although the research streams have remained surprisingly separate, cross-pollination is increasing among researchers, who are more likely to attend the same meetings.

Still, at the start of the twenty-first century, the CFS framework remains in the air. Patient groups focused on what they see as a problematic name have continued to lobby the CDC to revise it—preferred alternatives, according to a recent Internet vote circulated among patients, include CFIDS and "Cheney syndrome." The CDC has responded by organizing groups, including an external review committee with medical and patient representatives, to reexamine clinical and empirical knowledge concerning the definition of CFS and determine whether it needs to be reworked. The CFS Coordinating Committee at the U.S. Department of Health and Human Services is also working with all relevant government branches (the CDC, NIH, FDA, Social Security Administration) and activists to consider a name change. After a meeting of the CFS Coordinating Committee, members of the CFIDS Association of America complained: "We have participated in good faith, but to tell you frankly, we have met with great resistance. We have been shut out. We have been lied to. We have been treated with hostility. Worse yet, the efforts of this committee and the efforts of the represented agencies have brought us no closer to diagnostic markers or effective treatments. . . . The patient community, in fact, largely views this committee's meetings as pointless discussions that lead nowhere" (CFIDS Association of America 2000).

In 1999, patient advocates and members of the CDC agreed that a name change was in order, but only through scientific methods and with a name that would be scientifically descriptive and valid (CDC 1999). The U.S. Department of Health and Human Services appointed a Name Change Workgroup, made up of leading CFS experts and patient representatives, which conducted research to gauge opinions of various names (Jason, Eisele, and Taylor 2001). At its second Annual Case Definition Workshop on Chronic Fatigue Syndrome, in 2001, the CDC launched an international collaborative group to explore data for an empirically driven name change (CDC 2003). That same year, the American Association of CFS—an academic group that emerged in 1992 from a patient organization, the National CFS Advisory Counsel—sponsored a meeting to revisit the case definition in light of scientific advances, and the American Academy of Family Medicine began a study to assess the level of scientific evidence for diagnosis and treatment of CFS.

In 2002, the Name Change Workshop recommended a name that, if length means legitimacy, would provide heaps: chronic neuroendocrineimmune dysfunction syndrome, or CNDS. Rather than as a replacement for CFS, it was offered as an umbrella term, under which CFS as defined by the

CDC would be one subtype. The other subtypes would be myalgic encepha-lomyelitis (ME) as defined by British medical standards, a Canadian subtype referred to as CFS/ME, and Gulf War Syndrome (Name Change Workgroup 2002). The Name Change Workgroup soon shortened the name to NDS, removing the word "chronic," following criticism from activists that chronic conditions rarely contain the word in their name, as well as complaints that the pronunciation of CNDS as "sins" would result in further stigma. The response from sufferer-activists to NDS has been mixed. The CFIDS Association of America (2002) commended the workgroup's proposed solution and the idea of having an umbrella term rather than a "replacement" name, which could simply carry existing stigma to the new name. Others, such as the National CFIDS Foundation, Inc., have petitioned to remove all references to the diagnosis of CFS, which they claim has been irreparably sullied by bad press (McLaughlin 2002). At the time of the writing of this chapter, CFS has not been replaced, subsumed, or rejected. While the new name and definition await approval by the U.S. secretary of health, debates continue within activist groups seeking medical legitimacy for their suffering.

## Conclusion

Through the efforts of illness movements, fibromyalgia and chronic fatigue syndrome have become legitimate diagnoses with the advantages that come with this status. Both have succeeded in gaining respect by the medical establishment, measured by research in peer-reviewed journals and flows of state money to support research, as graphed in figure 6.1. Both have been adopted as "real" conditions by state bodies, including the World Health Organization.

However, debates continue about the definition and name of CFS, even as it reached a certain level of legitimacy, whereas contestation within the fibromyalgia movement is minimal. This difference can be explained, I argue, by the (a) social basis of illness movement activists and (b) the frameworks used to define each.

The medical community holds a near monopoly on the rendering of illness categories, or nosology, and both movements included physicians as activists. The rheumatologists who drove the fibromyalgia movement have enjoyed professional consolidation, network potential, and influential connections such as the American College of Rheumatology. In contrast, CFS has drawn support from a more eclectic group that includes marginalized physicians, others who do not conduct research, and researchers from a variety of fields, making networking more difficult. Activists with these two illnesses also have contrasting relationships with and approaches to the medical mainstream. Fibromyalgia groups were formed in hospitals by health care professionals; CFS groups emerged in living rooms, in response to media coverage of a mysterious epidemic. Fibromyalgia lay activists sought medical legitimacy every step of the way, whereas CFS activists bifurcated over their acceptance of mainstream approaches. Fibromyalgia groups networked with the Arthritis Foundation, an established medical organization, as their ally; vocal CFS activists modeled their strategies after AIDS activism, an outside group.

The two groups also diverge in their relationship with science and "scientific" frameworks. Fibromyalgia activists self-consciously committed to scientific protocols. Fibromyalgia conferences, organized by medical professionals, restricted patient involvement. As the fibromyalgia paradigm gained legitimacy, collaboration increased with patient groups that adhered to the scientific framework. In contrast, CFS researchers have not adopted this exclusionary tactic, even though some, accustomed to academic interchange, report unease presenting ideas to a vested lay audience. CFS conferences have been organized and attended primarily by sufferers.

Fibromyalgia movement proponents chose inductive science as a legitimizing strategy. Rheumatologists conducted a multicenter examination of fibromyalgia sufferers, knowing that the ACR would likely adopt their findings, whatever they were, as the condition's "gold standard." Their scientized presentation, not any findings themselves, provided legitimacy. CFS activists operated through a more defensive, deductive framework, aimed at proving the existence of a viral epidemic. In contrast to the nonfalsifiable fibromyalgia paradigm, this framework pegged CFS credibility, in part, on the findings of blood tests. CDC researchers, summoned to investigate, conducted epidemiological tests to determine whether patients had a distinct presentation. Their investigation came up inconclusive, reporting that only 10 percent of the population that was allegedly affected even fit their working definition. Virologists and infectious disease specialists proceeded to search for a viral cause. Because this research failed to establish a blood basis for CFS suffering, each rejected viral theory weakened the footing of the CFS framework.

The naming of the illness further affected the stability of the two conditions. Fibromyalgia emerged in scientific circles and remained out of the media limelight until after the completion of the diagnostic study. In contrast, the media reported on CFS before it was named or defined by the scientific community. Although this fostered early solidarity among sufferers, the media propagated information that was harmful to their movement. The media framed CFS alternatively as a frightening contagious epidemic and a scourge of the affluent. CFS activists assumed damage control tactics without the backing of scientific discourse. When the CDC devised a definition for CFS, derisive terms did not disappear from the media. Criticisms from within the CFS movement focus on the name and depiction of the condition. On the urging of lay activists, the CDC has convened numerous meetings to reconsider its name and working definition. Had fibromyalgia been called "achy muscle syndrome," it would likely have undergone more public scrutiny. Reactions would have also been different had media reports of the condition belittled suffering or associated it with an unpopular group. Instead, its difficult-to-pronounce, scientized name created an aura of authenticity.

In sum, the legitimacy of fibromyalgia and CFS rested on the reputation of their advocates, the extent to which "science" was available and drawn on for support, and how each was framed in terms of etiology and affected population. The histories of these conditions may have implications for other controversial diagnoses such as Gulf War syndrome, multiple chemical sensitivity, repetitive stress injury, and myofascial pain syndrome.

REFERENCES

Aaron, Leslie A., Mary M. Burke, and Dedra Buchwald. 2000. "Overlapping Conditions among Patients with Chronic Fatigue Syndrome, Fibromyalgia, and Temporomandibular Disorder." *Archives of Internal Medicine* 160:221–227.

Barke, Megan, Rebecca Fribush, and Peter N. Stearns. 2000. "Nervous Breakdown in Twentieth-Century American Culture." *Journal of Social History* 33(3): 565–584.

Beard, George M. 1869. "Neurasthenia, or Nervous Exhaustion." *Boston Medical Surgery Journal* 3:217–221.

Bennett, Robert M. 1981. "Fibrositis: Misnomer for a Common Rheumatic Disorder." *Western Journal of Medicine* 134(5): 405–413.

———. 1989. "Introduction." *Journal of Rheumatology* 16 (Suppl. 19): 1.

———. 2000. "Chronic Widespread Pain and the Fibromyalgia Construct." FAME 2000 International Fibromyalgia Conference; www.immunesupport.com/fame/selectlecture.cfm (viewed June, 20, 2000).

Bohr, T. W. 1995. "Fibromyalgia Syndrome and Myofascial Pain Syndrome: Do They Exist?" *Neurologic Clinics* 13(2): 365–384.

Campbell, S. M., S. Clark, E. A. Tindall, M. E. Forehand, and R. M. Bennett. 1983. "Clinical Characteristics of Fibrositis. I. A 'Blinded,' Controlled Study of Symptoms and Tender Points." *Arthritis and Rheumatism* 26(7): 817–824.

Caplan, Eric Michael. 1995. "Trains, Brains and Sprains: Railway Spine and the Origins of Psychoneurosis." *Bulletin of the History of Medicine* (The American Association for the History of Medicine) 69:387–419.

CDC. *See* U.S. Centers for Disease Control and Prevention.

CFIDS Association of America. 1997. "Name-Change Survey Results." *CFIDS Chronicle.* Summer.

———. 2000. "Statement of the Appointed Members of the DHHS Chronic Fatigue Syndrome Coordinating Committee." February 8.

———. 2002. "Results of Recent Poll on the Name Change Workgroup Proposal." March 12, 2002. See www.cfids.org/advocacy/c-act_03122002.asp.

Chalmers, Andrew, et al., eds. 1965. "Fibromyalgia, Chronic Fatigue Syndrome, and Repetitive Strain Injury: Current Concepts in Diagnosis, Management, Disability, and Health Economics." A compilation of papers presented at the Physical Medicine Research Foundation's Eighth Annual Symposium. New York: Haworth Medical Press.

"Consensus Document on Fibromyalgia: The Copenhagen Declaration." 1993. *Journal of Musculoskeletal Pain* 1:295–312.

Copeman, W. S. C., ed. 1969. *Textbook of the Rheumatic Diseases.* Edinburgh: E. and S. Livingstone.

Csillag, Claudio. 1992. "Fibromyalgia: The Copenhagen Declaration." *Lancet* 340: 663–664.

Daugherty, S. A., et al. 1991. "Chronic Fatigue Syndrome in Northern Nevada." *Reviews of Infectious Diseases* 13 (Suppl. 1) (January–February): S39–44.

De Ras, Marion, and Victoria Grace, eds. 1997. *Bodily Boundaries, Sexualised Genders and Medical Discourses.* Palmerston North, New Zealand: Dunmore Press.

DuBois, Richard E., et al. 1984. "Chronic Mononucleosis Syndrome." *Southern Medical Journal* 77(11): 1376–1382.

Ediger, Beth. 1991. *Coping with Fibromyalgia.* Toronto: LRH Publications.

Feiden, Karyn. 1990. *Hope and Help for Chronic Fatigue Syndrome: The Official Guide of the CFS/CFIDS Network.* New York: Prentice-Hall.

Folks, T. M., et al. 1993. "Investigation of Retroviral Involvement in Chronic Fatigue Syndrome." *Ciba Foundation Symposium* 173:160–166; discussion 166–175.

Foucault, Michel. 1973. *Birth of a Clinic: An Archaeology of Medical Perception.* New York: Vintage Books.

Friedberg, Fred, and Leonard A. Jason. 1998. *Understanding CFS: An Empirical Guide to Assessment and Treatment.* Washington, D.C.: American Psychological Association.

Fukuda, Keiji, Stephen Straus, Ian Hickie, Michael Sharpe, James Dobbins, Anthony Komaroff, and the International Chronic Fatigue Syndrome Study Group. 1994. "The Chronic Fatigue Syndrome: A Comprehensive Approach to Its Definition and Study." *Annals of Internal Medicine* 121:953–959.

Gow, J. W., et al. 1992. "Search for Retrovirus in the Chronic Fatigue Syndrome." *Journal of Clinical Pathology* 45(12): 1058–1061.

Hadler, Nortin. 1999. "The Dangers of the Diagnostic Process: Iatrogenic Labeling as in the Fibromyalgia Paralogism." In *Occupational Musculoskeletal Disorders* (2d ed.), 18–45. Philadelphia: Lippincott.

Halliday, J. L. 1941. "The Concept of Psychosomatic Rheumatism." *Annals of Internal Medicine* 15:666–677.

Heneine, W., T. C. Woods, S. D. Sinha, A. S. Khan, L. E. Chapman, L. B. Schonberger, and T. M. Folks. 1994. "Lack of Evidence for Infection with Known Human and Animal Retroviruses in Patients with Chronic Fatigue Syndrome." *Clinical Infectious Diseases* 18 (Suppl. 1): S121–125.

Henig, Robin Marantz. 1987. "Tired All the Time; Doctors Study Chronic Fatigue and Its Link to the Epstein-Barr Virus." *Washington Post,* June 30, Z8.

Holmes, Gary P., et al. 1988. "Chronic Fatigue Syndrome: A Working Case Definition." *Annals of Internal Medicine* 108:387–389.

Holmes, Gary P., John Kaplan, J. A. Stewart, B. Hunt, P. F. Pinsky, and L. B. Schonberger. 1987. "A Cluster of Patients with a Chronic Mononucleosis-like Syndrome: Is Epstein-Barr Virus the Cause?" *Journal of the American Medical Association* 257(17): 2297–2302.

Jacobson, David. 1989. "Chronic Fatigue Debate Still Going Strong." *Los Angeles Times,* December 25, 1989, B3.

Jason, L. A., H. Eisele, and R. R. Taylor. 2001. "Assessing Attitudes toward New Names for Chronic Fatigue Syndrome." *Evaluation and the Health Professions* 24:424–435.

Jenkins, Rachel. 1991. Introduction to *Post-viral Fatigue Syndrome,* ed. Rachel Jenkins and James F. Mowbray, 3–39. New York: Wiley and Sons.

Johnson, Hillary. 1996. *Osler's Web: Inside the Labyrinth of the Chronic Fatigue Syndrome Epidemic.* New York: Crown Publishers.

Jones, J. F., et al. 1985. "Evidence for Active Epstein-Barr Virus Infection in Patients with Persistent, Unexplained Illnesses: Elevated Anti-Early Antigen Antibodies." *Annals of Internal Medicine* 102:1–7.

Kinzie, J. D., and R. R. Goetz. 1996. "A Century of Controversy Surrounding Posttraumatic Stress Stress-Spectrum Syndromes: The Impact on DSM-III and DSM-IV." *Journal of Trauma and Stress* 9(2): 159–179.

Kroenke, K. 1991. "Chronic Fatigue Syndrome: Is It Real?" *Postgraduate Medicine* 15.90(7): 23–24.

Libman, Joan. 1989. "Fatigue: A Mystery Unravels." *Los Angeles Times,* April 18, 1989, 5:1.

Llewellyn, Richard, and A. B. Jones. 1915. *Fibrositis: Gouty, Infective, Traumatic: So-Called Chronic Rheumatism, Including Villous Synovitis of Knee and Hip and Sacro-iliac Relaxation.* London: W. Heineman.

May, Ulrike. 1999. "Freud's Early Clinical Theory (1894–1896): Outline and Context." *International Journal of Psycho-Analysis* 80(4): 769–781.

McCain, Glenn. 1989. "Summary." *Journal of Rheumatology* 16 (Suppl. 19): 192.

McLaughlin, Jill. 2002. "Chronic Fatigue Syndrome: Name Change Proposal." ImmuneSupport.com, May 13, 2002. See www.immunesupport.com/library/showarticle.cfm/ID/3581.

Meyer, John W. 1994. "Rationalized Environment." In *Institutional Environments and Organizations: Structural Complexity and Individualism,* edited by Richard W. Scott and John W. Meyer, 28–54. London: Sage.

Name Change Workgroup. 2002. Recommendations of the Name Change Workgroup to Be Presented to the DHHS Chronic Fatigue Syndrome Advisory Committee. January 21, 2003. Draft copy. See www.co-cure.org/NCR.htm.

National Institute of Arthritis and Musculoskeletal and Skin Diseases. 1999. Significant Items in House, Senate, and Conference Appropriations Committee Reports. FY 1998 House Appropriations Committee Report Language (HR 105-205).

NIAMS. *See* National Institute of Arthritis and Musculoskeletal and Skin Diseases.

Painter, Kim. 1989. "Shedding Light on an Incapacitating Mystery Illness." *USA Today,* April 13, 1989, 5D.

Patnaik, M., et al. 1995. "Prevalence of IgM Antibodies to Human Herpesvirus 6 Early Antigen (p41/38) in Patients with Chronic Fatigue Syndrome." *Journal of Infectious Diseases* 172(5): 1364–1367.

Rand, K. H. 1991. "Chronic Fatigue Syndrome: Fact or Fiction?" *Proceedings, Annual Meeting of the Medical Section of the American Council of Life Insurance,* 135–144.

Russell, I. Jon. 1993. "Editorial." *Journal of Musculoskeletal Pain* 1:1–7.

———. 1996a. "Preface: MYOPAIN '95." *Journal of Musculoskeletal Pain* 4:xi–xiii..

———. 1996b. "A Tribute to Harvey Moldofsky, MD." *Journal of Musculoskeletal Pain* 4:1–3.

Schluederberg, A., et al. 1992. "Chronic Fatigue Syndrome Research: Definition and Medical Outcome Assessment—NIH Conference." *Annals of Internal Medicine* 117:325–331.

Shorter, Edward. 1992. *From Paralysis to Fatigue: A History of Psychosomatic Illness in the Modern Era.* New York: Free Press.

Showalter, Elaine. 1997. *Hystories: Hysterical Epidemics and Modern Culture.* New York: Columbia University Press.

Simms, R. W., D. L. Goldenberg, D. T. Felson, and J. H. Mason. 1988. "Tenderness in 75 Anatomic Sites: Distinguishing Fibromyalgia Patients from Controls." *Arthritis and Rheumology* 31(2): 182–187.

Simons, David. 1990. "Muscular Pain Syndromes." *Advances in Pain Research and Therapy* 17:1–41.

Smythe, Hugh, and Harvey Moldofsky. 1978. "Two Contributions to Understanding of the 'Fibrositis' Syndrome." *Bulletin of Rheumatological Diseases* 28(1): 928–931.

Straus, Stephen E., et al. 1985. "Persisting Illness and Fatigue in Adults with Evidence of Epstein-Barr Virus Infection." *Annals of Internal Medicine* 102:7–16.

Telling, Maxwell. 1911. "'Nodular' Fibromyositis, an Everyday Affliction, and Its Identity with So-Called Chronic Muscular Rheumatism." *British Medical Journal* 1:154–158.

Tobi, M., and S. E. Straus. 1985. "Chronic Epstein-Barr Virus Disease: A Workshop Held by the National Institute for Allergy and Infectious Disease." *Annals of Internal Medicine* 103:951–953.

Todd, Alexandra Dundas. 1989. *Intimate Adversaries: Cultural Conflict between Doctors and Women Patients.* Philadelphia: University Of Pennsylvania Press.

U.S. Centers for Disease Control and Prevention. 1986. "Chronic Fatigue Possibly Related to Epstein-Barr Virus." *Morbidity and Mortality World Report* 35(21): 350–352.

———. 1993. "Inability of Retroviral Tests to Identify Persons with Chronic Fatigue Syndrome, 1992." *Journal of the American Medical Association* 269(14): 1779, 1782.

———. 1999. Summary Report. October 13, 1999. Meeting of CFS Patient Advocacy Group Representatives. See www.cdc.gov/ncidod/dieases/cfs/meetings/10.99_ update.htm.

———. 2001. National Center for Health Statistics. Office of the Center Director, Data Policy and Standards. "A Summary of Chronic Fatigue Syndrome and Its Classification in the International Classification of Diseases." March, 1–2.

———. 2003. Report of the Third Annual CFS Case Definition Workshop. See www.cdc.gov/ncidod/diseases/cfs/meetings/case_def_workshop.htm.

Voorhees, Paula. 1989. "Painful Disease Eludes Testing; Fibromyalgia Frequently Misdiagnosed, but as Many as 3 Million in U.S. Have It." *Los Angeles Times,* June 24, 9:6.

Wallace, Daniel, and Janice Brock Wallace. 1999. *Making Sense of Fibromyalgia: A Guide for Patients and Their Families.* New York: Oxford University Press.

Wesseley, Simon. 1990. "Old Wine in New Bottles: Neurasthenia and 'ME.'" *Psychology and Medicine* 20(1): 35–53.

WHO. *See* World Health Organization.

Wolfe, Frederick. 1989. "The Design of a Fibromyalgia Criteria Study." *Journal of Rheumatology* 19 (Suppl.): 180–184.

Wolfe, Frederick, and M. A. Cathey. 1985. "The Epidemiology of Tender Points: A Prospective Study of 1520 Patients." *Journal of Rheumatology* 12(6): 1164–1168.

Wolfe, Frederick, et al. 1990. "The American College of Rheumatology 1990 Criteria for the Classification of Fibromyalgia." Report of the Multicenter Criteria Committee. *Arthritis and Rheumatism* 33(2): 160–172.

World Health Organization. 1992–94. *ICD-10: International Statistical Classification of Diseases and Related Health Problems.* 10th rev. Geneva: World Health Organization.

Yalcin, S., et al. 1994. "Prevalence of Human Herpesvirus 6 Variants A and B in Patients with Chronic Fatigue Syndrome." *Microbiology and Immunology* 38(7): 587–590

Yunus, Muhammad B., et al. 1981. "Primary Fibromyalgia (Fibrositis): Clinical Study of 50 Patients with Matched Normal Controls." *Seminar on Arthritis and Rheumatism* 11(1): 151–171.

# CHAPTER 7

# The Newtown Florist Club and the Quest for Environmental Justice in Gainesville, Georgia

ELLEN GRIFFITH SPEARS

The efforts of communities of suffering to make their illnesses visible often involve struggles over what constitutes legitimate knowledge. This theme recurs in the cases of activism by communities of suffering examined in this volume. Persons with chronic fatigue syndrome and fibromyalgia often assert the value of their firsthand experience with fatigue and pain over the clinical knowledge of physicians and researchers. The office workers described in Michelle Murphy's chapter on sick building syndrome rejected conceptions of environmental exposure based on measuring the effects of specific toxic exposures. They argued instead for a concept of environmental risk that mirrored their work experience, emphasizing the need to assess the cumulative effect of multiple exposures. This concept runs counter to the epidemiological assumptions underlying most occupational and environmental health investigations.

Both the challenges of establishing environmental health effects and the divide between community-based versus "scientific" knowledge are particularly great in situations in which a significant power gradient exists between communities of suffering and the representatives of both local industries and state agencies responsible for monitoring environmental hazards. Such cases exist throughout the American South, where African American communities live near industrial waste sites. Ellen Spears, an environmental activist and academic, describes the experience of one such community, the African American neighborhood of Newtown, located in Gainesville, Georgia. The experience of Newtown residents raises important issues concerning the challenges facing communities of suffering. Spears locates her discussion of Newtown within the context of a broader discussion of the nature and value of "situated knowledge." She concludes that in situations such as that in Newtown investigators into environmental hazards need to be conscious of, and sensitive to, the kinds of understandings and knowledge that local residents acquire over time. She rejects an overly romantic valuation of such "indigenous knowledge," recognizing that it is historically constituted and is shaped by a long history of racial oppression. Nonetheless, she argues that investigators have much to gain from working with local communities and integrating them into their investigations. This allows local residents to understand the nature and limits of such inquiries. In addition, local communities can bring a longitudinal perspective to environmental investigations that is often not visible to investigators involved in short-term inquiries. Understanding the importance that situated knowledge has for communities of suffering is critical for efforts to understand and facilitate interactions between communities of suffering and representatives of public health and medical institutions.

At the same time, situated knowledge may lead the members of communities of suffering to discount external claims of illness causation. The rejection of psychogenic explanations may lead individuals with chronic fatigue syndrome and fibromyalgia to foreclose therapeutic options involving behavioral modification and stress reduction. In the case of Newtown, the possibility exists that some of the behavioral issues raised by state officials may contribute in some way to broader health problems within the community.

---

Negotiating justice in contested environmental situations is never simple. Basic facts are deeply disputed. The impact of environmental degradation on human health is often difficult to prove. Operating in a highly politicized context, polluting companies deny harm. Civil authorities decline to intervene. Adjudication in the courts requires often unattainable levels of proof. Residents are left with their fears and their concerns. Can there be a basis for justice in contested environmental situations when knowledge about cause and effect is incomplete, when scientific certainty about human health effects is not available, and the balance of power appears to reside with polluters?[1] How do these questions about knowledge affect the role of public health professionals and technical personnel in such disputed environments?

Environmental outcomes are rooted in politics, and the results of environmental disputes have much to do with who holds legitimate knowledge. To achieve just outcomes, a great deal of attention must be given to how communities can manage knowledge, especially how community residents handle scientific information. Justice requires valuing the unique knowledge that differently situated social groups bring to contested environmental circumstances. The public health professional, the hydrogeologist, the air-monitoring technician, even the environmental journalist often find themselves negotiating between the types of knowledge crucial to fair resolution of environmental justice struggles. Professional and technical people often occupy roles that require mediating between affected community residents and authorities charged with environmental regulation— roles that require valuing the kinds of knowledge community members bring. By valuing local knowledge and showing sensitivity to knowledge-related power dynamics, public health professionals and technical personnel can help achieve better outcomes for communities confronting environmental justice concerns.

Affected residents play a crucial role in discovering environmental hazards. "Popular epidemiology," that is, the participation of nonscientists in identifying patterns of disease, has come to describe the important role laypersons play in identifying toxic sites and causes. "The most significant force for detecting and ameliorating toxic disasters is the public that is affected by them," write Phil Brown and Edwin Mikkelson in *No Safe Place: Toxic Waste, Leukemia, and Community Action,* their book about the child leukemia cluster in Woburn, Massachusetts, that was the subject of the popular book and film *A Civil Action.* Although a much larger "public" is affected by toxic disasters than the "social movement" that brings toxic pollution to light, Brown and Mikkelson point out that "as in many other areas of medical knowledge, social movements play a crucial role in discovery and action."[2]

In the northern Georgia city of Gainesville, recognition of toxic problems began when members of a local community group identified multiple illnesses and toxic sites in Newtown, an African American neighborhood. As a result of the community's effort, even the business-boosting local paper, the *Times,* labeled Newtown an "industrial fallout zone" in an October 21, 1992, article. The battle over the environment in Newtown pitted a local organization of African American women against the largest privately held corporation in the United States, the global grain exporter Cargill. In doing so, the local group in Gainesville, the Newtown Florist Club, has challenged one of the world's top polluters, in a fight that is still unfolding.[3]

The Newtown Florist Club began in the 1950s as an informal women's association to care for the sick and assist in burial services. Through the civil rights years, the work of the club enlarged to include seeking access to the ballot, opposing unfair treatment of students in local schools, and protesting unequal recreation facilities available to African American youth. In the late 1980s, while carrying out their traditional mission, club members began to suspect that their neighbors experienced a high rate of certain illnesses like cancer, lupus, and respiratory ailments and an unusual number of deaths. "That put us on a wonder," said Mozetta Whelchel, one of the club's founders.[4] Another member of the Florist Club, Sara Nash, described what many residents felt: "It has to be a reason for that many clusters of diseases, that many clusters of illnesses, that many clusters of deaths within one area. I think that if it was in another area other than a minority area, I think it would be some investigation going on. I think some answers would be sought as to what is happening. It just bothers me that human life is no more important, that they seem to just ignore Newtown. Like it doesn't exist. They don't want it to exist."[5] The seventy-five well-tended Newtown homes built atop an old dump next to the railroad tracks are "surrounded by thirteen toxic industries, two identified potentially hazardous sites, numerous hazardous waste generators, and a rat-infested junkyard."[6] Residents of this community hypothesized that the illnesses and deaths were attributable to toxic pollution from surrounding industry. Upon investigation, they found that thirteen of the sixteen sites reporting toxic releases in Gainesville–Hall County in 1990 were located on the south side, where Newtown is located and the majority of the African American population resides.

In a study conducted that year, state Department of Human Resources officials acknowledged an elevated level of throat and mouth cancer.[7] Five years later, independent health professionals documented a higher than normal incidence of the immune system disease lupus.[8] Nevertheless, more than a decade after Newtown residents uncovered a pattern of fatal ailments, no comprehensive action has been taken by officials at any level to clean up a nearby hazardous waste dump site, restrict hazardous waste generators, or limit air or water pollution. "We know more about it, but not that much has been done," says Newtown Florist Club president Faye Bush.[9]

Newtown is not an isolated example; official inaction is common in such cases. Even when residents are successful in winning redress, the lag between discovery by residents and validation by court or settlement can be extremely long. In Fort Valley, Georgia, landowners near a Woolfolk Chemical plant that had produced agricultural pesticides since 1921 were notified of an environmental problem by the Environmental Protection Agency in

1991. Residents filed suit against Woolfolk and other chemical manufacturers in 1994. According to news accounts of statements by plaintiffs' lawyers at the trial, the owners and operators at Woolfolk had suspected since 1971, and known for a fact since 1984, that pesticides were leaking into surface and groundwater. In 1998, twenty-seven years after the first leakages were suspected by Woolfolk officials, the company agreed to compensate affected individuals. Chronologies like these raise particular problems about the management and ownership of knowledge about environmental pollution and its health effects.[10] The phenomenon extends well beyond a few local examples in Georgia. As Brown and Mikkelson wrote about the Woburn, Massachusetts, contamination, "The belated discovery of what residents knew long before is eery and infuriating—and, sadly it is common to many toxic waste sites."[11]

## Situated Knowledge

Residents cannot win remediation speedily–or at all—in part because their local knowledge is not valued. Environmental advocates contest the differences in the value attributed to lay knowledge and that attributed to scientific knowledge in environmental settings where sharp inequalities exist. The unequal relationship between lay and scientific knowledge is a function of power. "Images of science and rationality are embedded in structures and processes of authority," writes scientist-sociologist Brian Wynne.[12] Scientific knowledge becomes the privileged information. Rather than privileging one type of knowledge over the other, "we have to define conceptions of justice capable of operating across and through these multiple mediations," argues cultural geographer David Harvey.[13]

Harvey suggests in *Justice, Nature and the Geography of Difference* that the knowledge claims that become so important in identifying and resolving environmental battles are deeply embedded in the power relations between social groups. The concept of justice that currently operates in the courts in environmental cases requires clear evidence of quantifiable harm, as well as proof of a link between environmental degradation and human health effects as determined by the standards of positivist science. Harvey describes the conventional view of environmental management:

All state interventions, the logical tool of environmental management, are typically limited under the standard view by two important considerations. First, intervention should occur only when there is clear evidence of serious damage through market failure and preferably when that damage can be quantified (e.g., in money terms). This requires strong scientific evidence of connections. . . . And it also requires careful measurement of the costs of pollution and resource depletion because the second constraint is that there is thought to be a zero-sum trade-off between economic growth (capital accumulation) and environmental quality.[14]

This standard view rests strongly on Cartesian models that state that quantifiable evidence is the basis of scientific knowledge. The conventional view also holds that the capitalist virtues of economic growth weigh heavily in the balance against environmental quality. Harvey points out several problems with this "environmental management" view, including "after-the-event" approach, which assumes the rights of industry to operate without regard to environmental impact, in whatever way is deemed necessary, until ordered to do otherwise. This after-the-fact notion focuses on cleanup rather than prevention or other proactive strategies that would shift the burden of proof to corporations, which would have to demonstrate that their processes are safe.[15] Residents in communities affected by toxic pollution argue that justice is not served by the conventional view; the proof burdens are too great, and human health is damaged—sometimes lives are lost—as residents wait too long for remedial action by industry.

A reinvigorated conception of justice roots the challenge to the environmental degradation that is brought about by capitalist transformation within a worldview that respects differently situated knowledges. Cases such as Newtown demonstrate the need for a conception of justice that operates across the boundaries of situated knowledge and shows that knowledge and justice are constructed through social action.

"What is it that constitutes a privileged claim to knowledge," asks Harvey, "and how can we judge, understand, adjudicate, and perhaps negotiate through different knowledges constructed at very different levels of abstraction under radically different material conditions?"[16] As my use of the term already implies, "situatedness" is central to understanding this question. As described by feminist theorist Donna Haraway, situatedness emphasizes that all knowledge is partial, shaped by the perspective of the viewer. Haraway calls for "a deeper, broader, more open scientific literacy," which takes into account situated knowledges.[17]

### Knowledge and the Politics of Difference

The process of negotiating what is valued as knowledge produces new conditions for determining what is just, for all knowledges are not equally true. Justice is not simply relative. In seeking objective standards that provide a basis for action, it is important to recognize that those standards are constantly being changed and adapted through social action. These issues of the philosophy and sociology of knowledge are being addressed by feminist theorists, like Iris Marion Young, who are also, like Harvey, engaged in a critique of Cartesian scientific rationality. Young examines the politics of difference and how vantage point plays a defining role in what people "know."[18] She argues that situatedness is not reducible to biographical categories—black, white, Hispanic, male, female, working class, or agent of capital. Situatedness, or one's position in relation to knowledge, is not cast in simple biographical terms in place or history but, as Harvey describes it, "a dialectical power relation between oppressor and oppressed."[19]

The knowledge that arises from differently situated social groups is valuable. "Structured social differences of gender, race, age, or sexuality, as well as class," Young argues, "can serve as important resources for learning about what is needed and reasoning about what is fair."[20] But even though she finds much to agree with in Harvey, Young challenges what she sees as his notion of basing "appeals to justice on perceptions of similarity, because it invites us to deny that we have obligations to those whom we perceive as different."[21] Donna Haraway agrees and points out that appeals for a new conception of knowledge do not emerge spontaneously out of respect for difference but through determined social action by subjects who themselves occupy different positions in relation to knowledge and power: "Renegotiating what counts as knowledge, and as property, emerged not from spontaneous multicultural goodwill but from specific organization, articulation, and struggle by people locally and globally, in processes that have produced new kinds of indigenous subjects on the world stage as well as in national courts."[22]

### Situated Knowledge in Newtown

Although the situated knowledges operating in Newtown are numerous, a contrast between the views of some residents and the perspectives of some state health officials is instructive. Because of where it was situated and the nature of their role in caring for the community, the Newtown Florist Club spotted the health problems first. The club members conducted an informal house-to-house survey, identifying eighteen people of all ages with a variety of different illnesses, all residing in an area that included about seventy-five homes. Neighbors pressed the Georgia Department of Human Resources (DHR) to conduct a health study. DHR epidemiologist Dr. David Williams and Tom McKinley supervised the study, with the assistance of the Gainesville Public Health Department. As residents suspected, cancer levels were high. There were four cases of throat and mouth cancer where fewer than one case would be expected in a similar population. That result was not contested. What came to be disputed, and remains so to this day, was the conclusion of the DHR study that the elevated cancer levels were due to residents' "lifestyle." McKinley and Williams claim that the prevalence of the disease was due to residents' excessive use of tobacco and alcohol. The DHR study did not, however, quantify use of alcohol or tobacco products on the part of residents.[23]

Residents heard within the term *lifestyle* a strong, negative message about the race and class of the residents. Residents were outraged at the conclusion, which they saw as thinly veiled racism. Even much later, at public programs residents held about the situation, the lifestyle comment was raised and refuted. An excerpt from this 1994 interview with neighborhood resident Mae Catherine Wilmont, who has lupus, illustrates the point: "I just figure it's got to be the environment. It has to be. Just over the years, down through the years where I've breathed all this toxic, you know. A lot of people said it was our lifestyle over here. Our lifestyle over here is not half as bad as their lifestyle over yonder [in the white community] because they've got plenty that drink and smoke. Do you hear me? They can afford it. We can't afford it."[24] McKinley and Williams' conclusion clearly contradicted everything residents knew or believed about the situation. To residents, it revealed the bias of the researchers working for the state health department and undermined the scientists' credibility.[25]

The failure of state researchers to ask the right questions, quantify their findings, and base their conclusions on the available evidence reflected a devaluation of knowledge held by residents. As a result, as sociologist of science Brian Wynne suggests in *Risk Management and Hazardous Waste*, the scientists' expertise is questioned.

> What the conventional approach assumes to be lack of technical precision (scientific uncertainty) is often structural uncertainty or latent conflict between divergent perceptions and social rationalities. In its unawareness of these social dimensions framing rationality, we argue that "rational" methods in regulation and decision analysis effectively reject a necessary dimension of democratic negotiation between expertise and lay experience. By inadvertently dismissing a potential development of the relations of expertise with lay experience, they ultimately undermine the credibility and authority of expertise altogether.[26]

The 1990 study illustrated that scientific approaches themselves are place-based and culture-bound and function in a political world. As Brian Wynne noted in a case study of sheep farming in northern England after Chernobyl,

"Abstract scientific knowledge may seem universal, but in the real world, it is always integrated with supplementary assumptions that render it culture-bound and parochial. The validity of this supplementary knowledge crucially affects the overall credibility of 'science' or 'experts.'"[27] In their case, Newtown residents rejected the survey results but were aware of the power that those results wielded. Though angry, they were not surprised. Weighed against their own observations over time, and with the undisputed evidence of an elevated cancer rate, the skewed analysis of the DHR study reinforced residents' belief in and commitment to identifying environmental causes. The study's conclusions also reinforced mistrust in the epidemiologists, or any public officials, employed by the DHR.[28]

In addition to discounting residents' health concerns, the lifestyle conclusion gave local Gainesville authorities a way to place the blame outside the polluting industries. The illnesses could be projected as residents' own doing, and the responsibility for taking action shifted away from the polluters. McKinley and Williams' conclusion had a major impact in blocking action to clean up the environment. "State Official: Newtown Cancer Rate Is Normal," declared the headline in the local newspaper's account of the DHR results.[29] The conclusion allowed local government and state regulators to sidestep the environmental problems.

State officials occupy a situated position characterized by competing obligations. As James C. Cobb writes about the contradictory roles state officials play in preserving a favorable climate for business, "Preserving such a permissive atmosphere [lax environmental standards for industry] proved more difficult, however, when environmental problems emerged as a major public concern, because in addition to encouraging and facilitating industrial development, state government was given the responsibility of preventing incoming plants from releasing pollutants into the air and water."[30] Advocates see technicians as hopelessly imprisoned in bureaucracy, too beholden to existing power structures. Technicians often see communities as mired in unscientific reasoning, producing unsound explanations for the aggregate harms they face.

The scientist who studies a community during a three-week period takes a snapshot. The very distance that presumes to give the scientist objectivity works to prevent certain observations that residents can readily share. Although the scientist's knowledge is privileged in court and policy proceedings, that same scientist may have missed information that is available to residents who piece together their explanations over a period of years. The scientific studies conducted in Newtown, for example, have not been longitudinal; they have lasted a few weeks, at most a few months. But the residents have been observing the actions of industry—and the pattern of ailments and skin rashes, their neighbors' illnesses and deaths—over a period of years. In short, the locally collected information is more historically grounded.

Exploring similar questions in another, quite different environment, researchers in South Africa tested the validity of residents' recollections of soil conservation techniques as a way of rounding out the historical component of their environmental impact assessments. Even in evaluating environmental practices long past, Kate B. Showers and Gwendolyn M. Malahleha concluded that "local environmental knowledge has been shown to have an historical dimension which can be used in combination with formal historical materials to develop a picture of environmental events."[31] Though shaped at a local level, this situated knowledge collected over time remains of incomparable value in reconstructing historical events.

Even without questions about the lack of historicity that is a function of the researcher's situatedness, establishing a causal link is "a complicated and debatable exercise."[32] Quantifying harm and establishing scientific connections is no easy matter. As Gina Kolata wrote in 1999, examples of proven cause and effect are rare. The rare proven exceptions, "have two things in common: The chemical exposure was enormous, and the disease was extraordinarily rare."[33] Even when cause and effect can be determined, isolating those relationships is usually costly and time-consuming. Questions persist about issues such as sample size and how the boundaries around a cluster are drawn. The sample may be too small to be statistically significant. Within a half-mile radius, the number of cases of a disease might suggest a cluster, but the incidence may not be abnormally high if the boundary is a circle two miles wide. All in all, as Atul Gawande of the Harvard School of Public Health wrote in "The Cancer Cluster Myth," the numbers demonstrate how difficult it is to establish environmental factors as a cause of illness: "Raymond Richard Neutra, California's chief environmental health investigator and an expert on cancer clusters, points out that among hundreds of residential clusters in the United States, not one has convincingly identified an underlying environmental cause. Abroad, in only a handful of cases has a neighborhood cancer cluster been shown to arise from an environmental cause. And only one of these cases ended with the discovery of an unrecognized carcinogen."[34] Gawande perhaps overstates the case, but his assertion points out just how problematic establishing linkages remains.

Environmental advocates anticipated the limitation of the standard view on proving causation when drafting their principles at the 1992 International Summit on the Environment. Principle 15 of the Rio Declaration reads, "Where there are threats of serious or irreversible damage, lack of full scientific certainty shall not be used as a reason for postponing cost-effective measures to prevent environmental degradation."[35] Principle 15 invokes the precautionary principle, reflecting the recognition that justice may not be served by requiring proof of causation before recommending remediation.

Describing another community facing high levels of cancer, anthropologist Martha Balshem discusses how locally generated knowledge and scientific or medical knowledge are assigned different values. Based on her work as a public health educator in a European American working-class community in Philadelphia, Balshem's book, *Cancer in the Community: Class and Medical Authority,* describes how this unequal relationship factors into the debate about environmental versus lifestyle causes of cancer. She states, "Medical social scientists have described in elaborate detail the physician's power to confer or deny legitimacy to particular interpretations of patient sign, symptom, and behavior; charged that through the distinction between scientific and folk knowledge, *lay interpretations are cast as illegitimate and inconvenient counterpoints to real medical knowledge;* documented that the dialogue through which physicians decide whose knowledge is real is kept secret from patients."[36] Although Balshem's role was to educate the community about individual lifestyle choices (smoking, diet, etc.) that could reduce cancer, residents felt certain that environmental factors in the heavily industrial area were to blame. Residents understood how physicians and scientists viewed them.

> Community residents are aware that the worldview they express is regarded by science as a problem, as a view that needs changing. Scientists, in the residents' interpretation, see themselves as possessing the only valid authority, one that dictates a negating of the value of deep-seated wisdoms. The antagonism that the urban working class feels toward scientific authority is tied to a larger antagonism toward powerful forces from the outside that bear down upon the community like a juggernaut, causing social, economic, and health problems that are beyond local control. In this context, local traditions are held up, to self and other, as being of higher value than scientific knowledge. . . .
>
> Thus, the community expression of a counterpoint to science is tied to strong feelings about access to power in society.[37]

Like their Philadelphia counterparts, Newtown residents' skepticism toward experts was related to feelings about access to power. After the DHR report was published, they became *more* convinced of their belief in an environmental cause and less willing to trust the state's Environmental Protection

Division (EPD) scientists. "We don't have faith in EPD period," said Faye Bush.[38]

Understanding the power of scientific studies in assisting or preventing redress, residents sought to combat the conclusions of the DHR study with independent analyses of the state's data, as well as scientific investigations of their own. Although the state's unsupported interpretation was infuriating, the numbers did document what the residents had already observed. The patent inadequacy of the study boosted residents' confidence in their own convictions and their ability to manage scientific evidence. Neighborhood resident and club member Rose Johnson reports,

> After [the DHR's] study was complete, we took our time to analyze its contents. And we issued our own little report objecting to those things that we thought were inaccurate, still with no knowledge of those technical things but only our knowledge of what we knew to be the truth in terms of the cancers and where the clusters were and what [the state epidemiologists] did and did not do [in conducting the study].
>
> We were able to clarify from that point on what it was we thought we wanted in terms of research, what we wanted to know.[39]

"Never has it been more necessary for the public to be aware of technical and scientific issues," wrote Alan McGowan in *Environment* magazine, stressing the need for effective communication between lay people and scientists.[40] Residents in Newtown knew they needed more scientific knowledge. "With technical issues, like those associated with environmental justice, lay leaders face the daunting task of becoming educated about complex and unfamiliar topics."[41]

Newtown residents did not directly tackle the relative privilege of scientific over lay knowledge. While local organizing, which could have brought a shift in the power relations, continued, the club adopted a major focus on seeking independent scientific verification, in an attempt to compete with scientific knowledge on its own terms. Even with independent scientific work at their disposal, shifting the strategic focus from confronting local leaders to gathering scientific evidence limited the residents' gains. The independent lupus study conducted by Emory University School of Public Health Environmental and Occupational Health specialists Dr. Howard Frumkin and Tarik Kardestuncer in 1995 revealed a rate of occurrence of new cases of the immune system disease lupus nine times higher than would be expected. The findings suggested a link with "long-standing exposure to industrial emissions."[42] But residents did not define a strategy designed to challenge authorities directly by using the findings. A stream of other independent scientists came through Newtown, providing helpful but often inconclusive studies, and the power relations with industry and civic leadership in Gainesville remained largely unchanged.

## Constructing Knowledge through Social Action

Iris Marion Young argues that mediating between these knowledges is essential to finding just outcomes, but renegotiating the power relations is even more critical. The social movement in Newtown gained fresh momentum when the 1995 occurrence of two toxic releases shifted the focus away from blaming the victims, giving residents new knowledge as well as a further handle with which to mount opposition. On the evening of May 19, 1995, and again in early December, local emergency management personnel evacuated residents from two streets near Newtown, Black Drive and Cooley Drive, in response to an odor in the area. Early news reports stated that Cargill was suspected of releasing hexane during a cleaning process. Thirty people were examined at local hospitals, including one resident who was hospitalized for five days.[43] Although denying responsibility for the spill, Cargill sent letters to some residents offering to pay medical bills. The May 24, 1995, letters plant manager Catherine Hay sent to residents examined at the hospital read that though it is "not determined" what caused the odor, "Cargill has made arrangements to pay your medical expenses associated with this odor."[44]

The Georgia Environmental Protection Division did not take air samples on the night of the first evacuation, "since the material was no longer detectable." The EPD investigator also wrote that plant manager "Hay clearly stated Cargill's position that they did not feel they released hazardous materials, but that they had likely released an odor that passed through the community."[45] While not finding Cargill guilty, the EPD did issue two recommendations: first, that the EPD and Cargill identify better procedures for performing annual cleaning to minimize community odors, and second, that Hall County consider forming a local emergency planning committee (LEPC) to involve the entire community in emergency planning.[46]

Newtown residents had demanded the establishment of an emergency planning committee some years before, and after the 1995 crisis, the business community, city emergency management planners, and community representatives began to hold meetings. The new forum offered the possibility for a shift in the power relations between community and industry, but the Cargill plant manager was named to chair the LEPC. Belinda Dickey, a community representative who lives near Newtown, was appointed as vice chair, and Florist Club president Faye Bush also served on the committee.

Bush's assessment of the LEPC is that it became "bogged down."[47] Communication between these variously situated authorities, city emergency management personnel, company officials, state natural resources officials, and community residents cannot be effective without a radical renegotiation of power.

With knowledge of the spills, the residents sought to shift the balance of power by taking Cargill to court. Again, the knowledge that local residents had accumulated was quite different from the medical and scientific knowledge necessary to prove an environmental claim in court. In the course of challenging Cargill in a lawsuit, residents learned what had happened the night of the first spill. Supporting the argument that knowledge is constructed through social action, residents amplified their knowledge by challenging industry and the local authorities.

The environmental justice movement is skeptical of "experts" and "professional" knowledge and is, in Harvey's words, in "search for an alternative rationality" that values traditional ways of knowing.[48] Still, in the courtroom, the residents' observations are devalued; science is the privileged knowledge. "[Critics of science] are skeptical about the universal pretensions of scientific knowledge but lack alternative bases for cognitive authority," writes Stephen Yearley.[49]

Although the courts' exclusive reliance on scientific validation may limit attempts to achieve just solutions, this does not mean that neighborhood residents already know the full story and are just waiting for corroboration. Harvey warns against romanticized notions about intuitive understandings held by marginalized people.[50] Lay observations must be part of creating a full picture of possible hazards and potential causes; like the investigator's observations, they are developed within a bounded frame and are therefore only partial.

Harvey correctly points out some of the limits of knowledge too closely bounded by place. He urges that we not patronize local people in environmental struggles by assuming that, by virtue of their subjugated posture, they have some special understanding that science has yet to grasp. What people know locally out of their own experience may serve as an effective ground for resistance but is insufficient as the only ground for evaluating environmental harm. For example, Newtown residents view the 1936 dump atop which several homes were built as a source of toxins, even though independent experts volunteering for the residents have been unable to identify landfill remnants hazardous enough to cause current health effects. Some independent experts believe that the fact that the homes were built on a dump may hold far more symbolic power because of what it says about how people of color are valued. Here, as Harvey points out, there lies a "battle between different levels of abstractions, between distinctively understood particularities of places and the necessary abstractions required to take those understandings into a wider realm."[51]

## Implications for the Public Health Professional or Technician

If the power relations are more significant than scientific information in determining outcomes, where does this leave the technical professional or health expert in negotiating between differently situated knowledges? "More and more I find myself saying dispense with the technical people," one environmental and occupational health specialist who preferred not to be quoted by name half seriously says. Of course, professional and technical experts remain crucial. With an awareness of the underlying power dynamics surrounding scientific and lay knowledge, technical people can sensitively negotiate these dynamics, helping communities realize just outcomes.

Practical measures that can help ensure that local knowledge is valued include

1. *Making oral historical environmental assessment an element in any study.* A core tenet of the environmental justice movement is that community members speak for themselves. Seeking out community residents' stories becomes crucial to solving environmental problems.

2. *Shaping the research agenda in collaboration with local residents.* At the beginning of any research project, the limitations of scientific studies should be made clear. The certainty communities desire or authorities require is rarely achievable. As the Committee on Environmental Justice of the Institute of Medicine found, "If the residents do not appreciate the limitations [of a study] at the outset and the study gives a negative result, the community representatives may believe that the scientists are hiding something and that they purposefully used a method that could not find a cluster."[52] Residents may also be able to suggest design alterations more suited to the specific community.

3. *Actively aiding in the expansion of community residents' knowledge.* The institute's Committee on Environmental Justice also noted that "one important result of participatory research is the educational opportunity for those members of a community of concern who participate. By becoming involved in the development, execution, and analysis of research that addresses their health concerns . . . they will better understand the current state of the knowledge and its limits."[53]

4. *Recognizing the race, class, and gender dynamics.* Residents, researchers, and authorities who have different types of knowledge may often hold divergent views rooted in distinct experiences, all of which may be useful in achieving just outcomes.

5. *Involving local residents as researchers in health studies.* Cost considerations and the sheer scale of the problem in some situations have led researchers to involve community residents in collecting health data for environmental studies. Careful orientation is necessary, as it is with any researcher, but local people may be able to obtain more detail by more easily establishing rapport with interviewees.

6. *Acknowledging that outcomes rest on power dynamics.* Understand from the start the limitations of the technical professional's role. Informed public participation is key to winning any environmental justice battle.

The movement for environmental justice in Newtown is far from finished. But, as the work of the Newtown Florist Club suggests, the local knowledge of residents is key in identifying toxic exposures and achieving better outcomes for communities confronting environmental justice concerns.

## NOTES

1. On "scientific certainty," Sandra Harding wrote, "Scientific work—at least twentieth century scientific work—never claims to produce true statements, but only statements that are less likely to be false than the alternatives that have been considered." Harding, "Who Knows?," 112.

2. Brown and Mikkelson, *No Safe Place,* 3.

3. Council on Economic Priorities Corporate Environmental Data Clearinghouse, *Cargill,* 1. "Cargill reported the release of 10.2 million pounds of toxic chemicals in 1989, down from 13.4 million pounds in 1988. In terms of aggregate releases, the company reported the release of more toxic chemicals than all other companies in the food industry in 1988 and ranked second in 1989."

4. Mozetta Whelchel, interview with author, March 24, 1994.

5. Sara Nash, interview with author, October 31, 1993.

6. Spears, *Newtown Story,* 2.

7. McKinley and Williams, report, Newtown Neighborhood Cancer Investigation.

8. Frumkin and Kardestuncer, "Systemic Lupus Erythematosus," 85.

9. Spears, *Newtown Story,* 57.

10. This chapter argues for the importance of pushing for environmental solutions even when positivist science cannot determine cause and effect, recognizing that in many cases in which the facts are not in dispute, still no action is taken. Harvey cites an example, with lead paint in Baltimore: since the 1960s, the epidemiology, victims, health effects, and treatment for lead exposures have been known, but the local pow-

er structure has done nothing because of the race and class of the victims (*Justice*, 393). In the face of such recalcitrant local power structures, how can we expect a change when the evidence is less clear? Harvey's solution remains the same in both cases: empowerment of the poor and working classes (*Justice*, 395).

11. Brown and Mikkelson, *No Safe Place*, 144.

12. Wynne, *Risk Management and Hazardous Waste*, 397.

13. Harvey, *Justice*, 349.

14. Ibid., 374–75.

15. Ibid., 374.

16. Ibid., 23.

17. Haraway, *Modest_Witness@Second_Millennium*, 11.

18. "Critical theoretical accounts of instrumental reason, postmodernist critiques of humanism and of the Cartesian subject, and feminist critiques of the disembodied coldness of modern reason all converge on a similar project of puncturing the authority of modern scientific reason. Modern science and philosophy construct a specific account of the subject as knower, as a self-present origin standing outside of and opposed to objects of knowledge—autonomous, neutral, abstract, and purified of particularity." Young, *Justice and the Politics of Difference*, 125.

19. Harvey, *Justice*, 355.

20. Young, "Harvey's Complaint with Race and Gender Struggles," 41.

21. Ibid.

22. Haraway, *Modest_Witness@Second_Millennium*, 140–41.

23. In some ways the state epidemiologists' study is not a fair one to use to compare lay knowledge and scientific knowledge because it is so clearly not good science. Independent evaluators found numerous other problems with the DHR study in addition to the unsupported conclusions about "lifestyle." Stephen Lester, science director for the Citizens Clearinghouse for Hazardous Waste (CCHW), criticized the study for conducting the interviews by phone, not distinguishing people by how long they had lived in the community, and using inappropriate comparison groups and sample sizes. But the study has been influential, and the debate around it illustrates the different points of view. Letter in the possession of the author, April 5, 1993.

24. May Catherine Wilmont, interview with author, June 13, 1993.

25. "There are historic grounds among black citizens to at least suspect the government's credibility," wrote the *Montgomery Advertiser*, recalling the infamous Tuskegee syphilis experiment in an April 1, 1999, editorial about the Knollwood community in Montgomery County, Alabama, where residents fear an unusual number of cancer cases may be related to the water supply.

26. Wynne, *Risk Management and Hazardous Waste*, 7.

27. Wynne,"Sheepfarming after Chernobyl," 12.

28. Interviewed four years later, McKinley still was not self-critical about the method or conclusions of the study, though he admitted that "not much quantification" of tobacco or alcohol use had been done. Tom McKinley, phone interview with author, August 10, 1994.

29. Lambert, "State Official," 1.

30. Cobb, *Selling of the South,* 229.

31. Showers and Malahleha, "Oral Evidence in Historical Environmental Impact Assessment," 295.

32. Institute of Medicine, Committee on Environmental Justice, Health Sciences Policy Program, Health Sciences Section, *Toward Environmental Justice,* 16.

33. Kolata, "Probing Disease Clusters," 6.

34. Gawande, "Cancer Cluster Myth," 35.

35. *Rio Declaration on Environment and Development,* The United Nations Conference on Environmental and Development, June 1992. (The cost-effective clause in Principle 15 was the result of a compromise, but the principle nonetheless illustrated the recognition of the problems of science in validating environmental claims.)

36. Balshem, *Cancer in the Community,* 7; emphasis added.

37. Ibid., 85.

38. Faye Bush, conversation with author, April 7, 1999.

39. Rose Johnson, interview with author, Georgia Government Documentation Project, March 6, 1994.

40. McGowan, "Effective Communication," 2.

41. Institute of Medicine, Committee on Environmental Justice, Health Sciences Policy Program, Health Sciences Section, *Toward Environmental Justice,* 55.

42. Frumkin and Kardestuncer, "Systemic Lupus Erythematosus," 85

43. Not only may science and technical people not know what is causing a problem; sometimes what they think they know make things worse. It is possible that local emergency management personnel in Gainesville actually increased individuals' exposures by removing residents from their homes to higher ground down the street. This same issue is raised in Brian Wynne's study of sheep farmers in northern England, in a very different geographical space. "British government officials first reacted to the deposition in northern England of radioactive fallout from Chernobyl by ignoring or denying the facts and then by recommending impractical solutions stemming from their ignorance of hill sheepfarming." "Sheepfarming after Chernobyl," 34–37.

44. Catherine Hay to residents, May 24, 1995. In the possession of the author.

45. Report, Georgia Environmental Protection Division, 1995.

46. Local emergency planning committees have been required by the Emergency Planning and Community Right-to-Know Act (EPCRA) since the late 1980s, though the Gainesville LEPC was not established until after the 1995 spills.

47. Faye Bush, interview with author, April 25, 1997..

48. Harvey, *Justice,* 386.

49. Yearley,"Environmental Challenge to Science Studies," 469.

50. Harvey, *Justice,* 100.

51. Ibid., 34.

52. Institute of Medicine, Committee on Environmental Justice, Health Sciences Policy Program, Health Sciences Section, *Toward Environmental Justice,* 89.

53. Ibid., 55.

## REFERENCES

Balshem, Martha. *Cancer in the Community: Class and Medical Authority.* Washington: Smithsonian Institution Press, 1993.

Brown, Phil, and Edwin J. Mikkelson. *No Safe Place: Toxic Waste, Leukemia, and Community Action.* 1990. Berkeley: University of California Press, 1997.

Bullard, Robert D. *Dumping in Dixie.* Boulder, Colo.: Westview Press, 1990.

———, ed. *Confronting Environmental Racism: Voices from the Grassroots.* Boston: South End Press, 1993.

Cobb, James C. *The Selling of the South: The Southern Crusade for Industrial Development, 1936–1980.* Baton Rouge: Louisiana State University Press, 1982.

Council on Economic Priorities Corporate Environmental Data Clearinghouse. *Cargill: A Report on the Company's Environmental Policies and Practices.* New York: Council on Economic Priorities, 1992.

Crawford, Colin. *Uproar at Dancing Rabbit Creek: Battling over Race, Class, and the Environment.* Reading, Mass.: Addison-Wesley, 1996.

Frumkin, Howard, and Tarik Kardestuncer. "Systemic Lupus Erythematosus in Relation to Environmental Pollution: An Investigation in an African American Community in North Georgia." *Archives of Environmental Health* 52 (1997): 85–90.

Gawande, Atul. "The Cancer Cluster Myth." *New Yorker,* February 8, 1999, 34–37.

Gottlieb, Robert. "Beyond NEPA and Earth Day: Reconstructing the Past and Envisioning a Future for Environmentalism." *Environmental History Review* 19, no. 4 (1995): 1–13.

Haraway, Donna. *Modest_Witness@Second_Millennium. FemaleMan©_Meets_OncoMouse,™: Feminism and Technoscience.* New York: Routledge, 1997.

———. "Situated Knowledges: The Science Question in Feminism and the Privilege of Partial Perspective." In *Human Geography: An Essential Anthology,* edited by John Agnew, David N. Livingstone, and Alisdair Rogers, 108–28. Malden, Mass.: Blackwell Publishers, 1996.

Harding, Sandra. "Who Knows? Identities and Feminist Epistemology." In *(En)Gendering Knowledge: Feminists in Academe,* edited by Joan E. Hartman and Ellen Messer-Davidow, 100–115. Knoxville: University of Tennessee Press, 1991.

Harr, Jonathan. *A Civil Action.* New York: Vintage Books, 1996.

Harvey, David. *The Condition of Postmodernity: An Enquiry into the Origins of Cultural Change.* Malden, Mass.: Blackwell Publishers, 1990.

———. *Justice, Nature and the Geography of Difference.* Malden, Mass.: Blackwell Publishers, 1996.

———. *Social Justice and the City.* London: Johns Hopkins University Press, 1973.

Hayden, Dolores. *The Power of Place: Urban Landscapes as Public History.* Cambridge: MIT Press, 1997.

Institute of Medicine. Committee on Environmental Justice. Health Sciences Policy Program. Health Sciences Section. *Toward Environmental Justice: Research, Education, and Health Policy Needs.* Washington: National Academy Press, 1999.

Keller, Elizabeth Fox. "The Origin, History, and Politics of the Subject Called 'Gender and Science': A First Person Account." In *Handbook of Science and Technology Studies,* edited by Sheila Jasanoff, Gerald E. Markle, and James C. Petersen, 80–94. Society for Social Studies of Science. Thousand Oaks, Calif.: Sage Publications, 1995.

———. *Secrets of Life, Secrets of Death: Essays on Language, Gender and Science.* New York: Routledge, 1992.

Kolata, Gina. "Probing Disease Clusters: Easier to Spot Than Prove." *New York Times,* January 31, 1999, 6.

Lambert, Clay. "State Official: Newtown Cancer Rate Is Normal." *Gainesville (Ga.) Times,* June 19, 1990, 1.

McGowan, Alan. "Effective Communication: There Is Much to Learn." *Environment* 31, no. 2 (1989): 2.

McKinley, Thomas W., M.P.H., and David Williams, M.D. Report. Newtown Neighborhood Cancer Investigation. Georgia Department of Human Resources. June 7, 1990.

Melosi, Martin V. "Equity, Eco-racism and Environmental History." *Environmental History Review* 19, no. 3 (1995): 1–16.

Scott, Joan W. "The Evidence of Experience." In *The Lesbian and Gay Studies Reader,* edited by Henry Abelove, Michele Aina Barale, and David M. Halperin, 397–415. New York: Routledge, 1993.

Showers, Kate B., and Gwendolyn M. Malahleha. "Oral Evidence in Historical Environmental Impact Assessment: Soil Conservation in Lesotho in the 1930s and 1940s." *Journal of Southern African Studies* 18, no. 2 (1992): 277–96.

Spears, Ellen Griffith. *The Newtown Story: One Community's Fight for Environmental Justice.* Atlanta: Center for Democratic Renewal, 1998.

Watson-Verran, Helen, and David Turnbull. "Science and Other Indigenous Knowledge Systems." In *Handbook of Science and Technology Studies,* edited by Sheila Jasanoff, Gerald E. Markle, and James C. Petersen, 115–39. Society for Social Studies of Science. Thousand Oaks, Calif.: Sage Publications, 1995.

Wynne, Brian. *Risk Management and Hazardous Waste: Implementation and the Dialectics of Credibility.* Berlin: Springer-Verlag, 1987.

———. "Sheepfarming after Chernobyl: A Case Study in Communicating Scientific Information." *Environment* 31, no. 2 (1989): 10–39.

Yearley, Steven. "The Environmental Challenge to Science Studies." In *Handbook of Science and Technology Studies,* edited by Sheila Jasanoff, Gerald E. Markle, and James C. Petersen, 457–79. Society for Social Studies of Science. Thousand Oaks, Calif.: Sage Publications, 1995.

Young, Iris Marion. "Harvey's Complaint with Race and Gender Struggles: A Critical Response." *Antipode: A Radical Journal of Geography* 30, no. 1 (1998): 36–42.

———. *Justice and the Politics of Difference.* Princeton, N.J.: Princeton University Press, 1990.

CHAPTER 8

# Occupational Health from Below

## The Women Office Workers' Movement and the Hazardous Office

MICHELLE MURPHY

How did women office workers come to see their work environment as unhealthy? Historian Michelle Murphy argues in this chapter that the emergence of what became known as sick building syndrome resulted from the efforts of women office workers' movements to raise the consciousness of office workers. Occupational health, redefined in terms of the cumulative impact of office conditions on women's bodies rather than the dangers of a single toxic exposure, was a central instrument used by the organization 9to5 and other movements to raise worker consciousness. To achieve this consciousness, worker activists "rematerialized" the work environment, transforming it from a comfortable and efficient workspace to one characterized by multiple hazardous substances and conditions. Murphy's chapter raises several themes that are central to understanding the broader process of disease emergence and the role of activist groups in making illness visible.

The concept of "rematerialization" is useful for thinking about the work of all grassroots organizations concerned with exposing the health effects of environmental hazards. From Love Canal housewives, to the Florist Club in Newtown described by Spears, to Nigeria's Ogoni Peoples Survival Organization discussed during the Sawyer Seminar by Dr. Owen Wiwa, environmental activists must confront and recast materializations, or representations of the environment as benign that are produced by representatives of industry. They also need to challenge characterizations of their illnesses by state officials and others as the product of their personal failings. Thus, 9to5 fought to redefine stress as a biological response to social processes rather than the inability of individuals to cope with psychological stresses.

Residents of Newtown similarly challenged the conclusions of state environmental investigators who claimed that the poor health of Newtown residents reflected their bad personal habits including excessive consumption of alcohol and tobacco. It was the space the Newtown residents occupied, not the lives they led in that space, that was the source of their illnesses.

Rematerialization in all these cases is based on what Spears refers to in the previous chapter as "situated knowledge," that is, knowledge and understandings that are located within specific illness communities. Murphy's chapter, like Spears's, reminds us that situated knowledge is not organic, in the sense of being easily or naturally articulated by illness sufferers. The production and dissemination of situated knowledge require the accumulation of individual illness experiences by organizations such as the Newtown Florist Club or 9to5. In other cases, the cyberspace communities of suffering on the Internet fulfill this function. Situated knowledge is produced through the mobilization of sufferers. In the absence of organizations able to elicit and accumulate individual illness experiences, the collective understandings of the relationship between exposure and illness may remain invisible.

---

Office workers are not falling off tall buildings, emerging at 5 P.M. covered with soot, or getting their hands caught in dangerous machines. But as an understanding of chemical and psychological hazards has increased, we have learned that office workers are exposed to severe dangers, all the more severe because they are often invisible and unrecognized.
—ELLEN CASSEDY, 9to5, 1979

Imagine an office with carpeted floors and textiled cubicles containing plastic modular furniture, with the clicking of fingers on keyboards and the whirring of photocopiers, with computer screens more prevalent than windows. A woman working at her desk crowded in the corner, staring at a screen, rubs her irritated eyes. In a nearby cubicle, a second worker, also a woman, wipes her runny nose and pops an allergy pill. Standing at the photocopier, a third woman covers her mouth as she coughs. A fourth passes a lozenge to a fifth. A sixth begins to feel dizzy as the printer spews paper. A seventh, an eighth—crowd of corporeal complaints begin to form. Dispersed in far-flung corners of an office, these women, workers in the gender-segregated information economy of late capitalism, may never have thought twice about their irritations; never have connected their bodies' rebellions to the material conditions of their office. How did such women workers come to perceive connections between their bodies and the built environments they inhabited—offices, homes, and other spaces conventionally coded as ordinary, comfortable, and safe?

Black lung, brown lung, silicosis, radiation poisoning—these are the kinds of diseases that spring to mind when we think of occupational health in the twentieth-century United States.[1] Compared with the chemical and mechanical hazards of industrial work that led to loss of limb and life, the carpeted, air-conditioned, clean interiors of late capitalist office work appear benign refuges from occupational disease and explicit exploitation. Even the successful feminization of the lowest ranks of office work was dependent on this relative safety, which further combined with a genteel etiquette and a middle-class dress code that at first operated to cover over class stratifications and eventually became an expectation associated with the job.[2] Conditions that factory or mine workers regularly tolerated would be absolutely unacceptable to an office worker, regardless of her rank. Thus, the very conditions under which information work arose made it difficult to imagine—for occupational health experts, management, and even workers—that they might inflict themselves on the body.

This chapter is concerned with how offices workers came to perceive connections between their workplaces and their bodies in terms of occupational illness. The emergence of office occupational health problems was not accomplished by experts discovering a new disease that could be added according to an established formula to the list of already recognized occupational diseases. Rather, the articulation of office occupational health problems called into question the very way occupational illness was conventionally defined and detected. The set of historical contingencies—clinical methods, laboratory tests, and insurance policies, for example—that shape what is recognized as an occupational health problem also set up constraints that delineate other corporeal relationships with place as outside perception. One of the reasons office occupational health problems were so difficult to articulate or even to imagine was the history of workers' compensation in the twentieth-century United States.

At the turn of the century, the "nature" of occupational health was formulated largely through the initial establishment of state workers' compensation boards concerned exclusively with industrial accidents locatable in discrete time and place.[3] Accidents and their compensation were codified in schedules: lose an arm, receive X amount of money. In the 1930s workers' compensation boards began adding select occupational diseases to these schedules. Newly established industrial hygiene laboratories were now able to identify occupational diseases in terms amenable to such schedules: as a specific disease that an acute exposure to a specific chemical predictably caused, for example, silicosis and lead poisoning.[4] Such compensated occupational diseases were also locatable as a singularity: as the predictable and regular result of an exposure that could be pinpointed.[5] To render perceptible the causal pathway between the workplace exposure and its corresponding physiological expression, scientists used technologies such as blood tests, X-rays, air samplers, and exposure chambers. Enshrined in toxicological protocols, juridical precedents, and workers' compensation rules, the use of such technologies and the formulation of specific causality became a necessity for proving the existence of occupational illness. In this specific sense, however, there was no "disease" associated with office work; nor was there an acute chemical exposure to measure. Office workers instead experienced a multitude of often minor corporeal complaints in myriad and seemingly innocuous circumstances. If occupational illness was to be associated with office work, its "nature" had to defy causal specificity and the search for discrete disease entities.

The history of any particular occupational health problem—from black lung to sick building syndrome (SBS)—is more than the story of its successful medical codification and consequent compensation; it is often first a history of grassroots struggle to make sense of and bring attention to an occupational illness. Most incidents of occupational illness, community chemical exposures, and cancer clusters would not have become recognized without the efforts of workers and lay people, often called "popular epidemiology."[6] This chapter is concerned with the history of one such struggle, the efforts of the women office workers' movement of the 1970s and 1980s, epitomized by 9to5, National Organization of Women Office Workers. The grassroots methods and analyses used by the women office workers' movement, so this chapter argues, were crucial for rendering perceptible the corporeal effects of office work. This movement maintained that, despite appearances, the office can be perceived as containing a dispersed, insidious, and gendered form of oppression that is invisible because it successfully covers over its own exploitations. Moreover, it argued that in the office specific chemical causes typical of industrial settings were replaced by *nonspecific* assaults by seemingly banal technologies.

This nonspecific account of occupational health was to become more widespread. The two most prevalent and recognized occupational illnesses associated with office work in the 1990s were SBS and repetitive strain injury (RSI).[7] These occupational illnesses were surrounded by media coverage, a flood of workers' compensation claims, investigations by the National Institute for Occupational Safety and Health (NIOSH) and the Occupational Safety and Health Administration (OSHA), professional consultants, conferences, vendors of specialized projects, and medical specialists (though these last are often marginal within their profession). Unlike most recognized industrial diseases, neither SBS nor RSI is caused by an accident or acute injury occurring at a specific place and time. Rather, they are both nonspecific illnesses associated with an accumulation of small causes. RSI is affiliated with repetitive movements; SBS is a syndrome defined as a constellation of multiple symptoms experienced by workers in a building for which no single cause can be found.[8]

Today many occupational health practitioners now accept that seemingly ordinary environments, like homes and offices, can cause illness because they are filled with objects that release chemicals into the air. Yet when women office workers began organizing, such claims were usually considered laughable and even taken to be examples of a feminine predilection toward hysteria. Even today a debate over the reality and origins of SBS continues: Is it a physical or a psychological phenomenon, a cumulative chemical exposure, or a form of mass hysteria?

### Resistance and Materialization

Who opens the mail? Who types the memos? Who keeps their computer systems running? The vast majority of people working in these companies are women and without us these companies wouldn't be running at all. So if they have the power of money, then we have the power of woman!
—DARLENE STILLE of Women Employed, 1976

Where there is power, there is resistance.
—MICHEL FOUCAULT, *The History of Sexuality,* 1978

An office can be thought of as an elaborate apparatus for extracting time and labor from the bodies of workers.[9] The office apparatus has many components—it is an assemblage of management discourses, information technologies, and workers. It comprises office equipment, furniture, systems of surveillance, spatial and architectural arrangements, gender roles, and job taxonomies. It includes managers, secretaries, mailroom sorters, and machine operators. An office can be thought of as a tightly knit assemblage of material and social coercions, the exact characteristics of which have changed over time. The office apparatus of late capitalism holds within it three different phases of its history. In the Taylorist office of the 1920s and 1930s, office workers labored on light machinery—stenographs, comptometers, typewriters. The purpose of the office apparatus was to maximize the efficiency of the worker's corporeal gestures by rationalizing the stretch of the worker's arm, her posture in a chair, and the duration of each of her motions. In the modernist offices of the 1950s, new ventilation technologies facilitated the construction of large, suburban glass-and-steel sealed worlds organized into a rigid, gender-based hierarchy that determined one's task, size of desk, benefits, and degree of privacy. The object of efficiency analyses shifted from the gestures of workers' bodies to the organization of a self-sufficient "corporate machine." By the time of the women office workers' movement, a new set of coercions were inscribed into the office apparatus. In this cybernetic office, office work increasingly became computerized, which in turn made possible a new way to surveil workers, an "information panopticon" that could monitor speed, accuracy, and time away from workstations.[10] Within the cybernetic office, space was rearranged through the "open plan": there were no individual offices; instead, modular workstations were arranged in a large carpeted space according to cybernetic circuit diagrams of information flow between workers. By the 1980s, the open plan had devolved into the crowded warren of cubicles with which we are all so familiar. With the cubicles, computers, and carpets came a flood of new synthetic materials, mostly plastics.[11] For management, the office was a place where information circulated, and thus labor took the form of head and not hand work, and the physical circumstances of the office were neutral, if not pleasant, to the body.

How did the women office workers' movement of the 1970s and 1980s resist this office apparatus? Resistance was not simply in the negative, a "saying no" to the coercions of the office; nor was it a liberation from or escape out of office work or late capitalism. Resistance was in the production of an alternative way of apprehending office work that allowed workers to articulate the form of a previously unmarked oppression and then to perform acts of intervention. This resistance was made possible by a certain assemblage of tools and practices available to women through feminism and technoscientific culture. In this way, the feminist office workers' movement *rematerialized* the office. I mean this literally. The built environment of the office was not just a "thing" out there that workers simply responded to, were effected by, or understood more or less accurately. Rather, women office workers assembled a set of tools and practices through which the office acquired a certain constellation of characteristics. To be more precise, the women's office labor movement mobilized feminism and tools from science to rematerialize the office as an apparatus composed of a tide of small details, the microphysics of office work.

I use the term *materialization* here to underline the importance of furniture and efficiency analyses, meetings and surveys, technologies and practices in the process of apprehending the effects of office work on women's bodies. The office was not just an "it" described with different words by workers and management. I use the term *materialization* precisely to avoid the inference that what is at stake is only words and their meaning. Materialization is the process by which historically specific assemblages of technologies, people, and practices allow objects and events—in this case the office—to be concretely engaged with, manipulated, acted on. Materialization is the process by which objects and events are brought into being for us and acquire some qualities and not others. Thus, when I use the term *materialization* I am indicating not just what we can see about the world but also what work we can do with it.

Even though a given materialization can seem to be self-evident to us in our own time—the feeling, How could it be another way?—historicizing health in the way I am suggesting can allow us to see that there are always multiple materializations at any one time, each performing different kinds of actions with varying efficacy on ourselves, on bodies, and on objects around us. By tracing conflicting, yet coexisting, materializations, what before had seemed self-evident or natural can be shown to be only partial, constrained by the practical limits of what a particular constellation of instruments, methods, and styles of thinking can do together. Thus, while each materialization renders certain aspects of the world perceptible, intelligible, and thus also manipulable, it at the same time renders other aspects imperceptible, unintelligible, and thus outside our conscious intervention.

Whereas office management—through its theories, architecture, work protocols, and technologies–grasped the office as a site where work could be painlessly, pleasantly, and passively extracted from the bodies of "head" workers, the feminist office workers' movement used a different method of analysis (consciousness-raising) combined with a corporeal understanding of office work, combined with an interpellation of women as active agents who worked together to allow an apprehension of office work as harmful to women's bodies. A given materialization, then, is not singular but always coexists with and builds on other materializations. Thus, every materialization, whether explicitly fashioned as an act of resistance or not, is always to some degree a *rematerialization*. The assemblage of tools and practices that made up the feminist office workers' movement led to a way of apprehending office health that had to grapple with the way the office was already understood to affect bodies. Instead of a machine for extracting efficiency in a pleasant environment, the office was "rematerialized" as a site of oppression and pathology.

In the following pages, I argue that materializations from below—by office workers, feminists, and labor activists—are technoscientific actions requiring as much skill with bodies, language, and artifacts as the highest technoscientific materializations.[12] Concentrating on the work of 9to5's Project Health and Safety, I unpack how occupational health was rematerialized in terms of dispersed, nonspecific, low-level physical assaults that produced myriad symptoms and chronic nonphysical assaults that resulted in the corporeal expression of "stress." I also pay critical attention to the important role of "experience" as a category of knowledge within these methods. Finally, I map out some of the exclusions and limits produced by this particular articulation of office occupational health.

### Feminism in the Office

> The women's liberation movement appears to be on the threshold of a second phase—when feminist consciousness reaches beyond white, middle-class women. . . . [A] clear sign of the new awareness felt by the so-called typical woman is the way office workers are asserting themselves. There is nothing quite so typical as a woman who is a secretary, typist, clerk, or keypunch operator.
> —MARJORIE ALBERT, union organizer, 1973

Typists, clerks, and keypunch operators were not passive conduits in the communication network of the information economy; they were also women and, as women, open to the hail of feminism.[13] Feminism called upon women as political agents able to challenge a gender-based oppression and develop their own modes of analysis for uncovering that oppression. When the women office workers' movement split from the mainstream feminist movement, it brought that agency to the workplace. Feminism provided women office workers with a means to organize around gender instead of simply class.[14]

The women's movement of the 1960s and 1970s, though fractured within itself, tended to be critical of the role that technoscience, especially biomedicine, had played in maintaining women's oppression.[15] This was particularly true of the radical wing, which was also responsible for most of theoretical writings from this period. The women's health movement, itself influenced by radical feminism, pointed out that the medical profession not only was dominated by men but also produced sexist representations of women's bodies, misdiagnosed women, and worked for the maintenance of patriarchy. Patriarchal biomedicine could not be trusted to determine what constituted female pathology and had to be resisted, not simply by increasing women's access in the medical professions but through a feminist politics centered on reclaiming the body. In the words of feminist activist Claudia Dreifus, in 1977: "It is not factories or post offices that are being seized, but the limbs and organs of the human beings who own them."[16] Women's bodies were to be taken back from biomedicine by creating new accounts based on women's lived experiences.[17]

Like the broader feminist movement, the women office workers' movement that sprang from it was not a single coherent entity. It was made up of scattered cells that had independently sprung up in cities sprinkled around the nation. On International Women's Day in 1971, at a National Organization of Women (NOW) conference in San Francisco, a handful of working-class women banded together to form their own feminist organization, called Union WAGE. In Chicago, Women Employed was established; in Cleveland, Working Women; in New York, Women Office Workers. In Boston in 1972, two secretaries at Harvard's School of Education, Ellen Cassedy and Karen Nussbaum, formed 9to5, which began in a room at the Cambridge YWCA.[18] Similar grassroots organizations sprang up in other cities. *Ms.* magazine reported that women's caucuses were even appearing in companies like Polaroid, Blue Cross, General Electric, and AT&T.[19] Critiques of the technologies and scientific management techniques composing the late capitalist office were central to this fledgling grassroots movement.

The office workers' movement was both a new kind of feminism and a new kind of labor movement. First, its founders envisioned their efforts as a practical extension of the larger women's movement into the lives and concerns of working-class women. Second, the movement was formed of extraunion organizations independent of the male-dominated labor unions. The founders considered their efforts an innovation within the labor movement, providing women, many of whom had misgivings about unionism, an alternative to unionism for responding to inequalities in the workplace. The organization 9to5, for example, saw its role within the labor movement as "working on issues of discrimination and fair employment in general, developing activists in the cause of working women, and serving as a lightning rod for the expression of problems and concerns."[20] Feminism provided an inroad for organizing women across the nation around broad issues, even when their individual workplaces were not amenable to unionization. Although the founders of the office workers' movement were feminists and the membership almost entirely female, not all the women 9to5 tried to organize would have called themselves "feminists." Yet these women had been affected by the ideas of the women's movement, believing that they were due equal pay and respect. In response to many office workers' ambivalent relationship to feminism, organizers within the women office workers' movement exclusively focused on issues pertaining to working women, declining to take positions on such issues as abortion and sexuality.[21] Moreover, the women office workers' movement organized only those women who worked with information technologies, entirely neglecting other groups of women who worked in office buildings—cleaners and custodial workers. These workers, often new immigrants or women of color, remained invisible and excluded from the movement, despite the occupational health hazards they faced.

The women office workers' movement was not by any means antiunion. However, the movement did feel a need to develop techniques outside the traditional union drive that would speak to women who, though perhaps not feminists themselves, were living in the context of feminism. To this end, the women office workers' movement both drew on the agency feminism had granted women and imported tactics from feminism for its own grassroots organizing. In particular, the technique of consciousness-raising became a crucial tool of the movement, used to convince women office workers that they deserved rights and respect in the workplace, thereby paving the way for possible unionization.

## Consciousness-Raising

Consciousness-raising, which had begun among radical feminists in New York in the late 1960s, spread rapidly across America.[22] By the end of 1970, every major city in the country had consciousness-raising groups—New York City alone had hundreds.[23] Even liberal feminist organizations like NOW were founding consciousness-raising groups. Consciousness-raising was a discursive instrument: small groups of women got together, rapped, shared their "experiences," found commonalties, and began to analyze them in a discussion. It was a powerful technique for translating seemingly idiosyncratic personal events and emotions into a gender-based experience.

Consciousness-raising can be thought of as a technique for unsaying what has already been said (the office works efficiently and safely) and saying what was unsaid (the office is an exploitative and harmful technology for extracting labor from bodies). It was a technique for undoing the self-evidence of what would otherwise appear to be a fixed or natural social structure. It was also a technique that operated by collecting variation and turning it into commonality—it took accounts of different women's lives and abstracted a common womanness or class membership. This ability to produce an analysis based in commonality was later critiqued by feminists of color who pointed out the false universality of "woman" and called for methods that allowed for, rather than covered over, differences among women. The use of consciousness-raising by the women office workers' movement, however, depended precisely on this problematic ability to transform difference into commonality.

Women Office Workers (WOW) of New York was founded through consciousness-raising techniques in the summer of 1973: "Several office workers got together to rap about what could be done to improve the life of clerical workers. Some of us were active in the women's organizations and some in unions. . . . We also believe[d] that office workers are not immune to the consciousness-raising that has been changing the lives of professional and middle-class women. What was missing was a vehicle for bringing together women office workers who often felt alienated from what went on at typical women's movement meetings."[24] WOW advertised its consciousness-raising sessions in its newsletter: "If you want to find out more about WOW, we can come to your home to talk to you and your co-workers. It's sort of a Tupperware party idea, except we're not selling pans—we're exchanging experiences and ideas."[25]

Similarly, Karen Nussbaum advised women in *9 to 5: The Working Woman's Guide to Office Survival:* "Suppose the office workers at your company have no organized group and you would like to start one. . . . Where do you begin? If your office chair is uncomfortable and the woman at the next desk has complained that her chair is also uncomfortable, then chances are you have already begun. Organizing begins when two or more people exchange common needs or grievances."[26]

The successful Hollywood movie *9 to 5* (1980) even brought consciousness-raising to the masses. Its stars, Jane Fonda, Lily Tomlin, and Dolly Parton, play disgruntled office workers; in a pivotal scene, the women, encouraged by smoking marijuana one evening, begin to exchange experiences in a consciousness-raising session, realize their common oppression, and plot comic revenge.[27]

Through consciousness-raising, experience played a critical role in feminist analysis and was more generally an important referent in many political movements of the 1960s and the 1970s; it was mobilized as a counter-knowledge that could set into question and even replace other types of expertise.[28] Experience was considered by radical feminists to be an alternate and more accurate source of knowledge about the status of women than already existent, male-produced knowledge. Kathie Sarachild, the member of New York Radical Women who coined the term *consciousness-raising,* argued that by basing their methods in experience, they "were in effect repeating the 17th century challenge of science to scholasticism: 'study nature, not books,' and put all theories to the test of living practice and action."[29] Within the feminist women's health movement, experience was a type of personal knowledge often originating in the body itself, and women were encouraged to trust their own corporeal sensations over medical authority. Thus, feminists tended to claim an epistemic privilege for experiential knowledge. It was possible for some feminists to argue that the nature of oppression can be fully comprehended only by the oppressed (and thus no one else can refute their assertions). As members of a cultural elite, male scientists, sociologists, and managers were alienated from this knowledge gained through life under oppression. Thus, not only did appeals to experience bring political struggles down to the level of the body, but they also asserted that oppression and suffering themselves provided special access to a type of knowledge that other expertise overlooked.

"Experience" is a category of knowledge that is just as historical as other forms of knowledge. The claim being made here is not as straightforward as asserting that our experiences of the world are shaped by culture and history. Rather, I am drawing attention to a stronger point: it is only through particular methods rooted historically in time and space that experience became a kind of evidence imbued with certain truth-telling qualities. Put more specifically, with the method of consciousness-raising, experience became a kind of evidence that involved the analysis of the intimate and minute details of one's personal life, and even one's body, that could reveal more general gendered phenomena. The persuasive force of the evidence of experience came from being marked as (rather than self-evidently being) more "natural" or "authentic" than expert knowledge. In short, in the practice of consciousness-raising, experience was not just collected but was *generated* as a kind of evidence.

### Oppression in the Details

With the aid of consciousness-raising, feminists of the 1970s transformed individual women's experiences into an analysis of the accumulation of small, trivial, day-to-day behaviors that made patriarchy possible and that, in turn, patriarchy constituted as invisible. It was this apprehension of how power operates through the details that provided the underlying grammar for the women office workers' rematerialization of the office. Feminist labor activists argued that oppression in the office wasn't necessarily obvious— women worked in air-conditioned rooms with potted plants and modern technological conveniences; rather, oppression took place in the seemingly trivial details of day-to-day interactions.[30]

Whereas designers promoted the plush carpets and modern furniture of the information economy as "humanity in an age of machines,"[31] feminist workers identified these same attributes as microtactics: "An older man sitting behind a huge oak desk in a large room with pictures on the wall of him shaking hands with President Nixon or John F. Kennedy, and his secretary at his side, uses all of that to intimidate. . . . One of these guys *starts off* by referring to you as girls, by intentionally forgetting your name and making sure that you know his. People now understand these things as *tactics* and they are willing to develop their own tactics to counterbalance his."[32] The subtle signals and daily minuscule humiliations—the way yellow and orange walls were meant to make workers feel "up," the way partitions isolated workers from one another, the way plastic modular furniture systems were advertised as a means to "solve your people problems,"[33] the way women had to fetch coffee—all these tiny details as construed by consciousness-raising constituted the dispersed operation of office oppression.[34]

In 1977 the dispersed pockets of women activists joined together to become a national group based in Cleveland called Working Women and renamed 9to5 in 1983. By the early 1980s the movement had locals in Los Angeles, Baltimore, Boston, Minneapolis, Atlanta, Philadelphia, Washington, D.C., Pittsburgh, Cincinnati, Cleveland, Dayton, Rhode Island, Seattle, San Francisco, and New York, with a total of more than ten thousand members. As the movement grew and became more institutionalized, the grassroots technique of consciousness-raising fell away, but the materialization of office oppression as occurring in the details remained. Following feminism's lead, the office workers' movement began to ask how the dispersed quality of office oppression might inscribe itself directly upon the body itself, producing material effects that could be called occupational illness.

## The Toxic Office

From its inception, the women office workers' movement was critical of the effects new digital technologies and the "automation" of the office might inflict on workers.[35] Although desktop computers were not part of the office landscape until the 1980s, many back office workers were already working on manual tasks affiliated with the computerization of office work. In the 1970s, an office worker might, for example, enter data on a video display terminal (VDT). An office worker might also work with a photocopier, which by the 1970s was standard office equipment. Not all new technologies were machines: carbonless paper and correction fluid had become common parts of office work, as had furniture made out of plastic, synthetic textiles, and chipboard. At the same time, the budding environmental movement of the 1970s, especially in relation to incidents like Love Canal in which supposedly ordinary neighborhoods were found to be toxic, provided a new critical way of looking at everyday technologies and environments. Although the feminist office workers' movement made no explicit political alignment with the environmental movement, it is not surprising that office workers in the 1970s began to examine with suspicion the new technologies that made up the ordinary environment of the office: "No one knows much about the new machines being introduced as large offices automate or the chemical products that have replaced the old rubber eraser and typewriter brush. They smell bad, but just how much harm will they do over a period of years?"[36] Much suspicion was concentrated on the VDT: not only was it tied to concerns about automation and routinization, but workers who spent hours staring at a small green-on-black screen also ended up with eyestrain, headaches, and cramped necks. Sitting inches away from a terminal, workers began to wonder if it emitted some kind of low-level radiation that might cause miscarriages or otherwise damage their reproductive abilities.[37]

Jeanne Stellman was one of the few occupational health researchers in this period to be interested in the idea that the office could be hazardous to workers' health. With a Ph.D. in physical chemistry (1972) from the City University of New York, she began working on occupational health issues with industrial unions. Before the early 1970s, very little attention was given to the study of women and occupational illnesses, and Stellman was part of the vanguard of women in the 1970s who defined this area of study.[38] In 1978, she became the scientific consultant for the Coalition of Labor Union Women and published the influential book *Women's Work, Women's Health,* which included a section on the toxic and material dangers of office work, as well as a more general chapter on stress.[39] At conferences and meetings held by the feminist office workers' movement, Stellman was often the lone speaker on occupational health.

In 1978 Stellman founded the Women's Occupational Health Resource Center in Brooklyn, New York, which served as a clearinghouse for information on women's occupational health issues and for a time became the center of feminist occupational health efforts. Between 1981 and 1983 the resource center held training sessions for more than four thousand workers, published a newsletter, and produced fact sheets (many based on Stellman's book), some of which covered the dangers of photocopiers, indoor air, VDT work, and other occupational hazards besieging office workers. Fact sheets typically listed a hazard and suggested a means of prevention. For example, the general fact sheet on "clerical work" published in 1980 listed excessive sitting, fatigue, stress, noise, poor office design, ozone, and organic solvents as the most prevalent office hazards. Above the list was the following caveat:

"The exact nature and extent of office hazards are not known. They vary from office to office."[40] Without scientific or epidemiological studies that rigorously connected a hazard with a specific occupational illness, Stellman could only use her fact sheets as a tool for consciousness-raising and make speculative suggestions about individual exposures based on chemical ingredients known to have a harmful effect in industrial settings. With her background in chemistry, Stellman concentrated on mechanical injuries and chemical exposures rather than VDT emissions.[41] She justified her extrapolation from industrial exposures to office settings with her observation that "today all workers have—in one way or another—become chemical workers and everybody is exposed to chemicals in the workplace."[42]

Encouraged by the suggestive statements in Stellman's work, the women office workers' movement more carefully cast a suspicious eye at office technologies and quickly found that workers were surrounded by products with potentially toxic chemicals as their ingredients. Although these chemicals were usually present in minute quantities, no one had investigated their cumulative effect. Framing the effects of the office in terms of occupational health created a vehicle for connecting "the body"—the pivotal site of feminist politics—to office work. By 1980 occupational health became a core issue for 9to5, and Project Health and Safety was founded.

Project Health and Safety began by researching some of the chemical ingredients composing office technologies and building materials. Project members found reports that photocopier toner was linked to cancer.[43] They cited the testimony of one woman who had successfully filed a workers' compensation claim for exposure to photocopier exhaust.[44] They unearthed a handful of instances in which office workers had suffered from formaldehyde poisoning, which they in turn linked to federal consideration of a urea formaldehyde foam insulation ban.[45] Correction fluid also fell under their suspicion after newspapers reported that a Texas teenager died by using it as an inhalant.

Despite this research, members of Project Health and Safety did not collect enough evidence to launch a campaign around any specific technology, specific chemical, or specific illness associated with office work. Rather, what they had assembled was a sense that possible toxic exposures lurked everywhere, even in the most seemingly banal of office technologies. The barely discernible quality of these potential hazards made them not more benign but, as Ellen Cassedy explained, "all the more severe because they are often invisible and unrecognized."[46] This perception of insidious hazards lurking in the most unsuspecting places was aptly captured in the following speech by Karen Nussbaum, quoted here:

Let me give you a guided tour of the hazards in just sending out one letter:

Alice prepares to type a letter for Mr. Big. The carbonless typing paper she uses is made with abietic acid to fill the pores, and PCB's—polychlorinated byphenyls. Abietic acid has been found to cause dermatitis and PCB's are extremely toxic, causing irritation to eyes, skin, nose and throat, can cause sever liver damage, and are suspected carcinogens. The typing ribbon she uses also contains PCB's.

To correct an error she uses a correction fluid containing trichloroethylene—TCE. In high doses, TCE can have a depressing effect on the central nervous system and can cause liver damage and lung dysfunction. . . . Alice goes to make a copy of the letter on the photocopy machine, which may emit ozone, a deadly substance. In poorly ventilated areas it's not hard to raise ozone to at least twice the federal standard. That black powder in the machine—the toner—may have nitropyrene or TNF, trinitroflourenone–suspected mutagens. . . . While in the copying room, Alice breathes methanol from the duplicating machine, which can cause liver damage. She ends her hazardous journey by filing Mr. Big's copy of the letter in a plastic file containing polyvinyl chloride, which can cause skin lesions and dermatitis.[47]

In other words, like office oppression, toxicity was in the details. To complicate matters further, these details varied from office to office and their health effects from person to person.

The women at Project Health and Safety understood that their conception of office occupational health was problematic. Medical diagnosis, workers' compensation standards, and juridical precedent, for one, demanded more specific and acute causal explanations. As Stellman explained, "There is no disease currently known as officeitis, nor is there likely to be one. One can explain this by considering that each of the occupational health hazards . . . usually produce a slow, subtle, insidious effect over time for which cause and effect may not be discernible. . . . In essence, the development of similar symptoms from many different low-level causes makes all chronic disease particularly difficult to understand."[48] The obstacles barring the recognition of nonspecific forms of occupational health problems at a clinical level were numerous. Most physicians were trained to search for discrete disease entities, and there was no "officeitis." Further, doctors diagnosed workers one at a time, making it difficult for them to perceive any buildingwide pattern that consciousness-raising might have revealed. Finally, the vague and often mild symptoms women workers presented could be explained by other diagnoses medicine had developed long ago— "dysmenorrhea," "hysteria," or "psychosomatism"—and then treated with pharmaceuticals. In contrast to feminism, medicine did not find it imperative to consider women accurate registers of their own bodily experiences. Additionally, occupational health investigators at NIOSH were not particularly concerned with office occupational health issues; the conditions of office work paled next to those industrial workers endured. When members of NIOSH did investigate office conditions, they usually came up empty handed—their instruments, designed to register high levels of toxic chemicals in industrial settings, found no significant exposures to record in the office.[49] Not only was a physical phenomenon imperceptible through their techniques, but also the nonspecific phenomenon officer workers complained of failed to fit the toxicological model of specific and acute chemical exposure and thus seemed outside the realm of possibility. Investigators at NIOSH found themselves turning to psychosomatic explanations, such as mass hysteria and mass psychogenic illness, to make sense of the variety and nonspecificity of women office workers' complaints.[50] Perhaps, some investigators suggested, such symptoms were a gendered psychological response to life stresses.

### Stressed-Out

Project Health and Safety's portrait of "toxicity in the details" became further complicated by its desire to expand its materialization of workplace assaults from chemical exposures to include nonphysical chronic social assaults. The term *stress* encapsulated this vision of the dispersed and chronic character of office exploitation manifesting itself through bodily symptoms. Stress, however, was a polyvalent term whose meaning the movement could not control. Project Health and Safety had to negotiate carefully the quagmire surrounding articulations of stress, for "hysteria" and "psychosomaticism" were psychological explanations closely tied to it. "Stress" was in danger of being swallowed by "hysteria" and thus being used to explain away often vague physical symptoms as simply the result of a gender-based psychological disposition, perhaps even rooted in menstruation, rather than adverse work conditions.[51]

Instead of granting stress a psychological foundation, feminist office activists wanted to couple stress tightly with the rise of the late capitalist information age. "If loss of limb and back pain are the characteristic hazards of the industrial age," explained Karen Nussbaum, "then job stress is the characteristic hazard of the computer age. It's an insidious hazard, because it's hard to identify and as Americans we frown on what we consider to be a personality weakness. But stress is not in your head, it's in the office. And it could be a blight on society."[52] Stress still proved a slippery subject to define. Union WAGE described stress in vague terms as something that "cannot be measured the way the noise level or temperature can be and is connected to workers having some control over their working conditions."[53] How could one measure stress? How to avoid its dismissal as an individual's psychological inability to cope with the pressures of the workplace? The link between stress and office work was confirmed for activists in a series of epidemiological studies that lent their claims scientific credence. In 1975 NIOSH reported that office workers had the second highest incidence of stress-induced diseases of any occupational group and in 1980 that VDT workers had the highest stress levels ever recorded.[54] The Framingham Heart Study provided further evidence, calculating that women office workers developed heart disease at a rate two times that of other women workers.[55] Project Health and Safety was willing to point to these studies to lend credence to its claims, but it was not willing to accept the often accompanying analysis that stress was rooted in a gendered psychology. Office activists mobilized a two-pronged strategy for explaining stress in nonpsychological terms: stress was a *biological* reaction to *social* conditions.

By pointing to social causes, office activists were using stress to argue for the necessity of a labor movement that would fight to restructure office work. The members of the short-lived, humorously named Nasty Secretaries Liberation Front (who later published *Processed World*) costumed themselves as computer terminals while handing out leaflets on stress to office workers in San Francisco. The pamphlet insisted that "stress is a social disease and it has a social cure. . . . Stress is not a result of individual failings. It is the result of an irrational and inhumane society. The solution to stress will not be found in any special seminar or in any special meditation or exercise techniques. . . . It will take a deliberate restructuring of the social order to reduce it in any real sense."[56] What elements of the "social order" should be restructured? The litany of social and work factors contributing to stress in the office were endless and diffuse. They covered every facet of workplace conditions: from environmental quality issues (temperature and overcrowding), to labor relations (lack of respect and no promotions), to job design (constant sitting and repetitive work). Stress was the expression of the dispersed quality of office oppression, and thus its presence condemned the social order, not the worker.

The proof of "stress," for feminist office activists, was in workers' bodies: shaky hands, headaches, nervous stomach, high blood pressure, ulcers, and menstrual cycle irregularities, any of which led many women to turn to pharmaceuticals to get through the day. This corporeal evidence was accounted for by pointing to an underlying nonpathological biological mechanism that translated adverse social conditions into biological expressions; borrowing from the work of Dr. Hans Selye, one of the fathers of modern stress theory, the feminist office movement called upon a biologically based "General Adaptation" response.[57] Jeanne Stellman, for example, explained stress in the following fashion: "We can think of the stress response as a mechanism for adapting to the environment. In fact, some scientists call the stress response the general adaptation syndrome . . . the body responds to different inputs in an identical manner."[58] In other words, the causes of stress were nonspecific, but their effects on the body accumulated in a general way and could be seen in the bodily symptoms of stress. The organization 9to5 described stress similarly in its newsletter: stress is "the body's way of protecting itself from physical, mental, or emotional strain. Heart and breathing rates increase, muscles tighten, and stomach acid and adrenaline are released. After a brief stressful incident, the body quickly returns to a balanced state. But constant stressful demands keep the body off balance, creating symptoms such as headaches, nervousness and fatigue. Prolonged stress can cause serious illnesses like high blood pressure and coronary heart disease."[59] Through such descriptions, feminist labor activists attempted to render stress a normal biological reaction, rather than a pathological psychological reaction: stress was not the result of an abnormal individual's psyche but the normal response to unhealthful conditions.

The meanings circulating under the term *stress,* however, were impossible to fix within this two-pronged strategy. In the late 1970s and 1980s, the term was used widely to describe the anxiety of living in the late twentieth century.[60] Popularly, and in contradistinction to the NIOSH studies, stress was most often associated with high-pressure jobs: executives, careerists, and professionals. A stress industry blossomed catering to the stressed-out, high-income market, providing hordes of advice, self-help books, exercise, hobbies, diets, and biofeedback therapies. The cottage industry that developed around stress focused on changing the individual rather than social conditions.

Even feminist labor activists themselves had difficulty entirely avoiding individualistic accounts of stress. Their own advice had a tendency to fade into platitudes indistinguishable from popular rhetoric. Even though 9to5's *Working Woman's Guide to Office Survival* noted that the only real solution to stress was to eliminate its sources through workplace organizing, the majority of its advice and column inches were dedicated to visualizing, stretching, dieting, hobbies, and expressing oneself.[61] Office-related stress was even included in a chapter of *Jane Fonda's Workout Book*—a succinct expression of feminism as self-discipline rather than social change.[62]

Stress remained a danger-fraught diagnosis. Yet the claim that the dispersed quality of office oppression was materially revealed in the biology of office workers was a radical twist.[63] Bodies reacted not only to low-level chemical exposures but also to unjust social factors. How, then, to capture these bodily reactions while still leaving their causes dispersed and open? The women office workers' movement still needed to devise a vehicle through which its formulation of nonspecific occupational health could be rendered politically effective, despite its slipperiness, in the context of compensation claims and government agencies.

### Surveys and Experience

Project Health and Safety needed a tool more suited to its new institutional form, one that could do some of the work consciousness-raising had done when the movement was more grassroots. The project members needed a tool that could gather into a single event an unwieldy constellation of health effects caused by both physical and nonphysical assaults; a tool that was affordable and easy to use without experts; a tool, unlike consciousness-raising itself, that would transport their controversial vision of occupational health beyond themselves into political action. Here they turned to the survey, which, unlike consciousness-raising, was technique imported from the social sciences, rather than from feminism.

As 9to5 increasingly left one-to-one organizing to union drives in the 1980s, it instead focused on articulating—to workers, government, and the media—the broad issues that women office workers shared. Likewise, Project Health and Safety became concerned with occupational health, not to organize individual workplaces but to educate workers and authorities about the broader issue. The survey came to join the grassroots feminist methods that had given birth to the movement. The power of the survey was that it performed several functions. For 9to5, it was important that the survey work as an extension of consciousness-raising by other means. It had to compel workers to ask probing questions about their own bodies and surroundings and continue the work of assembling dispersed experiences into a pattern of endemic oppression.[64] Of equal importance was the survey's ability to transport this apprehension to media, government, and expert audiences. Although the survey performed these functions well, it also came into conflict with the way the women office workers' movement had previously appealed to experience as a counterknowledge that could challenge expert knowledge and reductive science. The survey came with a price—it changed the status of "experience" into evidence to be resourced.

This change in the place of "experience" partially resulted from the institutional character of 9to5, which now had an executive board composed of members who chaired committees, a staff director who in turn hired other staff, and an executive director, Karen Nussbaum. No longer was the movement formed of office workers getting together at informal gatherings likened to "Tupperware parties." Instead, it was run by professional activists, mostly white and college-educated, who found themselves lobbying politicians, giving lecture tours, and testifying at congressional hearings.[65] The members of Project Health and Safety were no longer directly compelled by their own firsthand encounters of working in the office. For the career activist, the survey gathered the "experience" of office work at a distance. The social science survey already had a long tradition of such use by educated feminist social workers. In the opening decades of the twentieth century, progressive social activists like Alice Hamilton and other women at Hull House had innovatively used the social science survey in their work.[66] Surveys, administered by social workers, doctors, or union organizers, were a means to find out about the people one meant to help. Thus the survey was a technology middle-class activists used to assess the conditions of others, especially the working class, from afar.

The survey, however, was not simply a vehicle for gathering information; it was a tool that 9to5 used to construct a nonspecific health event by virtue of gathering that information. In other words, Project Health and Safety already apprehended the nonspecificity of office occupational health, and the survey was a means not to create but to make available that apprehension to others. The surveys Project Health and Safety used came in a variety of forms: some were large and others small; some were used to gather statistical data, and others were sent out for their consciousness-raising effects alone; some were administered by Project Health and Safety, and others were designed for workers to administer to one another in their own office buildings. The first large survey conducted by Project Health and Safety was funded by an OSHA grant in 1981; distributed to eight thousand workers in the Boston and Cleveland areas, it gathered thirteen hundred responses. The survey asked a hodgepodge of questions, covering such topics as stress levels, air quality, office environment, and office machines.[67] Its intent was not to uncover a particular disease or common problem underneath the variety of symptoms. Rather, the survey was used to constitute the "nature" of office occupational health—its diverse symptoms and varying conditions—by rendering nonspecificity perceptible. Thus surveys became not simply a way of gathering a diversity of information but a script through which diversity was arranged to illustrate a certain way of thinking about occupational health.

In 1982, Project Health and Safety administered an even larger "Stress

Test," distributing more than forty thousand surveys largely through women's magazines.[68] The Stress Test, like 9to5's first survey, asked many questions, listing more than thirty-four possible sources of stress (e.g., rate of promotion, deadlines, keystroke quotas, isolation, sexual remarks) and signs of stress (high blood pressure, smoking, heart disease, miscarriages, nerves, etc). Again, the survey was not intended to whittle stress down to a few well-delimited expressions. Rather, it operated to spread the effects of stress and its sources as widely as possible, thereby constituting the capillary nature of office occupational health. These large surveys were then supplemented by small, self-administered surveys—at meetings, in the project's newsletter, and in women's magazines—intended to elicit a consciousness-raising effect. On the heels of Project Health and Safety's work followed the publication of several popular books on office hazards, each appended with a comprehensive survey that workers could use to assess their own work conditions.[69] A survey distributed in a particular office building was sometimes used explicitly to organize workers around the conditions found there, sparking what came to be called "sick building" episodes.

Although a survey could be used to render perceptible and then organize workers around a sick building syndrome episode, its power to transport "experience" outside the movement came from its ability to objectify and quantify.[70] With the survey, workers no longer had to actually talk to one another about their experiences; instead, they could anonymously check off predetermined categories on a sheet of paper. These anonymous responses were then tabulated to create a statistical profile. The survey need not uphold strict standards of scientific rigor to be effective; the statistical format of the survey was usually sufficient to signal a scientistic "objective" measure. By objectifying and quantifying, surveys translated the nonspecific nature of office pathology into a scientistic format and bureaucratic language socially invested with the power to measure and give shape to the material. With the survey, "experience" was turned from a naturalized "counterknowledge" into evidence to be analyzed with quantitative methods. In the process, the voices of workers were transformed, as was their claim to a fuller comprehension of oppression, to be replaced by statistics more palatable to the bureaucrat and which could in turn better mobilize government resources. Thus, its reliance on the survey moved 9to5 further away from the grassroots techniques of person-to-person communication that had originally given birth to the women office workers' movement.

Paradoxically, this distancing from "experience" was what made the survey successful. By translating "experience" into statistics, surveys allowed

9to5 to transport its perception of office occupational health to government agencies and expert audiences in a language invested with the power to persuade. Through surveys a host of impressive numbers were now at their disposable. A congressman on the committee for indoor air could be told that "70% of the office workers surveyed complained of severe indoor pollution, poor circulation, and irritating fumes."[71] Or a press release might announce "4 out of 5 workers describe their jobs as somewhat or very stressful."[72] Project Health and Safety widely publicized the results of its survey on the radio, in newspapers, at congressional hearings, and at medical conferences. At the level of the building, a survey could be used to provoke a NIOSH Health Hazard Investigation.

Thus, the survey became a powerful technology for making nonspecific office occupational health perceptible to both workers and experts. It was a technology—unlike the air samplers, X-rays, and other instruments used by conventional occupational health investigations—able to capture a phenomenon that was nonspecific and only discernible within a group. And it was able to transport this "countermaterialization" outside the movement. Yet the survey was not a savior; like any technology, it had limitations. First, it was not a "bottom-up" method and thus could only gather "experience" in the form of evidence. Second, it did not provide the kind of causal proofs workers' compensation boards demanded. Surveys could only materialize the contours of an occupational health phenomenon, not the causes behind it. Thus, the survey left the lingering question of the psychological or material origins of office occupational health unresolved.

## Conclusion

There is still no single "disease" associated with office work. Instead of finding a shared affliction such as black lung or silicosis, the women office workers' movement materialized a nonspecific phenomenon that clashed with juridical, medical, and compensatory institutional demands for proof of linear causality. The women office workers' movement was only able to capture the symptomatic *expression* of workplace conditions by simultaneously leaving the question of their causal root uncertain. Thus, the history of office occupational health raises questions about the efficacy of articulating occupational health problems in a form that does not fit with the criteria of established institutions. I would argue that, despite this dilemma, the women office workers' movement was efficacious in rendering perceptible a new kind of occupational health event that workers could, and would, organize around later as sick building syndrome.

SBS is not a "disease" at all but is defined by occupational health experts as a phenomenon that manifests as a multitude of symptoms, mostly minor, associated with a particular office building and for which no single cause can be found. The symptoms associated with SBS vary not only from episode to episode but also from worker to worker within a single episode. Moreover, the proof of SBS is not in a blood test, air monitor, petri dish, or other form of "objective" scientific measure but in the density of worker complaints, first as expressed through worker protests and later as captured by experts' surveys. It has been the numerous and increasing grassroots protests around individual problem buildings that have kept the perception of office occupational health in the limelight.

SBS, however, has acquired circumstances quite different from those generated by the women office workers' movement. With the dramatic rise of unionism in the public sector and a corresponding greater willingness to organize women among established organizations like the AFL-CIO, SBS has become untangled from the extraunion women office workers' movement and its feminist character. Karen Nussbaum became exemplary of the changed status of women office workers in unions—head of the Women's Bureau in the U.S. Department of Labor in 1993 and then director of the Working Women's Department at the AFL-CIO, she is now the highest-ranking female official in the U.S. labor movement.[73] Although the rank and file of clerical unions may be female, SBS is rarely figured as a women's issue, even though the role of gender in SBS remains heatedly debated among occupational health professionals and in courtrooms. Why did episodes primarily affect women? Did the predominance of women indicate a gendered psychological cause? Is SBS caused by cumulative chemical exposure or mass hysteria?[74]

The emergence of office occupational health problems cannot be simply understood as the result of newly pathological workplace conditions or as the discovery of a previously hidden problem by experts or even by workers. These problems did not emerge as such but were materialized by a set of methods, tools, and actions assembled by office workers that together made possible a kind of occupational health event otherwise excluded from perceptibility. Yet this materialization, like all others, carried its own set of constraints, for the assemblage of tools and practices the feminist office workers' movement used performed only in certain ways. On the one hand, although they were successful in articulating nonspecific health events, they could not connect bodies and work conditions in the terms of causality and specificity that health institutions conventionally demanded. On the other hand, by connecting office work to bodies primarily as health events, rather than as acts of oppression, the feminist office workers' movement unwittingly set the stage for the depoliticization of office occupational health events through the later diagnosis of SBS.

## NOTES

I would like to thank 9to5 and Jeanne Stellman for permission to use their papers, housed at the Schlesinger Archive, Radcliffe College, Harvard University. Revised version reprinted by permission from "Toxicity in the Details: The History of the Women's Office Worker Movement and Occupational Health in the Late Capitalist Office" by M. Murphy, *Labor History* 41, no. 2 (2002). www.tandf.co.uk/journals/titles/002365x.html

1. Robert Botsch, *Organizing the Breathless: Cotton Dust, Southern Politics and the Brown Lung Association* (Lexington: University Press of Kentucky, 1993); Claudia Clark, *Radium Girls, Women, and Industrial Health Reform, 1910–1935* (Chapel Hill: University of North Carolina Press, 1997); Bennett Judkins, *We Offer Ourselves as Evidence: Toward Workers' Control of Occupational Health* (New York: Greenwood Press, 1986); David Rosner and Gerald Markowitz, *Deadly Dust: Silicosis and the Politics of Occupational Disease in Twentieth-Century America* (Princeton, N.J.: Princeton University Press, 1991); Barbara Ellen Smith, *Digging Our Own Graves: Coal Miners and the Struggle over Black Lung Disease* (Philadelphia: Temple University Press, 1987).

2. On the history of gender and office work, see Cindy Sondik Aron, *Ladies and Gentlemen of the Civil Service: Middle Class Workers in Victorian America* (New York: Oxford University Press, 1987); Margery Davies, *Women's Place Is at the Typewriter: Office Work and Office Workers, 1870–1930* (Philadelphia: Temple University Press, 1982); Ileen Devault, *Sons and Daughters of Labor: Class and Clerical Work in Turn-of-the-Century Pittsburgh* (Ithaca, N.Y.: Cornell University Press, 1990); Lisa Fine, *Souls of the Skyscraper: Female Clerical Workers in Chicago, 1870–1930* (Philadelphia: Temple University Press, 1990); Kenneth Lipartito, "When Women Were Switches: Technology, Work, and Gender in the Telephone Industry, 1890–1920," *American Historical Review* 99.4 (1994): 1075–1111; Ellen Lupton, *Mechanical Brides: Women and Machines from Home to Office* (Princeton, N.J.: Princeton Architectural Press, 1993); Massachusetts History Workshop, *They Can't Run the Office without Us: Sixty Years of Clerical Work* (Cambridge: Massachusetts History Workshop, 1985); Stephen Norwood, *Labor's Flaming Youth: Telephone Operators and Worker Militancy, 1878–1923* (Urbana: University of Illinois Press, 1990); Elyce Rotella, *From Home to Office: U.S. Women at Work, 1870–1930* (Ann Arbor: UMI Research Press, 1981); Sharon Hartman Strom, *Beyond the Typewriter: Gender, Class, and the Origins of Modern American Office Work, 1900–1930* (Urbana: University of Illinois Press, 1992).

3. R. Rudy Higgens-Evenson, "From Industrial Police to Workmen's Compensation: Public Policy and Industrial Accidents in New York, 1880–1910," *Labor History* 39.4 (1998): 365–80.

4. Christopher Sellers, *Hazards of the Job: From Industrial Disease to Environmental Health Science* (Chapel Hill: University of North Carolina Press, 1997).

5. What diseases to include on these schedules became a source of contention between workers and state compensation systems, as expressed in the black and brown lung movements of the 1960s and 1970s. In the 1970s a federal report critiqued such systems as out of date (National Commission on State Workmen's Compensation Laws, *The Report of the National Commission on State Workmen's Compensation Laws* [Washington, D.C.: Government Printing Office, 1972]). State after state amended its laws to include *any* occupational disease that could be demonstrated to be caused, aggravated, or hastened by the workplace. Yet a worker must still document that he or she is suffering from a particular *disease* and that this disease is considered by occupational health professionals to be caused by conditions found at the person's workplace. Although it is technically possible to get compensation for a chronic injury, such as a back injury, the burden of proof is on the worker. Allard E. Dembe, *Occupation and Disease: How Social Factors Affect the Conception of Work-Related Disorders* (New Haven: Yale University Press, 1996).

6. Phil Brown and Ewin Mikkelsen, *No Safe Space: Toxic Waste, Leukemia, and Community Action* (Berkeley: University of California Press, 1990).

7. On RSI, see Dembe, *Occupation and Disease*; R. Dennis Hayes, "Digital Palsy: RSI and Restructuring Capital," in *Resisting the Virtual Life: The Culture and Politics of Information,* ed. James Brook and Iain A. Boal (San Francisco: City Lights, 1995), 173–80. On sick building syndrome, see Michelle Murphy, "Sick Buildings and Sick Bodies: The Materialization of an Occupational Illness in Late Capitalism" (Ph.D. diss., Harvard University, 1998).

8. I want to make clear that this picture of nonspecific causality and nonspecific reaction should be differentiated from accounts of multifactor causality. In an instance of multifactor causality, the combination of causes A and B and C, for example, causes an occupational illness. This picture also differs from accounts in which A or B or C can cause a given illness. In SBS, an unpredictable combination of possible causes A through Z creates a multitude of individual reactions 1 though 100. Both the cause and effect are nonspecific.

9. In *Discipline and Punish: The Birth of the Prison,* trans. Alan Sheridan (New York: Vintage Books, 1979), Michel Foucault uses this notion of the apparatus to analyze the prison as a historically specific technology of discipline and subjection. Foucault defines the term in the following way, "What I am trying to pick out with this term [apparatus] is, firstly, a thoroughly heterogeneous ensemble consisting of discourses, institutions, architectural forms, regulatory decisions, laws, administrative measures, scientific statements, philosophical, moral and philanthropic propositions—in short, *the said as much as the unsaid.* Such are the elements of the apparatus. The apparatus itself is the system of relations that can be established between these elements" (194–95; italics mine) My analysis of the office apparatus, however, differs from Foucault's analysis of the prison apparatus. Foucault does a marvelous job at drawing his topography in terms of the knowledges and material technologies that main-

tained subjection, and thus the prisoner tended to be represented as the outcome or effect of the orderly workings of the prison. I am attempting to draw a topography of the material technologies of resistance that interfered with and rematerialized the apparatus.

10. Shoshana Zuboff, *The Age of the Smart Machine: The Future of Work and Power* (New York: Basic Books, 1988). On the changes the computer brought to office organization, see Joan Greenbaum, *Windows on the Workplace: Computers, Jobs, and the Organization of Office Work in the Late Twentieth Century* (New York: Monthly Review Press, 1995). On gender, computer work, and capitalism, see Evelyn Nakano Glenn and Roslyn Feldberg, "Proletarianizing Clerical Work: Technology and Organizational Control in the Office," in *Case Studies on the Labor Process,* ed. Andrew Zimbalist (New York: Monthly Review Press, 1979), 51–72; Rachel Grossman, "Women's Place in the Integrated Circuit," *Radical America* 14 (1980): 29–49; Heidi Hartmann, ed., *Computer Chips and Papers Clips: Technology and Women's Employment* (Washington, D.C.: National Academy Press, 1987).

11. On the history of plastic products, see Jeffrey Meikle, *American Plastic: A Cultural History* (New Brunswick, N.J.: Rutgers University Press, 1995).

12. This is a paraphrase from Donna Haraway, "Situated Knowledges: The Science Question in Feminism and the Privilege of Partial Perspective," in *Simians, Cyborgs and Women* (New York: Routledge, 1991), 191.

13. On the relationship between feminism and labor movements, see Deborah Bell, "Unionized Women in State and Local Government," in *Women, Work and Protest: A Century of US Women's Labor History,* ed. Ruth Milkman (Boston: Routledge and Keegan Paul, 1985), 280–99; Linda Briskin and Patricia McDermott, eds., *Women Challenging Unions: Feminism, Democracy, and Militancy* (Toronto: University of Toronto Press, 1993); Roslyn Feldberg, "'Union Fever': Organizing among Clerical Workers, 1900–1930," *Radical America* 14.3 (1980): 53–70; Roberta Goldberg, *Organizing Women Office Workers: Dissatisfaction, Consciousness and Action* (New York: Praeger, 1983); Carol Kates, "Working Class Feminism and Feminist Unions: Title VII, the UAW and NOW," *Labor Studies Journal* 14 (1989): 28–45; Gail Gregory Sansbury, "'Now What's the Matter with You Girls?': Clerical Workers Organize," *Radical America* 14 (1980): 67–75; Nancy Seifer and Barbara Wertheimer, "New Approaches to Collective Power: Four Working Women's Organizations," *Women Organizing: An Anthology* (Metuchen, N.J.: Scarecrow Press, 1979), 152–83.

14. This, however, was not the first time women office workers had heard the call of feminism or even of labor organizing. Feldberg, "'Union Fever'"; Sharon Hartman Strom, "'We're No Kitty Foyles': Organizing Office Workers for the Congress of Industrial Organizations, 1937–1950," in Milkman, *Women, Work and Protest,* 206–34.

15. See, for example, Suzanne Arms, *Immaculate Deception: A New Look at Women and Childbirth in America* (San Francisco: San Francisco Book Co., 1975); Barbara Ehrenreich and Deirdre English, *Complaints and Disorders: The Sexual Politics of Sickness* (Old Westbury, N.Y.: Feminist Press, 1973); Barbara Ehrenreich and Deirdre English, *Witches, Midwives, and Nurses: A History of Women's Healers* (Old Westbury, N.Y.: Fem-

inist Press, 1973); Donna Haraway, "Animal Sociology and a Natural Economy of the Body Politic, Part I and II," *Signs* 4.1 (1978): 21–60; Donna Haraway, "The Biological Enterprise: Sex, Mind, and Profit from Human Engineering to Sociobiology," *Radical History Review* 20 (1979): 206–37; Ruth Hubbard, Mary Sue Henifin, and Barbara Fried, eds., *Women Look at Biology Looking at Women: A Collection of Feminist Critiques* (Boston: G. K. Hall and Co., 1979); Carolyn Merchant, *The Death of Nature: Women, Ecology, and the Scientific Revolution* (New York: Harper and Row, 1980); Barbara Seaman and Gideon Seaman, *Women and the Crisis in Sex Hormones* (New York: Rawson Associates, 1977). By the 1980s diverse writings about science and medicine by feminist historians, philosophers, sociologists, and biologists formed a new field of academic inquiry, "feminist science studies," from which this article descends.

16. Claudia Dreifus, ed., *Seizing Our Bodies: The Politics of Women's Health* (New York: Vintage Books, 1977): xxxi.

17. By affirming "experience" as a legitimate form of knowledge just as valuable as scientific knowledge, the women's health movement challenged the scientization of clinical medicine. Over the course of the twentieth century, doctors had moved away from using their patients' accounts of their illness as the primary source of information for making their medical diagnosis. With the success of bacteriology and the rise of scientific biomedicine, doctors turned increasingly to laboratory tests and instruments that could make graphical representations of bodily functions, such as heart function, blood pressure, or brain waves. Further, the rising dominance of medical insurance led to the standardization of physical exams and history-taking techniques (Angela Nugent, "Fit for Work: The Introduction of Physical Examinations in Industry," *Bulletin of the History of Medicine* 57 (1983): 578–95). The purpose of the clinical encounter became to extract data from the patient's body by using instruments that measured and quantified bodily functions and by employing an interview technique that uncovered information that would otherwise be obscured by the patient's subjective account of illness. The role of the doctor was to gather "objective" data and then interpret them using expertise unavailable to the patient. On the women's health movement, see Sheryl Ruzek, *The Women's Health Movement: Women's Alternatives to Medical Control* (New York: Praeger, 1978).

18. Jean Tepperman documented the feminist clerical movement in the Boston area. She published two books and interviewed many clerical workers in 1974–75. Jean Tepperman, *Sixty Words a Minute, and What Do You Get? Clerical Workers Today* (Somerville, Mass.: New England Free Press, 1972) and *Not Servants, Not Machines: Office Workers Speak Out* (Boston: Beacon Press, 1976). The interviews are housed at Schlesinger Archive, Radcliffe College, Harvard University.

19. Susan Davis, "Organizing from Within," *Ms.* 1 (1972): 92–99.

20. David Plotke and Karen Nussbaum, "Women Clerical Workers and Trade Unionism: Interview with Karen Nussbaum," *Socialist Review* 49.10 (1980): 153.

21. Ibid., 154–55.

22. Alice Echols, *Daring to Be Bad: Radical Feminism in America, 1967–75.* (Minneapolis: University of Minnesota Press, 1989).

23. Ellen Messer Davidow, "Acting Otherwise," in *Provoking Agents: Gender and Agency in Theory and Practice,* ed. Judith Keegan Gardner (Urbana: University of Illinois Press, 1995): 36–37; Anita Shreve, *Women Together, Women Alone: The Legacy of the Consciousness-Raising Movement* (New York: Viking, 1989), 11.

24. *Women Office Workers News,* January/February 1974.

25. *Women Office Workers News,* April/May 1974.

26. Ellen Cassedy and Karen Nussbaum, *9 to 5: The Working Woman's Guide to Office Survival* (New York: Penguin, 1983), 158.

27. The film was used as a vehicle to promote the women office workers' movement. Jane Fonda went on a national promotional tour with it. The movie's success spawned a hit single, Parton's "Working 9 to 5," and a short-lived television series in 1982.

28. On experience, see Elizabeth Bellamy and Artemis Leontis, "A Genealogy of Experience: From Epistemology to Politics," *Yale Journal of Criticism* 6.1 (1993): 163–84; Chandra Mohanty, "Feminist Encounters: Locating the Politics of Experience," *Copyright* 1 (1987): 30–44; Joan W. Scott, "'Experience,'" in *Feminists Theorize the Political,* ed. Joan W. Scott and Judith Butler (New York: Routledge, 1995), 22–40.

29. Kathie Sarachild, "Consciousness-Raising: A Radical Weapon," in *Feminist Revolution,* ed. Redstockings (New York: Random House, 1975), 132.

30. This is not to argue that all feminists avoided pointing to large structural inequalities. In particular, feminists building on Marxian analyses tended to argue for an underlying "dialectics of reproduction" at the root of women's oppression. See, for example, Shilamuth Firestone, *The Dialectic of Sex* (New York: William Morrow and Co., 1970), and Mary O'Brien, *The Politics of Reproduction* (Boston: Routledge and Kegan Paul, 1983).

31. Herman Miller Office Furniture advertisement, *Wall Street Journal,* November 14, 1974.

32. Karen Nussbaum quoted in Tepperman, *Not Servants, Not Machines,* 89.

33. Herman Miller Office Furniture advertisement, *Wall Street Journal,* November 14, 1974.

34. It is interesting to note the striking similarity between this analysis and that developed roughly contemporaneously by Michel Foucault, who showed how the operation of any modern institutional apparatus, be it a prison or an office, is exercised in the abundance of small details, of daily gestures, bodily disciplines, habits, schedules, spatial demarcations, and repetitions—what he called the "microphysics" of power or the "capillary" operation of power. Foucault, *Discipline and Punish,* 26.

35. 9to5, *Hidden Victims: Clerical Workers, Automation, and the Changing Economy* (Cleveland: Working Women, 1985); 9to5, *VDT Syndrome: The Physical and Mental Trauma of Computer Work* (Cleveland: Working Women, 1988); Daniel Armschall and Judith Gregory, eds., *Office Automation, Jekyll or Hyde? Highlights of the International Conference on Office Work and New Technology* (Cleveland: Working Women Education Fund, 1983); Tepperman, *Sixty Words;* Tepperman, *Not Servants, Not Machines;* Working Women, ed., *Race against Time: Automation of the Office* (Cleveland: National Association of Office Workers, 1980).

36. "Health and Safety: Writing and Negotiating Your Union Contract," *Union WAGE,* March/April 1975.

37. Although no proof exists that VDTs are linked to reproductive health problems, some activists remain concerned that low-level radiation causes health effects.

38. However, there were specific instances of famous occupational diseases that had affected women, for example, the radiation poisoning of the radium dial painters and phosphorus poisoning of women who worked in match factories. David Moss, "Kindling a Flame under Federalism: Progressive Reformers, Corporate Elites, and the Phosphorus Poisoning Campaign of 1909-12," *Business History Review* 68 (1994): 244-75; Clark, *Radium Girls.*

39. See "Work, Stress, and Health," and pp. 103-18 in Jeanne Stellman, *Women's Work, Women's Health: Myths and Realities* (New York: Pantheon Books, 1977).

40. Women's Occupational Health Resource Center, "Clerical Workers: Fact Sheet," box 2, unlabeled black folder, Jeanne Stellman Papers, Schlesinger Archive, Radcliffe College, Harvard University.

41. Stellman, unlike the women at Project Health and Safety, did not believe that the evidence supported the claim that VDTs emitted radiation.

42. Jeanne Stellman, speech, Institute on Safety and Health, Columbus, Ohio. Quoted in Margaret Banning, "Workplace Chemicals Put the Emphasis on Occupational Health," *Monitor* (May 1984): 16-17.

43. Photocopiers had been maligned in the press in a series of articles with titles such as "Cancer from Photocopiers?," *Mother Jones,* December 1980, 8; G. Lofroth, E. Hefner, I. Alfhelm, and M. Moller, "Mutagenic Activity in Photocopies," *Science* 209 (August 1980): 1037-39; and "IBM Brass Knew TNF 'Possible' Carcinogen More Than a Decade Ago," *Computerworld* 8 (September 1980): 1.

44. Testimony by Anita Reber, House Subcommittee on Health, "Women's Health Issues," 97th Cong., 1st sess., February 4, 1981.

45. "Alert Sheet: Urea Formaldehyde Foam Insulation," Consumer Product Safety Commission, Washington, D.C., December 1980.

46. Ellen Cassedy to Karen Nussbaum, memo, October 9, 1979, 9to5 Papers, box 18, folder 1199, Schlesinger Archive, Radcliffe College, Harvard University.

47. Karen Nussbaum, speech, Women's Occupational Health Resource Center, New York, October 1981. Quoted in "Sending a Letter: PCBs and Other Office Hazards," *Working Women News,* November/December 1981.

48. Jeanne Stellman and Mary Sue Henifin, "Health Hazards in the Computerized Office," in Armschall and Gregory, *Office Automation, Jekyll or Hyde?,* 63.

49. On the history of NIOSH investigations of office buildings, see Murphy, "Sick Buildings and Sick Bodies."

50. Michael Colligan, James Pennebaker, and Lawrence Murphy, eds., *Mass Psychogenic Illness* (Hillsdale, N.J.: Lawrence Erlbaum Associates, 1982); Neal Schmitt, Michael Colligan, and Michael Fitzgerald, "Unexplained Physical Symptoms in Eight Organizations: Individual and Organizational Analyses," *Journal of Occupational Psychology* 53.4 (1980): 305-17.

51. Jerome Singer, Carlene S. Baum, Andrew Baum, and Brenda D. Thew, "Mass Psychogenic Illness: The Case for Social Comparison," in Colligan, Pennebaker, and Murphy, *Mass Psychogenic Illness,* 166. On the history of menstruation and work, see Susan Cayleff, "She Was Rendered Incapacitated by Menstrual Difficulties: Historical Perspectives on Perceived Intellectual and Physiological Impairment among Menstruating Women," in *Menstrual Health in Women's Lives,* ed. Alice Dan and Linda Lewis (Urbana: University of Illinois Press, 1992), 229–35; Sioban Harlow, "Function and Dysfunction: A Historical Critique of the Literature on Menstruation and Work," in *Culture, Society, and Menstruation,* ed. Virginia Olesen and Nancy Fugate Woods (Washington: Hemisphere Publishing Co., 1986), 39–51; Emily Martin, *The Woman in the Body* (Boston: Beacon Press, 1989), 92–138; Anne Walker, "A History of Menstrual Psychology," in *The Menstrual Cycle* (London: Routledge, 1997), 30–58.

52. Karen Nussbaum, "Office Automation: Jekyll or Hyde?," in Armschall and Gregory, *Office Automation, Jekyll or Hyde?,* 19.

53. "Clerical Workers Tell All," *Union WAGE,* July/August 1979, 3.

54. M. G. Smith, "Potential Health Hazards of Video Display Terminals," NIOSH, Cincinnati, June 1981.

55. S. Haynes and M. Feinleib, "Women, Work and Coronary Heart Disease: Prospective Findings from the Framingham Heart Study," *American Journal of Public Health* 70.2 (1980): 133–41.

56. "Stress Is a Social Disease," 1973. Reprinted in *Processed World,* Fall 1987, 41.

57. Hans Selye, *The Stress of Life* (New York: McGraw-Hill, 1965).

58. Stellman, *Women's Work, Women's Health.*

59. "Stress: The Office Worker's Illness," *9to5 Newsletter,* September/October 1993, 3.

60. Neurasthenia, hysteria, and nerves, like stress, were also thought to result from the "new" anxieties of their moment. Robert Kugelmann, *Stress: The Nature and History of Engineered Grief* (Westport, Conn.: Praeger, 1992).

61. Nussbaum, "Office Automation," 19.

62. Jane Fonda, *Jane Fonda's Workout Book* (New York: Simon and Schuster, 1981).

63. In this same period psychosomatic explanations split into two politically divergent camps. In the first, and more politically conservative, form, physical illness with no known biological underpinning is explained, usually by a process of elimination, as a somatic expression of an underlying gendered psychological disorder or disposition. In its second, and more politically radical, form, psychosomaticism is understood as a biological expression caused by racial, sexual, or other kinds of oppression. Thus, there is a recent scientifically rigorous epidemiological literature on the relationship between racism and hypertension. For a historical account of this association, see Edwin J. Greenlee, "Biomedicine and Ideology: A Social History of the Conceptualization and Treatment of Essential Hypertension in the United States" (Ph.D. diss., Temple University, 1989).

64. Davis, "Organizing from Within," 92–99.

65. In fact, some of the volunteers in Project Health and Safety critiqued the way the organization of the committee reflected office work itself. As one volunteer

explained in her resignation letter, "In our questionnaires, we inquired about the issue of lack of control as being a contributory factor causing stress in the workplace. I feel a lack of control and input into committee decisions that is as devastating as that which I currently experience in my workplace. It is as if the decisions for the Committee have already been made by 9 to 5 and Working Women in advance and we are simply to fulfill these constant requests. The members do not appear to be an integral part of the decision-making process." Letter of resignation to Health and Safety Committee, May 6, 1981, box 5, folder 150, 9to5 Papers. The committee was also critiqued for not representing the diversity of office workers (committee members tended to be young, college-educated, white, and single). Self-evaluation of the Committee on Health and Safety, July 7, 1981, box 5, folder 151, 9to5 Papers.

66. Linda Gordon, *Pitied but Not Entitled: Single Mothers and the History of Welfare, 1890–1935* (Cambridge: Harvard University, 1994); Helene Silverberg, ed., *Gender and American Social Science* (Princeton, N.J.: Princeton University Press, 1998); Kathryn Kish Sklar, "Hull-House Maps and Papers: Social Science as Women's Work in the 1890s," in *The Social Survey in Historical Perspective, 1880–1940,* ed. Martin Bulmer, Kevin Bales, and Kathryn Kish Sklar (Cambridge: Cambridge University Press, 1991), 111–47.

67. According to the survey, 70% of women clerical workers experienced an inadequate supply of fresh air, two-thirds reported air circulation problems, and 25% reported irritating fumes. Stress was also widely reported, particularly among office workers who used VDTs.

68. 9to5, *The 9to5 National Survey on Women and Stress* (Cleveland: 9to5, National Association of Working Women, 1984).

69. Joel Makower, *Office Hazards: How Your Job Can Make You Sick* (Washington, D.C.: Tilden Press, 1981); Stellman, *Women's Work, Women's Health;* Jeanne Stellman and Mary Sue Henifin, *Office Work Can Be Dangerous to Your Health* (New York: Pantheon, 1983); Working Women, *Warning: Health Hazards in the Office* (Cleveland: Working Women, 1981).

70. On the history of objectification through quantification, see Ian Hacking, *The Taming of Chance* (Cambridge: Cambridge University Press, 1990); Theodore M. Porter, *Trust in Numbers: The Pursuit of Objectivity in Science and Public Life* (Princeton, N.J.: Princeton University Press, 1995).

71. Lisa Fein, 9to5's Project Health and Safety, to Rep. T. Moffet, U.S. House of Representatives. April 24, 1981, box 5, folder 150, 9to5 Papers.

72. Mary Mitchell, speech, April 22, 1981, box 5, folder 150, 9to5 Papers.

73. Before holding this post, Nussbaum was appointed to head the Women's Bureau of the Labor Department during Bill Clinton's first term. On the recent increase of women in unions, both in the rank and file and in positions of power, see Julie Kosterlitz, "Luring Women to Labor's Ranks," *National Journal,* March 15, 1997, 541; Amy Waldman, "Labor's New Face," *Nation,* September 22, 1997, 11–16. This increase in unionism among women has come hand in hand with a return to organizing techniques that involve one-to-one contact. See, for example, John Hoerr, *We*

*Can't Eat Prestige: The Women Who Organized Harvard* (Philadelphia: Temple University Press, 1997).

74. See, for example, Peter Boxer, "Occupational Mass Psychogenic Illness," *Journal of Occupational Medicine* 27.12 (1985): 867–72; Halley Faust and Lawrence Brilliant, "Is the Diagnosis of 'Mass Hysteria' an Excuse for Incomplete Investigation of Low-Level Environmental Contamination?," *Journal of Occupational Medicine* 23.1 (1981): 11–16; Christopher Ryan and Lisa Morrow, "Dysfunctional Buildings or Dysfunctional People: An Examination of the Sick Building Syndrome and Allied Disorders," *Journal of Consulting and Clinical Psychology* 60.2 (1992): 220–24; Michael Colligan, "The Psychological Effects of Indoor Air Pollution," *Bulletin of the New York Academy of Medicine* 57.10 (1981): 1014–26.

# PART II // INSTITUTIONAL RESPONSES TO EMERGING ILLNESSES

CHAPTER 9

# "Always with Us"

## Childhood Lead Poisoning

## as an Emerging Illness

CHRISTIAN WARREN

What determines how public health institutions respond to emerging illnesses? The chapters in this section of the volume explore the range of political, economic, and cultural factors that shape institutional responses to disease.

The response of public health institutions to childhood lead poisoning, described in this chapter by Christian Warren, and tuberculosis, discussed in the next chapter by Sandy Smith-Nonini, reveals the ways in which institutional responses to a health problem may be driven as much by changes in the wider political environment as by the epidemiology of the problem.

Warren's chapter illustrates two other themes that run through the cases in this volume. First, the history of childhood lead poisoning reinforces the importance of social class in shaping public health responses. For most of the twentieth century, lead poisoning was perceived to be a problem of the urban poor, living in houses containing lead paint. As such, it drew little public health attention outside local health departments and urban hospitals. Childhood lead poisoning became a national public health issue only when it became part of a larger social movement—the War on Poverty—in the 1960s or when, in the 1970s and 1980s, it was seen as a problem of atmospheric pollution associated with automobile emissions and thus a threat to all Americans. When the War on Poverty ended, replaced by more conservative political agendas, and when lead was removed from gasoline, lead poisoning resettled among the poor and lost much of its visibility.

Second, the history of childhood lead poisoning, like the history of Lyme disease, described in Chapter 11, demonstrates that efforts to understand the public health responses to emerging illnesses require an examination of the interplay of numerous institutions. Although the Centers for Disease Control and Prevention played a central role in combating childhood lead poisoning, efforts to respond to the problem also involved the Environmental Protection Agency, the Department of Housing and Urban Development (HUD), medical associations, local public health offices, and the lead-manufacturing industries.

---

The history of lead poisoning might make it an odd candidate for inclusion in a volume on emerging illnesses. After all, observers since antiquity have described lead's symptoms, noted epidemics of occupational lead poisoning, and warned of the dangers of lead-adulterated foods and beverages.[1] In the early twentieth century, occupational lead poisoning frequently provoked public furor and reformers' outrage. Lead poisoning, it would seem, finished emerging long ago.

On the other hand, the recent history of childhood lead poisoning reveals much common ground with other emerging illnesses: for many years it remained practically invisible, its diagnosis difficult and its definition contested, the sickness and death it caused attributed to other diseases. Presumed to be a disease of poverty and of race, childhood lead poisoning remained stigmatized— and underresearched. This combination of indeterminate diagnosis and stigma delayed appropriate institutional responses until the late 1960s, when community activists mobilized local resources and forged alliances with medical professionals to make childhood lead poisoning, which they christened "the Silent Epidemic of the Slums," a major thrust of community action programs. Often they had to overcome indifference and ossification at city hall and hospitals. Always they faced critical shortages of money and social capital, even after the early 1970s, when the federal government made lead poisoning a health priority.

Although the "silent epidemic" can be studied historically as an illness that successfully "emerged," the definitions of childhood lead poisoning changed radically in the late 1970s, prompting a reemergence of childhood lead poisoning in a markedly different form—one that threatened children from all classes and ethnicities. This reconceived threat demanded action on two fronts: reducing universal exposures and implementing universal screening. Ironically, the campaign for universal screening foundered on the success of programs to reduce exposure. By the end of the 1990s, childhood lead poisoning was once again widely portrayed as an exclusively inner-city problem.

This chapter examines the very different institutional responses these two episodes of childhood lead poisoning provoked. Lead's ubiquity in consumer products and the environment eventually required the efforts of many institutions not chiefly concerned with health, and for this reason agencies such as HUD and the EPA appear in this chapter. However, its chief subjects are those institutions that most directly addressed the problems of childhood lead poisoning, especially local public health departments, the CDC and its federal predecessors, hospitals and medical associations, antilead activist organizations, and the lead industry itself. Examining this shifting cast of institutional characters throughout the second half of the twentieth century reveals something of the dynamic interrelations that shape the natural histories of emerging illnesses—interrelations among medical and epidemiological expertise, activism, governmental authority and will, and public perception of both the nature of the threat and the population most "at risk."

At the turn of the century, lead poisoning was considered exclusively an occupational hazard. Childhood lead poisoning was all but invisible, its toll in morbidity and mortality ascribed to a menagerie of infectious "diseases of filth." As lead-using industries adopted occupational hygiene programs, rates of occupational lead poisoning fatalities plummeted. In a few cities, interaction between occupational researchers, public health departments, and pediatricians increased awareness of childhood lead poisoning.[2] Until the 1930s, children accounted for fewer than 10 percent of reported deaths from lead poisoning (but bear in mind that these statistics gauge only perceptions, not extensive epidemiological data). This percentage doubled during the Great Depression, and by the postwar years children under five accounted for about 30 percent of reported lead poisoning deaths.[3] The public and professional perceptions of the day defined childhood lead poisoning as a disease of poverty, another troubling problem among the many that blighted the nation's growing population of urban poor.

As American physicians began finding American children dying from lead paint, it became obvious that childhood lead poisoning was not uncommon. Public health officials redefined childhood lead poisoning as a social problem caused by bad housing, "backward" children, and their "ignorant" parents: no one ate paint unless he or she was starved owing to poverty or deficient in culture, either of which might predispose to the *real* cause of childhood lead poisoning—the eating disorder called pica, after the Latin name for the omnivorous magpie. The strong association between poverty and plumbism meant that detecting and preventing lead poisoning would come to be part of the mandate of those agencies whose primary duties involved practical action on the problems of the poor: city health departments and large urban hospitals where lead-poisoned children showed up every year. Decades before the federal government began coordinating screening and treatment programs, local institutions made considerable progress in identifying cases and reducing the rates of fatal lead poisoning.[4]

Since the early 1930s, Baltimore led the nation in locating, studying, and combating childhood lead poisoning. Through the activism of Huntington Williams, Baltimore's commissioner of health, the Baltimore Health Department developed strong ties with researchers at the Johns Hopkins University and the University of Maryland, where pediatricians had demonstrated continuing interest in lead poisoning since about 1910.[5] As early as 1936, the health department routinely sent blood samples from children to Maryland's Bureau of Occupational Diseases for lead testing. In most years, the ratio of confirmed cases to patients tested remained close to 25 percent. In 1942, Baltimore case-finding efforts turned up nearly one-quarter of the childhood lead poisoning cases reported in the United States, earning Baltimore the erroneous distinction as the nation's capital for childhood lead poisoning.[6]

Baltimore's efforts quickly caught the attention of the Lead Industries Association (LIA), a voluntary trade association comprising most of the nation's lead producers and industrial consumers. The LIA's health-related activities had traditionally centered around occupational hazards, but concerns with pediatric plumbism had increased over the 1930s and 1940s. In 1949, the LIA contracted with the Johns Hopkins Department of Physiological Hygiene to analyze all the reported cases of childhood lead poisoning in Baltimore between 1948 and 1949. The study confirmed that the sixty cases included nine "probables" and fifty-one "positive" cases; eight children in the latter group had died. Lead paint was clearly implicated in all the cases.[7] If the LIA intended this study to fend off unfavorable publicity, its plan backfired. The association's director of health and safety, Manfred Bowditch, wrote to Harvard physiologist Joseph Aub for help: "I hope you may find the time to discuss with me some of the matters referred to. . . . These young Baltimore paint eaters are a real headache and I need all the advice I can get."[8]

A second study coming out of Baltimore's program would have an even more significant impact on doctors' perceptions of the scope of the problem. A 1956 study of children with no reported symptoms of lead poisoning revealed that more than 40 percent had blood leads in excess of 50 micrograms of lead per deciliter of blood (μg/dL). Although this level did not put these children in the "positive" category in the mid-1950s, twenty-one of the children had blood leads above the industry standard of 80 μg/dL, and eight developed acute symptoms requiring hospitalization.[9] If two-fifths of "well babies" in Baltimore had this level of lead absorption, what was the cost in chronic illness?

Following Baltimore's lead, many eastern cities established lead poisoning detection and prevention programs. St. Louis instituted free blood-lead testing in 1946, driving its reported cases up by an order of magnitude over the next decade. New York City instituted a small but aggressive case-finding program in 1951, which it reinforced in 1955. In 1958 the city health department began intensive training of hospital staffs and continued to improve case finding, despite limited funding and competition from other pressing public health issues such as drug addiction, venereal disease, and tuberculosis.[10] Chicago began actively searching for lead poisoning in 1953.[11] In Cincinnati, the Kettering Laboratory, funded by the leaded gasoline industry, provided the facilities and expertise to local childhood lead poisoning efforts.[12] In Philadelphia, where the American lead paint industry began in 1804, case finding limped along with little in the way of a concerted effort until 1959, when the Children's Hospital of Philadelphia established screening protocols and initiated a two-year study.[13]

Several factors hampered local health departments' efforts to control childhood lead poisoning. There was no effective treatment prior to the mid-1940s, and accurate diagnosis was costly and invasive; both factors discouraged widespread screening. The best preventive measure—environmental cleansing—was far beyond the scope of city health departments. In 1941 Baltimore's housing ordinance authorized the health department to make landlords abate hazardous conditions in their properties, although landlord resistance made this ordinance essentially unenforceable. In the first year of the new ordinance, the health department ordered one hundred landlords to remove flaking paint from apartments. All the landlords complied with this modest demand, although two did so only after being taken before a judge and fined. The city council diluted the ordinance in 1958, so that it now required lead abatement only in homes in which a child had been poisoned.[14] In 1959 the city of New York banned the sale or application of any paint containing more than 1 percent lead but made little effort to enforce the ban.[15]

Without the regulatory power to remove lead paint from tenement walls, local programs could not hope to eliminate lead poisoning. But those cities with aggressive case-finding programs significantly reduced the number of children who died from it. New treatments developed in the years just after World War II multiplied the benefits of case-finding programs. As each city began looking, the same pattern developed. Initial case finding detected only the most serious cases, giving the impression of an acute but limited problem. As the program reached deeper into the affected community, it identified greater numbers of children who were amenable to treatment. As a result, the fatality rate among lead poisoning cases fell, often dramatically.

Some health practitioners in the late 1950s were prepared to declare victory over lead poisoning. Recalling pediatrician John Ruddock's assertion in 1924 that "the child lives in a lead world," radiologist Paul Woolley summed up progress over the intervening years: "A final phase in the story might be the observation that Ruddock's lead world has been shrinking until now it is almost a microcosm composed of neglected surfaces from which paint is crumbling and peeling." Lead poisoning remained a persistent problem, to be sure, but one for which adequate measures had been taken. Lead poisoning was simply one of many problems of the ghetto and was likely to get the same attention that other "ghetto problems" received. As one researcher had concluded in 1940, "like the poor, lead poisoning is always with us."[16]

Many pediatricians who began their careers in the 1960s learned their trade in hospitals in which, year after year, doctors and nurses fought—often in vain—to save lead-poisoned children. These doctors would forge strong alliances with politicians, lawyers, and the communities whose knowledge of lead poisoning came from personal experience—usually of loss. Lead poisoning remained a "disease of poverty," but in the years of civil rights, the Great Society, the Occupational Safety and Health Administration (OSHA), and Rachel Carson, the fact that most lead victims were poor or nonwhite or laborers would be an increasingly inadequate excuse for neglect.

In these times, lead poisoning was seen less as the result of a sick or backward child with pica than as direct proof that poverty kills. The lead-poisoned child became a sympathetic figure, a capacitor for collecting a charge of moral outrage. It is critical to note that this change was not accompanied by significant improvements in treatment or better statistics about the number of poisoned children or their demographics. The same children as before were presumed to be at risk: poor children, usually children of color, living in terrible housing; but the nation's attitude toward those children had changed, as had the attitudes of the communities suffering the most from lead poisoning.

The War on Poverty and Great Society programs directed unprecedented resources at the problems of poverty and poor housing, and lead poisoning came to be seen as a fundamental but treatable symptom of poverty. But lead poisoning control was not a Great Society program; to suggest so reverses the flow of influence. Federal money eventually sustained the movement, but the impetus for change ran from the community to the city and beyond. This local initiative stemmed from the interactions of several distinct groups, each working on the problem from its own direction but in remarkable concert. Local public health officials, often portrayed at the time as entrenched or insensitive bureaucrats, were not deaf to the call to reduce lead poisoning. As a consequence of the growing poison control movement, most city public health departments had poison control officers, who naturally played an important role in administering lead poisoning programs. Critical in devising these plans was a new generation of doctors who often trained in city hospitals or took jobs in federally funded community health centers where there was plenty of exposure to childhood lead poisoning.

Two groups of professionals outside the realm of public health also supplied needed expertise and political energy. Scientific and medical advocacy groups such as the Scientists' Institute for Public Information and Physicians for Social Reform became deeply involved at both the local and national level. Hundreds of lawyers, working alone or in groups like the Welfare Rights Organization, Community Legal Services, and the Massachusetts Advocacy Center, helped individual families fight for safer housing and pushed for state and federal legislation to promote abatement and prevention.[17]

Lead poisoning would not long have remained a compelling issue for all these lawyers, scientists, physicians, and public health officials if not for the new resolve on the part of members of the communities ravaged by childhood plumbism. As one housing activist put it: "We have already been told by the Health Department that no money can be found for a testing program until the black community begins yelling 'Murder.'" And it did. Lead became a focus of inner-city activism, a symbol for all that was wrong with housing and medical care in America's largest cities from Baltimore to Chicago.[18]

Chicago was the second major city, after Baltimore, to institute a broad screening program. A number of researchers there had published statistics on lead poisoning from the late 1950s, and Chicago's hospitals were reasonably alert to symptomatic lead poisoning. Each year between 1959 and 1965, the board of health reported between 154 and 218 cases, but before 1966, the city did little more than collect data. In the summer of 1963, after Chicago hospitals had reported fourteen lead poisoning deaths, a child advocate group called on the governor to declare a state of emergency. Chicago's commissioner of health responded by sending thirty building inspectors to check the homes in which poisoned children lived; landlords were to repaint any homes found to contain lead paint.[19]

Community activists on Chicago's West Side began agitating for an end to this paltry response during the summer of 1964. Representatives from a broad range of Chicago's church and social agencies formed the Citizens Committee to End Lead Poisoning (CCELP).[20] The new organization secured the Chicago Department of Health's cooperation with a trial screening program. The following fall, the cooperative program began, as "a dedicated group of teenagers" canvassed Garfield Park, collecting almost six hundred urine samples. The city tested the samples and administered additional tests to children who tested positive for lead. The program identified four children for chelation therapy and reported their addresses to the Building Commission for inspection. When the trial program ended in November, it had demonstrated that untrained workers could help run a screening program. The board of health planned to employ War on Poverty workers in a citywide screening program beginning in spring of 1966. From November 1965 until the next May, the CCELP continued its program "as a reminder to the city."[21]

The CCELP stopped its operations in the spring of 1966, when Chicago began what would be the largest screening program in any city to date. Over the next twelve months, the Chicago Board of Health tested more than forty thousand urine specimens, and in October 1966 it augmented urine tests with a finger-prick blood test. More than seven hundred children exhibiting no clinical signs of plumbism were identified for treatment.[22] Between 1966 and 1971, Chicago tested more than a quarter of a million children from the city's most distressed areas, and its new lead clinic evaluated or treated almost ten thousand children. Just as case-finding programs brought increased awareness among medical professionals and reduced lead poisoning deaths, screening well children for lead reduced the average blood leads in the screened population. Chicago's case rate plummeted, from 8.5 percent of those screened in 1967 having blood leads more than 50 µg/dL to around 2 percent by the third year.[23]

Although Chicago's screening program was much larger than that any other city had undertaken, it still served only one-fourth of the city's children considered to be at risk.[24] In areas untouched by these programs, lead-poisoned children continued to turn up in emergency rooms with severe headaches or constipation, vomiting, and slipping into convulsive unconsciousness. A second limitation was the board of health's lethargic approach toward eliminating the source of these poisonings. As late as 1972, Chicago had no laws outlawing lead paint, and enforcement of hazardous housing codes was notoriously lax. Community activists and residents would have to rise to form new alliances in the 1970s to take on this very different issue, to treat lead poisoning "as a social disorder rather than a disease which will be remedied through . . . medical care."[25]

In the late 1960s, similar scenarios played out in New York City, Washington, Boston, and other northeastern cities, as activist groups pushed for local action. By 1970, Congress was hammering out the final version of a national law to promote case finding and abatement. But it does not appear that national publicity, or nationally coordinated activism, galvanized support for federal action.[26] Childhood lead poisoning was a local story repeated in many cities across the nation rather than a national issue playing out at the local level. Local media covered lead poisoning, but they emphasized local political issues and the responses of scandalized city bureaucrats. Local action on lead poisoning made good headlines for domestic consumption. But lead poisoning itself, as New York journalist Jack Newfield correctly observed, was a hard sell: "How do you show a *process*, how do you show indifference, how do you show invisible, institutionalized injustice in two minutes on Huntley-Brinkley?"[27]

The path to a federal commitment to fight childhood lead poisoning resembles that taken by local public health agencies. For many years, a small number of specialists within the U.S. Public Health Service had promoted lead programs. In the late 1960s, Jane S. Lin-Fu, a Children's Bureau pediatrician, directed much of the government's efforts, focusing on educating doctors, health departments, and parents. In 1967 the Children's Bureau published Lin-Fu's booklet *Lead Poisoning in Children*. Between 1967 and 1970, the bureau distributed close to thirty thousand copies.[28] Prior to 1971, federal health administrators also set standards for diagnosing and treating childhood lead poisoning, reducing the uncertainty resulting from local authorities adopting their own case definitions. An ad hoc committee convened in 1970 by Surgeon General Jesse Steinfeld established a three-tier standard for diagnosis. Children whose blood leads exceeded 79 µg/dL should be considered "as unequivocal cases of lead poisoning." Since at this level severe symptoms could materialize at any moment, these children should be "handled as medical emergencies" whether or not they showed symptoms. Children whose blood leads were between 50 and 79 µg/dL were to receive further medical evaluation. The standard recommended monthly blood tests and an evaluation of the child's environment. Third, a blood lead between 40 and 60 µg/dL was considered proof of "undue absorption of lead."[29]

Initially, this third level was omitted from the standard. But Lin-Fu lobbied the committee to recognize the growing research literature dealing with the effects of chronic blood leads in this "high normal" range. Opponents on the surgeon general's committee argued that because in some cities almost half the children screened had blood leads in this range, "this new standard would frustrate public health officials with an overwhelming caseload." According to Lin-Fu, the committee chairman reluctantly agreed to adopt this definition in the surgeon general's statement only after Lin-Fu promised "to respond to all the letters of complaint that might be written about the new limit."[30]

On November 7, 1970, the surgeon general released an official statement on childhood lead poisoning, which began: "The U.S. Public Health Service recommends that screening programs for the prevention and treatment of lead poisoning (plumbism) in children include all those who are 1 to 6 years of age and living in old, poorly maintained houses." But Steinfeld offered little hope that federal funding might help local communities implement his recommendations. Indeed, although the draft of a Department of Health, Education, and Welfare (DHEW) booklet to help cities develop programs listed potential sources of federal funds, it cautioned: "The term 'potential sources' cannot be stressed enough. . . . In none of these programs are there specific allocations of monies either for medical case finding or treatment."[31]

Federal categorical funds became available with the Lead-Based Paint Poisoning Prevention Act (LBPPPA) of 1971, legislation introduced by lawmakers from New York City, Philadelphia, and Boston—three cities with strong, vocal antilead activists.[32] The act created three programs. DHEW was to make grants to cities establishing lead paint abatement programs, screening, and treatment programs. HUD was to survey the scope of the lead paint hazard and establish methods for abatement. Finally, the act empowered DHEW to prohibit the use of "lead-based paint" in federally constructed or rehabilitated housing.

Not that DHEW wanted this power; in fact, the department opposed the LBPPPA. Under President Richard Nixon's "New Federalism," the department was not about to support a bill that mandated top-loaded categorical grants to be administered from Washington. The U.S. Public Health Service claimed that it already had the authority to award the grants the new bill mandated. The bill was unneeded, "since a successful program is dependent upon the resolve of States and localities to solve a local problem."[33] Massachusetts senator Edward Kennedy agreed that the USPHS had the authority. "But there is no evidence to indicate that the Public Health Service is prepared to expend any resources to fight this problem." Documents supplied to Kennedy's committee revealed that though the service had spent $45,850 to fund a few demonstration programs in recent years, twenty-seven state, local, and private institutions had requested funds totaling $33 million to establish or enhance lead poisoning programs. "To date," one report concluded, "the Department has been unable to fund any of these applications."[34]

It is important to note that the LBPPPA delegated primary responsibility to DHEW, even that pertaining directly to housing. The House version of the bill divided the tasks more evenly between DHEW and HUD, but Kennedy convinced Congress that his version would produce greater accountability.[35] Nonetheless, DHEW and HUD retained an often unproductive mix of autonomy and overlapping jurisdiction between housing and health issues, creating at the federal level a dynamic similar to that which caused so much friction at the city level, where health and housing departments jostled for tight funding and blamed each other for permitting the *real* cause of lead poisoning to go unchecked. As the federal government assumed greater responsibility for implementing lead poisoning programs, representatives of DHEW, HUD and the EPA would make similar accusations.

Funding the LBPPPA remained a serious problem. At first the Nixon administration ignored it; in 1971, DHEW asked for none of the $10 million authorized.[36] During the next year, when authorizations under the act doubled to $20 million, Nixon appropriated only $6.5 million. By the end of the decade, authorizations and appropriations reached something approaching a balance; but in 1982, the LBPPPA's appropriations were subsumed under the Maternal and Child Health block grant, part of President Ronald Reagan's Omnibus Budget Reconciliation Act.[37]

Still, as was true at the city level, increasing the resources available to lead programs made a tremendous difference. Despite the LBPPPA's glaring defects, it ended decades of halfhearted efforts on the part of federal health agencies that seldom produced more than "little pamphlets" or funds for basic research. In its first decade, the federal government disbursed more than $80 million in LBPPPA funds to local public health agencies for lead poisoning prevention and hazard abatement programs. By the early 1980s, federal funds had established more than 100 lead-testing laboratories; 4 million children had been screened; 250,000 at-risk children were identified for treatment; and 112,000 homes were cleared of toxic paints and plaster.[38] The act's ban on "lead-based" paints had an impact far beyond manufacturers of paints used in government-funded housing. Amendments in 1973 and 1975 lowered permissible lead contents, first from 1 percent to 0.5 percent lead in the dried film, then to 0.06 percent. In 1977, after lengthy legal, political, and scientific conflicts, the Consumer Product Safety Commission (CPSC) extended the ban to interstate commerce.[39]

Despite these successes, federal lead poison control programs served but a fraction of the children at risk. By 1981 only twenty-five states and the District of Columbia were drawing federal funds under the LBPPPA. Cities that did have screening programs concentrated their efforts in the oldest, most economically stressed communities. Although plentiful evidence showed that significant percentages of children living outside large cities had high blood-lead levels, suburban and rural children fell beyond the scope of most screening programs, as did children outside the industrial Northeast. East coast and midwestern states accounted for more than half of the states using federal funds for lead screening in 1981. Southern and western states were particularly slow to institute screening programs.[40] Given this often incomplete effort, it is puzzling that in the decade following the onset of federal lead poisoning programs, fatal childhood plumbism all but disappeared, and the number of children with high blood levels dropped dramatically.

It is now clear that the elimination of leaded gasoline from American cars was of signal importance. Leaded gasoline was introduced in the 1920s to increase fuel efficiency and automotive power.[41] Beginning in the mid-1960s, awareness of the tremendous scale and possible dangers of environmental lead pollution increased, and by 1980, federal restrictions reduced the amount of lead consumed annually in gasoline to the lowest level since 1951.[42] The Second National Health and Nutrition Examination Survey (NHANES II), conducted between 1976 and 1980, provided graphic evidence of the potential health benefits of deleading gasoline. The survey found that average blood-lead levels had declined by nearly 40 percent over the entire period, from 14.6 to 9.2 μg/dL. No other factor, such as reduced lead in tin cans and pesticides, improved screening and treatment, or lead paint abatement, bore any significant correlation to this nationwide detoxification.[43]

The elimination of leaded gasoline produced a healthy margin of safety in preventing "lead paint poisoning," at least as the standards of the early 1960s defined it. Of course, by the late 1970s those standards were no longer in place. As the amount of lead in the environment and in human blood decreased, clinicians found more evidence of the effects of low-level exposure. Faith in the old thresholds eroded, and with each reduction in "acceptable" blood leads, further studies revealed ever more subtle and widespread damage below the new threshold, eventually calling into question the very notion of a threshold for damage from lead.[44]

J. Julian Chisolm, a pediatrician at the Johns Hopkins Hospital, was the foremost childhood lead poisoning researcher in the early 1970s.[45] Chisolm and others were very interested in the effects of subacute, or asymptomatic, lead absorption. Their research found consistent dose-response relationships between blood damage and blood-lead levels well below those associated with clinical symptoms, prompting the first two downward revisions of the USPHS's definitions of "undue lead absorption," from 60 µg/dL to 40 in 1971, and to 30 µg/dL in 1975.[46]

While Chisolm's work focused on the physiological markers of low-level lead exposure, Pittsburgh psychiatrist Herbert Needleman exposed the mental and psychological costs of that exposure. In 1979 Needleman published a study implicating subclinical lead exposure in diminished IQs and school performance among Boston-area children. A follow-up study eleven years later retested 132 of the original subjects, finding among those whose childhood lead burden had been the highest "a seven-fold increase in failure to graduate from high school, lower class standing, greater absenteeism, impairment of reading skills . . . and deficits in vocabulary, fine motor skills, reaction time and hand-eye coordination."[47]

From an earlier year's perspective, the NHANES II findings that very few Americans had blood leads higher than 70 µg/dL and that average blood-lead levels for children were "only" 16 µg/dL suggested that lead poisoning would soon be vanquished. But according to the standards of the day, NHANES II conveyed some very bad news indeed. The study estimated that 4 percent of American children between the ages of six months and five years —approximately 675,000 children—had blood lead levels higher than 30 µg/dL. This estimate far exceeded projections from local screening programs.[48] The children whose blood leads exceeded the old standards best fit the conventional image of a poor child ingesting old flaking paint in substandard housing. But under the new categories, they were merely the worst off of a vast population of lead-poisoned children whose members could be found almost anywhere.[49]

As lead removal programs proliferated throughout the 1970s, so too did concern over the health consequences of the dusty work of abatement. "Urban lead miners" frequently exhibited sharp increases in blood-lead levels, and since no one suspected workers of nibbling on paint chips as they worked, lead dust must have been the cause of the workers' plumbism. Other researchers investigated the lead levels in children living in or near homes undergoing lead abatement.[50] The amount of lead in the dust on children's hands had been worrisome—but not alarming—when authorities labeled a child's blood lead of 39 µg/dL "below clinical concern." As low-level lead exposure became a concern, however, the fine particles of dust assumed gargantuan significance.[51]

Reconfiguring the threat from lead—from flakes ingested by children with "perverted appetites" to deadly dust breathed by innocents—created an entirely different moralization of childhood lead poisoning. For most Americans, assigning blame for ghetto children eating the paint they picked off baseboards and windows had been relatively easy. They could cite villainous landlords, "defective" children, parents mired in a "culture of poverty," or urban blight, factors that seemed safely remote from them. Dust was different. For one thing, exposure to lead dust required no dietary or behavioral pathology. More significant, for a generation of Americans raised on the hygienic visions of June Cleaver and Mr. Clean, dust was more than merely a vexing nuisance; the new allegations that this dust threatened their children's mental health only multiplied parents' worries and guilt.[52] Whereas middle-class professionals previously might have lent moral or electoral support to fight a killer of ghetto children, the reassessed risks of low-level exposure cast lead's threatening shadow over their bright, healthy offspring, making the push for screening and abatement a personal fight.

This new image of childhood lead poisoning should have nearly assured adoption of universal screening. In 1982, the USPHS recommended that all community child-health programs screen all children from one to five years of age in conjunction with other health screening activities.[53] The CDC's policy in 1985 recognized universal screening as an ideal goal, to be met as part of a proposed screening program for iron deficiency; however, the policy stated that resources should be targeted toward enrolling "the maximum number of children in high-risk groups." By 1991, the CDC's goals all but embraced universal screening: "Our goal is that all children should be screened, unless"—and here was the condition that more than any factor has prevented universal screening—"it can be shown that the community in which these children live does not have a childhood lead poisoning problem."[54]

Unfortunately, in the 1980s, national lead prevention policy and national health care politics were moving in opposite directions. Universal lead screening would require millions of federal dollars designated exclusively for such a program—the very type of "categorical" program to be eliminated under Reagan's campaign to return spending authority to the states. Instead, federal money was to be allocated through block grants, which each state could spend as local exigencies demanded. In 1981 Congress placed lead prevention programs under the new Maternal and Child Health block grant. The funding cuts accompanying block grants threatened the survival of existing state and local lead-screening programs, even in states with large "at-risk" populations. In a typical example, St. Louis—ranked third among the ten cities with the worst childhood lead poisoning problems—fired half of its lead control staff in 1982, reducing the number of children screened by 29 percent, four thousand fewer than in 1981. If block grants squeezed the budgets of existing programs, they discouraged other cities from starting new ones, eliminated incentive for rural states to embark on lead poisoning prevention, and made universal screening a near impossible undertaking.[55]

Still, lead's opponents pressed on. The reconstitution of childhood lead exposure as a pandemic thief of intelligence demanded their persistent efforts, no matter the political climate. And the ubiquity of the environmental hazard demanded more than one line of attack. Noting the tremendous health benefits resulting from the leaded gasoline phasedown, these activists successfully employed the new definitions of lead hazards to further reduce additional sources of bioavailable lead in the environment. In 1986, the EPA lowered the permissible lead contamination of drinking water by a factor of ten, prompting passage of federal laws sharply curtailing lead in solder, plumbing fixtures, and drinking fountains. The amount of lead in foods canned in the United States plummeted in the early 1990s, after the FDA secured a voluntary industrywide ban on lead solder for cans.[56]

Ironically, reductions in new sources of environmental lead made the push for universal screening even more difficult. Although both the CDC and the American Academy of Pediatrics (APP) endorsed universal screening in the early 1990s, opponents began publishing their doubts about low-level toxicity and the cost-effectiveness of testing all children. With reductions in bioavailable lead, and with average blood leads falling, the rhetorical power of lead as a universal menace began to wane.[57]

Much of the criticism of universal screening came from health professionals in western states. Oakland, California, pediatrician Edgar Schoen, of the Kaiser Permanente Medical Care Program (then the nation's largest HMO), assailed the CDC's strategic plan for universal screening, treatment, and abatement as too costly and questioned the evidence of neurological effects at low levels of lead exposure.[58] Others argued that universal screening was a waste of money in regions, such as much of the West, where housing stock was relatively new. George Gellert of the Orange County, California, Health Care Agency reported that of 5,115 low-income children tested there, only 371 (7.25%) had blood-lead levels above 10 μg/dL, and only 6 (0.12%) had blood leads higher than 25 μg/dL. Gellert estimated the costs of locating the six children with the highest blood levels at $19,139 per child, money he argued would be better spent on "more compelling child health problems, such as immunization, nutrition, or injury."[59]

Schoen's and Gellert's views on screening probably resembled those of most American pediatricians, who, if they did screen their patients for lead, found few children with elevated levels and worried that the low risks did not warrant the anxiety, pain, and expense of the venipuncture required for the most reliable lead test. In early 1994, the American Academy of Pediatrics published a forum on recent lead poisoning risk assessment studies in its journal, *Pediatrics.* In his concluding commentary, Stanford University Pediatrics chief Birt Harvey declared that it was "a time for action," but by this he meant that it was time to revise the 1991 CDC screening guidelines before more precious funds were wasted. "With government and industry demanding limitations on the cost of health care," he warned, "expenditures for new or expanded services must be justified, and necessary services must be provided in the most cost-effective manner."[60] Simply asking parents the right questions was as effective as the frequently inaccurate finger-prick test for identifying those children whose blood should be tested further. And why spend millions of dollars to label as "poisoned" thousands of children with blood-lead levels between 10 and 25 µg/dL when the medical and environmental interventions for such low levels were of such questionable value?

In 1994, the National Center for Health Statistics and the CDC released the blood-lead findings from the Third National Health and Nutritional Examination (NHANES III). The new data revealed that the low prevalence of elevated blood leads found in western cities more closely resembled national norms than the epidemiological studies from eastern lead-belt cities. A decade of aggressive restrictions on new sources of bioavailable lead had brought a 78 percent drop in average blood-lead levels since the midpoint of the NHANES II study, from 12.8 to 2.8 µg/dL. The change in childhood exposure was even more remarkable: NHANES II had estimated that almost 90 percent of American children had blood leads above 10 µg/dL; by the late 1980s and early 1990s, the percentage had fallen to under 9 percent.[61] Voices quickly rose to declare victory and cast off the heavy veil of "plumbophobia." In a 1995 issue of *Atlantic Monthly,* science journalist Ellen Ruppel Shell hailed America's "triumph over lead," calling it "a stunning example of the strength of activism over vested interests."[62]

The heartening news in the NHANES III reports all but guaranteed that the CDC's 1991 call for universal screening would fail. In 1995, the CDC began reviewing its recommendations for screening and abatement, bowing to public pressure, economic realities, and a shifting consensus within the agency regarding whether universal screening would—or should—ever be implemented. Late in 1997, the CDC published new guidelines, calling for "targeted" screening. Case finding and screening should focus on children in high-risk areas, children in high-risk groups, or children identified as at risk "by means of a personal risk questionnaire." Local and state authorities, not CDC officials, were to set priorities and criteria for blood screening.[63] The CDC asserted that the guidelines were in no way a retreat from screening children, that in fact, screening would increase if localities followed the new guidelines.[64]

The new CDC guidelines are sensible, sound, and well reasoned, based as they are on credible epidemiological data, humane principals, and sophisticated cost-benefit analyses. Unfortunately, their particularist view of the problem, abetted by the CDC's determination to turn authority over to local agencies, threatens to turn the clock back to the early 1980s, when only knowledgeable professionals and poor people living in enlightened towns got their children tested as a preventive measure. Even if the guidelines are perfectly implemented, many children will fall through the cracks: the child who moves to a "lead-free" town from a region with older housing or other sources of high exposure levels; the child whose family uses lead-containing folk remedies or cooks with improperly fired lead-glazed ceramics; children from towns where tight pockets or widely scattered sprinklings of older housing exist within "safe" areas. Presumably, these are acceptable losses, their risks balanced against the far greater cost-effectiveness of targeted screening.

Far greater is the risk that the new guidelines' abandonment of universalism and their acceptance of race specificity will perpetuate the status quo—that is, that the number of screening programs nationwide will stay roughly the same. Without a federal commitment to finding every lead-poisoned child, beleaguered communities with the worst problems will face the same old challenge. Even in cities like Newark, which has one of the highest percentages of lead-poisoned children in the country, letting down the federal guard may hamper local efforts at sustaining, let alone enhancing, much needed programs. In all probability, childhood lead poisoning will remain a defining environmental justice issue for the foreseeable future.

In 1994, EPA toxicologist Lynn Goldman and Joseph Carra wrote that "the US strategy must begin to focus more than ever on poor, nonwhite, and inner-city children."[65] But though treating lead poisoning as a disease of the poor makes good epidemiological sense, historical precedent suggests it is a dangerous choice. Substantial progress in reducing lead poisoning has been made only when lead was considered a common risk. Seventy years ago, when conventional wisdom held that lead poisoning was confined to the workplace, "universal" blood testing within lead factories prompted simple but effective environmental improvements, improvements few had been willing to make when only the "wops and hunkies" in a factory were thought to get poisoned.[66] The concept of lead as a "ubiquitous threat," as posited by geochemist Clair Patterson in the 1960s and borne out by Herbert Needleman and others since the 1970s, prompted the dramatic reductions in lead-contaminated air, food, and water that brought average blood-lead levels in the United States to their historic low.[67]

Although it is true that poor children shared in the universal reductions in environmental lead over the past twenty years, they remain at greatest risk of exposure to the lead that remains. And with no comparable improvement in universal access to health care, a return to particularism amplifies the dangers of that exposure. This brief survey of institutional responses to childhood lead poisoning makes it clear that without centralized programs for case finding and eradication—not merely management—environmental illnesses, like other emerging and reemerging diseases, will recede to the local pockets where they were first discovered. Their burden will fall invisibly upon the poor. Sixty years ago, a researcher said, "Like the poor, lead poisoning is always with us." Perhaps she was right—especially if current policies act to silence childhood lead poisoning once more.

## NOTES

This chapter is based on material in my book *Brush with Death: A Social History of Lead Poisoning* (Baltimore: Johns Hopkins University Press, 2000).

1. Jerome O. Nriagu, *Lead and Lead Poisoning in Antiquity*, Environmental Science and Technology Series, ed. Robert L. Metcalf and Werner Stumm (New York: John Wiley and Sons, 1983). In 1786, Benjamin Franklin remarked, "The opinion of this mischievous Effect from lead is at least above Sixty Years old, and you will observe with Concern how long a useful truth may be known and exist, before it is generally receiv'd an practis'd on." Franklin to Benjamin Vaughan, July 31, 1786, quoted in Carey P. McCord, "Lead and Lead Poisoning in Early America: Benjamin Franklin and Lead Poisoning," *Industrial Medicine and Surgery* 22 (September 1958): 393–99.

2. On lead poisoning in the occupational setting, see Christian Warren, *Brush with Death: A Social History of Lead Poisoning* (Baltimore: Johns Hopkins University Press, 2000), and Christopher Sellers, *Hazards of the Job: From Industrial Disease to Environmental Health Science* (Chapel Hill: University of North Carolina Press, 1997); on early city public health responses to childhood lead poisoning, see Elizabeth Fee, "Public Health in Practice: An Early Confrontation with the 'Silent Epidemic' of Childhood Lead Paint Poisoning," *Journal of the History of Medicine* 45 (1990): 570–606; on the early federal response, see Warren, *Brush with Death*, 134–51; on the lead industry's campaigns to control the scientific and regulatory processes involving lead, see Gerald Markowitz and David Rosner, *Deceit and Denial: The Deadly Politics of Industrial Pollution* (Berkeley: University of California Press, 2002).

3. Department of Commerce, Bureau of the Census, *Mortality Statistics*, 1931–50 (Washington, D.C.: GPO, 1935–50).

4. See Warren, *Brush with Death*, 152–77.

5. See Fee, "Public Health in Practice"; Kenneth Daniel Blackfan, "Lead Poisoning in Children with Especial Reference to Lead as a Cause of Convulsions," *American Journal of the Medical Sciences* 153 (June 1917): 877–87.

6. Baltimore, *Annual Report of the Department of Health* (Baltimore: n.p., 1941–51); John McDonald and Emanuel Kaplan, "Incidence of Lead Poisoning in the City of Baltimore," *JAMA* 119 (1942): 870–72, which reported forty-nine fatal pediatric lead poisonings between 1931 and 1940, approximately 25% of those reported in the nation.

7. Anna M. Baetjer to Manfred Bowditch, October 19, 1949, Joseph C. Aub, Office Files, box 2: Benison-Bowman Folder "Bowditch, Manfred [folder 3], 1948–1960" (hereafter cited as "Bowditch folder"), Harvard Medical School Archives GA-4, Harvard Medical School, Boston, Mass.

8. Manfred Bowditch to Joseph Aub, December 21, 1949, "Bowditch folder."

9. J. Edmund Bradley, Albert E. Powell, William Niermann, Kathleen R. McGrady, and Emanuel Kaplan, "The Incidence of Abnormal Blood Levels of Lead in a Metropolitan Pediatric Clinic," *Journal of Pediatrics* 49 (1956): 1–6, data from table, p. 5.

10. For St. Louis, see J. Earl Smith, B. W. Lewis, and Herbert Wilson, "Lead Poisoning: A Case Finding Program," *American Journal of Public Health* 42 (April 1952): 417–21, and Benjamin Lewis, Richard Collins, and Herbert Wilson, "Seasonal Incidence of Lead Poisoning in Children in St. Louis," *Southern Medical Journal* 48 (March 1955): 298–301. For New York City, see Mary Culhane McLaughlin, "Lead Poisoning in Children in New York City, 1950–1954," *New York State Journal of Medicine* 56 (December 1, 1956): 711–14, and "Two Health Problems, One Solution," *Bulletin of the New York Academy of Medicine* 46 (June 1970): 454–56; Harold Jacobziner, "Lead Poisoning in Children: Epidemiology, Manifestations, and Prevention," *Clinical Pediatrics* 5 (May 1966): 277–86; Gary Eidsvold, Anthony Mustalish, and Lloyd Novick, "The New York City Department of Health: Lessons in a Lead Poisoning Control Program," *American Journal of Public Health* 64 (October 1974): 956–62.

11. See David Jenkins and Robert Mellins, "Lead Poisoning in Children: A Study of Forty-Six Cases," *Archives of Neurology and Psychiatry* 77 (1957): 70–78, and Joseph Christian, Bohdan Celewycz, and Samuel Andelman, "A Three-Year Study of Lead Poisoning in Chicago," *American Journal of Public Health* 54 (August 1964): 1241–51.

12. Records of many of these analyses appear in box 19, folder "Distribution—Children," Robert A. Kehoe Archives, Cincinnati Medical Heritage Center, Cincinnati.

13. "Study Project, the Detection, Treatment, and Prevention of Lead Poisoning in a Children's Hospital Out-patient Department," Joseph Stokes Jr. Papers, Archive B:St65p, American Philosophical Society Library, Philadelphia.

14. Fee, "Public Health in Practice," 586–87.

15. Prepared statement by Vincent Guinee, director, Bureau of Lead Poisoning Control, Department of Health New York City, U.S. Congress, Senate, Committee on

Labor and Human Resources, *Lead-Based Paint Poisoning: Hearing before the Subcommittee on Health*, 91st Cong., 2d sess., November 23, 1970, p. 277; David Bird, "Test of Paint Finds 10% Has Illegal Lead Content," *New York Times*, July 24, 1971, 1.

16. John C. Ruddock, "Lead Poisoning in Children with Special Reference to Pica," *JAMA* 82 (1924): 1682; Paul Woolley, "Lead Poisoning during Infancy and Early Childhood," *American Journal of Roentgenology* 78, no. 3 (September 1957): 548; Nell Conway, "Lead Poisoning—from Unusual Causes," *Industrial Medicine* 9 (September 1940): 471.

17. Barbara Berney, "Round and Round It Goes: The Epidemiology of Childhood Lead Poisoning, 1950–1990," *Milbank Quarterly* 71 (1993): 13–14; for the Scientists' Institute for Public Information, see *Scientist and Citizen* 10 (April 1968); Richard W. Clapp, "The Massachusetts Childhood Lead-Poisoning Prevention Program," in Herbert Needleman, ed., *Low Level Lead Exposure: The Clinical Implications of Current Research* (New York: Raven Press, 1980), 285–92.

18. Paul Du Brul, quoted in Diana R. Gordon, *City Limits: Barriers to Change in Urban Government* (New York: Charterhouse, 1973), 28–29; Jonathan M. Stein, "An Overview of the Lead Abatement Program Response to the Silent Epidemic," in Needleman, *Low Level Lead Exposure*, 279–84.

19. David Elwyn, "Childhood Lead Poisoning," *Scientist and Citizen* 10 (April 1968): 53–57. The trial screening is described in Christian, Celewycz, and Andelman, "Three-Year Study of Lead Poisoning in Chicago"; "Lead Paint in Chicago," *Time*, August 9, 1963, 36–37.

20. Ann Koppelman Simon, "Citizens vs. Lead in Three Communities: 1. Chicago," *Scientist and Citizen* 10 (April 1968): 58–59.

21. Simon, "Citizens vs. Lead," 59; Berney, "Round and Round It Goes," 12; Simon, "Citizens vs. Lead," 59; Elwyn, "Childhood Lead Poisoning," 56.

22. Nancy Hicks, "Drive to Stop Lead Poisoning Begins," *New York Times*, October 10, 1970, 9.

23. Henrietta K. Sachs, "Effect of a Screening Program on Changing Patterns of Lead Poisoning," *Environmental Health Perspectives* 7 (May 1974): 41–45.

24. Phillip R. Fine, Craig W. Thomas, Richard H. Suhs, Rosellen E. Cohnberg, and Bruce A. Flaschner, "Pediatric Blood Lead Levels: A Study in 14 Illinois Cities of Intermediate Population," *JAMA* 221 (1972): 1475–79; Den Elger, "7 Suburbs Listed as Having Many with Lead Poisoning," *Chicago Tribune*, May 25, 1972.

25. "Lead Paint Law Delay Denied," *Chicago Tribune*, June 28, 1972, sec. 1, p. 3; "Two City Aides Defend Lead Detection System," *Chicago Tribune*, August 31, 1972, sec. 4a (N), p. 10. Marcia Opp, "Increase of Lead Poison Laid to Collusion of Slum Landlords," *Chicago Tribune*, May 4, 1972, sec. 2, p. 7.

26. The *Reader's Guide to Periodical Literature* listed only eleven feature stories on pediatric lead paint poisoning in the two years prior to the enactment of federal legislation.

27. Jack Newfield, "Silent Epidemic in the Slums," *Village Voice*, September 18, 1969, 3.

28. The government's distribution of Lin-Fu's tract paled next to the efforts of the Lead Industries Association, which distributed sixty-one thousand copies of the pamphlet as part of its free booklet *Facts about Lead and Pediatrics*. U.S. Department of Health, Education, and Welfare, Social and Rehabilitation Service, Children's Bureau, *Lead Poisoning in Children*, by Jane S. Lin-Fu, Children's Bureau Publication no. 452 (Washington, D.C.: GPO, 1967); distribution figures for both the department and the LIA appear in U.S. Congress, Senate, Committee on Labor and Human Resources, *Lead-Based Paint Poisoning*, 188–89.

29. For example, New York and Baltimore defined a "case" of lead poisoning as a child whose blood lead exceeded 60 µg/dL, Chicago reported children with 50 µg/dL as positive cases; others probably clung to the traditional industrial standard of 80 µg/dL; U.S. Department of Health, Education, and Welfare, "Medical Aspects of Lead Poisoning," Statement of the Surgeon General, November 8, 1970, in U.S. Congress, Senate, Committee on Labor and Human Resources, *Lead-Based Paint Poisoning*, 45–49.

30. Hoping to fend off professional opposition, Lin-Fu wrote a review article in the *New England Journal of Medicine*, and the Surgeon General's Office never received "a single letter of complaint" (Jane S. Lin-Fu, "Modern History of Lead Poisoning: A Century of Discovery and Rediscovery," in Herbert Needleman, ed., *Human Lead Exposure* [Boca Raton, Fla.: CRC Press, 1992], 34–43). For the state of early research into chronic "low-level" lead exposure, see J. Julian Chisolm, "Chronic Lead Exposure in Children," *Developmental Medicine and Child Neurology* 7 (1965): 529–36, and "Lead Poisoning," *Scientific American* 224 (February 1971): 15–23; Lin-Fu, "Modern History of Lead Poisoning," 34–43.

31. U.S. Department of Health, Education, and Welfare, "Medical Aspects of Lead Poisoning"; U. S. Department of Health, Education, and Welfare, Public Health Service, Environmental Health Service, Environmental Control Administration, Bureau of Community Environmental Management, "Control of Lead Poisoning in Children," Prepublication draft, July 1970, p. VI-1, printed in U.S. Congress, Senate, Committee on Labor and Human Resources, *Lead-Based Paint Poisoning*, 51–174, p. 135.

32. The lead poisoning bills' sponsors were William Ryan, representative from Manhattan; William Barrett, representative from Philadelphia; Massachusetts senator Edward Kennedy; and Pennsylvania senator Richard Schweiker; Susan Bailey, "Legislative History, P.L. 91–695," *United States Code: Congressional and Administrative News*, 91st Cong., 2d sess., 1970 (St. Paul, Minn.: West Publishing, 1971), 3:6130–39; *Congressional Quarterly Almanac, 1970* (Washington, D.C.: Congressional Quarterly, 1971), 590; for hearings on the bill, see U.S. Congress, Senate, Committee on Labor and Human Resources, *Lead-Based Paint Poisoning*; On Nixon signing bill, see *New York Times,* January 15, 1971, 10.

33. John Hanlon, deputy administrator of the Public Health Service's Environmental Health Service, testimony in U.S. Congress, Senate, Committee on Labor and Human Resources, *Lead-Based Paint Poisoning*, 188.

34. Ibid.; Kennedy quote on p. 175. The documents also noted a number of research

programs funded by the Bureau of Occupational Safety and Health and animal and clinical studies conducted by the NIH and other research bodies (p. 179). A Boston public health official testified that when he discussed his department's plans with a DHEW administrator, the department was discouraged from applying because of "the stringency of funding." Dr. Jonathan Fine, deputy commissioner, Boston Department of Health and Hospitals, testimony in ibid., 237.

35. *Congressional Quarterly Almanac, 1970,* 590; U.S. Congress, House, "Conference Report, Lead-Based Paint Poisoning Prevention Act," Report no. 91-1802, 91st Cong., 2d sess.

36. Jack Newfield, "Let Them Eat Paint," *New York Times,* June 1, 1971, 45.

37. "Legislative History of the Lead-Based Paint Poisoning Prevention Program," in U.S. Congress, House, Committee on Energy and Commerce, *Lead Poisoning and Children: Hearing before the Subcommittee on Health and the Environment,* 97th Cong., 2d sess., December 2, 1982, 3–12.

38. Lin-Fu, "Undue Lead Absorption and Lead Poisoning in Children—An Overview," in *Proceedings of the International Conference on Heavy Metals in the Environment, Toronto, Canada* (Toronto: Public Institute of Environmental Studies, 1975), 3:29–52; George Hardy, assistant director, Centers for Disease Control, U.S. Congress, House, Committee on Energy and Commerce, *Lead Poisoning and Children,* 28.

39. The 1973 amendment to the LBPPPA (PL 93-151) lowered the permissible content of lead in dried films to 0.5% and assigned the CPSC to determine whether a there was a safe level somewhere below 0.5% but above 0.06% (the lowest level test equipment could detect). The commissioner announced that 0.5% was safe, but the Consumer's Union and the Philadelphia Citywide Coalition Against Childhood Lead Poisoning sued the CPSC to force a reevaluation. Before the case was settled, another amendment to the LBPPPA (via PL 94-317) mandated a full review by the CPSC. This time the commission had to prove that any amount of lead added constituted no "unreasonable risk." Put this way, the National Academy of Science study commissioned by the CPSC was "unable to determine that 0.5 percent lead in paint is safe." The ban was promulgated in late 1977 and went into effect six months later. *Federal Register* 42 (1977): 9405–6, 16445; "Lead Poisoning from Paint: The End of the Road?" *Consumer Reports* 42 (March 1977): 124.

40. Based on DHEW reports of grants. Their data do not cover abatement programs that HUD might have funded; Susan Bailey, "Legislative History of the Lead-Based Paint Poisoning Prevention Program," in U.S. Congress, House, Committee on Energy and Commerce, *Lead Poisoning and Children,* 12; Douglass W. Green, "The Saturnine Curse: A History of Lead Poisoning," *Southern Medical Journal* 78 (January 1985): 48–51.

41. About 160 tons of lead went into gasoline in 1959, as compared with 95 tons in 1949; U.S. Department of the Interior, Bureau of Mines, "Lead," *Minerals Yearbook,* Series: 1920–88 (Washington, D.C.: GPO, 1922–90); for a scathing overview of the history of leaded gasoline, see Jamie Lincoln Kitman, "The Secret History of Lead," *Nation,* March 20, 2000, 11–44.

42. U.S. Department of the Interior, Bureau of Mines, "Lead"; Jack Lewis, "Lead Poisoning: A Historic Perspective," *EPA Journal* 11 (May 1985): 16.

43. NHANES II was conducted by the National Center for Health Statistics. Detailed medical information was collected from a sample population of about twenty-eight thousand; see Kathryn R. Mahaffey, Joseph L. Annest, Jean Roberts, and Robert S. Murphy, "National Estimates of Blood Lead Levels: United States, 1976–1980," *New England Journal of Medicine* 307 (1982): 573–79.

44. Before 1970, many doctors and hospitals used the "threshold" for toxicity established for diagnosing lead poisoning in adult lead workers. This standard defined blood-lead levels higher than 80 μg/dL as dangerous. As late as the early 1950s, Baltimore's lead screening program defined as "possible" lead poisoning cases in which the blood-lead level exceeded 70 μg/dL. On the gradual reduction in threshold values, see Karen L. Florini, George Krumbaar, and Ellen K. Silbergeld, *Legacy of Lead: America's Continuing Epidemic of Childhood Lead Poisoning* (Washington, D.C.: Environmental Defense Fund, 1990), 11–12; U.S. Department of Health and Human Services, Public Health Service, Centers for Disease Control, *Strategic Plan for the Elimination of Childhood Lead Poisoning* (Atlanta: U.S. Department of Health and Human Services, Public Health Service, 1991).

45. Since the early 1950s, James Julian Chisolm (1924–) has been a pediatrician at the Johns Hopkins University, where he received his M.D. in 1946. He has served as a consultant to the EPA since 1976 and on many government-sponsored review panels on lead poisoning.

46. Florini, Krumbaar, and Silbergeld, *Legacy of Lead*, 12; these levels defined only "undue lead exposure." In 1975 the threshold for "lead poisoning" remained set at 80 μg/dL.

47. Herbert L. Needleman et al., "Deficits in Psychologic and Classroom Performance of Children with Elevated Dentine Lead Levels," *New England Journal of Medicine* 300 (1979): 689–95; Herbert Needleman, Alan Schell, David Bellinger, Alan Leviton, and Elizabeth Allred, "The Long-Term Effects of Exposure to Low Doses of Lead in Childhood: An 11-Year Follow-up Report," *New England Journal of Medicine* 322 (1990): 83–88; on the unfriendly reception of this article, see Herbert Needleman, "Salem Comes to the National Institutes of Health: Notes from Inside the Crucible of Scientific Integrity," *Pediatrics* 90 (December 1992): 977–81.

48. J. L. Annest and K. Mahaffey, "Blood Lead Levels for Persons Ages 6 Months–74 Years, United States, 1976–1980," *Vital and Health Statistics,* ser. 11, no. 233. U.S. Department of Health and Human Services, Public Health Service, Publication No. (PHS) 84-1683 (Washington, D.C.: GPO, 1984), 3.

49. To a great extent, the NHANES II findings supported the conventional image of lead poisoning as a problem chiefly affecting poor blacks. The percentage of poor (household incomes under $6,000) black children with elevated blood leads was twenty-six times higher than that of whites with family incomes over $15,000 (18.5% of poor black children [i.e., from families with an annual income under $6,000] had blood leads over 30 μg/dL, compared with 0.7% of white children with family

incomes above $15,000: 18.5/0.7 = 26.43). Inner-city blacks were more than fifteen times as likely as rural whites to have elevated blood levels (18.6%, compared with 1.2% for rural whites). The findings also convey how the conventional view had determined screening efforts: almost one-fourth of inner-city black children had already been tested for lead prior to their NHANES II interviews; one-eighth of all black children had been tested, compared with fewer than 3% of white children. Computed from Annest and Mahaffey, "Blood Lead Levels," table B, p. 10

50. Robert Feldman, "Urban Lead Mining: Lead Intoxication among Deleaders." *New England Journal of Medicine* 298 (1978): 1143–45.; Muriel D. Wolf, "Lead Poisoning from Restoration of Old Homes," *JAMA* 225 (1973): 175–76; James W. Sayre, Evan Charney, Jaroslav Vostal, and I. Barry Pless, "House and Hand Dust as a Potential Source of Childhood Lead Exposure," *American Journal of Diseases of Children* 127 (1974): 167–70.

51. The comparative toxicity of paint chips and lead dust is a clear case in which size matters: lead in finely ground dust is far more bioavailable than that in intact chips; Sayre et al., "House and Hand Dust." Mark Farfel and J. Julian Chisolm Jr., "Health and Environmental Outcomes of Traditional and Modified Practices for Abatement of Residential Lead-Based Paint," *American Journal of Public Health* 80 (1990): 1240–45. See also E. Charney, B. Kessler, M. Farfel, and D. Jackson, "Childhood Lead Poisoning: A Controlled Trial of the Effect of Dust-Control Measures on Blood Lead Levels," *New England Journal of Medicine* 309 (1983): 1089–93.

52. On cleanliness in the postwar era, see Suellen Hoy, *Chasing Dirt: The American Pursuit of Cleanliness* (New York: Oxford University Press, 1995); Ruth Schwartz Cowan, *More Work for Mother: The Ironies of Household Technology from the Open Hearth to the Microwave* (New York: Basic Books, 1983); and Stephanie Coontz, *The Way We Never Were: America's Families and the Nostalgia Trap* (New York: Basic Books, 1992). All of these historians make it clear, as Coontz's title states explicitly, that the baby boomers were seeking in vain to re-create a world that never was, but the historical realities seldom slowed the guilt trip.

53. Jane Lin-Fu, "Children and Lead: New Findings and Concerns," *New England Journal of Medicine* 307 (1982): 615–17.

54. U.S. Department of Health and Human Services, Public Health Service, Centers for Disease Control, *Preventing Lead Poisoning in Young Children* (Washington, D.C.: GPO, 1985), 8, and *Preventing Lead Poisoning in Young Children* (Washington, D.C.: GPO, 1991), 39.

55. National Coalition for Lead Control, *Children, Lead Poisoning, and Block Grants: A Year-End Review of How Block Grants Have Affected the Nation's Ten Most Crucial Lead Screening Programs* (Washington, D.C.: Center for Science in the Public Interest, 1982), reprinted in U.S. Congress, House, Committee on Energy and Commerce, *Lead Poisoning and Children*.

56. In 1986, the EPA reduced permissible lead in drinking water from 50 parts per billion to 10; Ellen K. Silbergeld, "Preventing Lead Poisoning in Children," *Annual Review of Public Health* 18 (1997): 187–210. Consequently, Congress amended the

Clean Water Act to ban lead fixtures and lead solder in new water supply systems. In 1988, Congress passed the Lead Contamination Control Act (P.L. 100-572) to ban the use of lead in the manufacture of drinking fountains and to help local schools eliminate lead in existing fountains. Unfortunately, participation in the programs established under this act has been spotty; see Richard M. Stapleton, *Lead Is a Silent Hazard* (New York: Walker, 1994), 129-30.

57. U.S. Department of Health and Human Services, Public Health Service, Centers for Disease Control, *Preventing Lead Poisoning in Young Children* (1991), 39; Howard Pearson [president, AAP], "Stepped-up Lead Screenings Urged," *AAP News* 9 (April 1993): 1.

58. Schoen called the CDC's 1991 revised definitions "epidemic by edict" (Edgar J. Schoen, "Lead Toxicity in the 21st Century: Will We Still Be Treating It?," *Pediatrics* 90 [September 1992]: 481-82). Alluding to the investigation into Needleman's research then under way in Pittsburgh, Schoen asserted that research into low-level lead effects was plagued by "faulty methodology, flawed statistics, and overstated conclusions." Although Needleman's work had had "great influence on other investigators in the field," Schoen did not mention that those investigators, by and large, had corroborated Needleman's initial findings (Edgar J. Schoen, "Childhood Lead Poisoning: Definitions and Priorities," *Pediatrics* 91 [February 1993]: 504-5).

59. George Gellert, Gerald Wagner, Roberta Maxwell, Douglas Moore, and Len Foster, "Lead Poisoning among Low-Income Children in Orange County, California," *JAMA* 270 (1993): 69-71. Researchers at the University of Utah conducted a similar but smaller screening study of poor children in Salt Lake City. None of the 261 children screened had blood leads above 15 μg/dL, and only 4.2% measured above 10 μg/dL. The authors called for "public policy to reflect regional priorities to avoid divisive arguments within the community of pediatricians." William Banner, Barbara Vugnier, and Jannette Pappas, "Mythology of Lead Poisoning" (letter to the editor), *Pediatrics* 91 (January 1993): 161.

60. Birt Harvey, "Should Lead Screening Recommendations Be Revised?," *Pediatrics* 93 (February 1994): 201-4.

61. Debra Brody et al., "Blood Lead Levels in the US Population: Phase 1 of the Third National Health and Nutrition Examination Survey (NHANES III, 1988 to 1991)," *JAMA* 272 (1994): 277-83; James Pirkle et al., "The Decline in Blood Lead Levels in the United States: The National Health and Nutrition Examination Surveys (NHANES)," *JAMA* 272 (1994): 284-92.

62. Ellen Ruppel Shell, "An Element of Doubt: Disinterested Research Casts Doubt on Claims That Lead Poisoning from Paint Is Widespread among American Children," *Atlantic Monthly* 276 (December 1995): 24-39; see also Peter Samuel, "Lead Hype," *National Review* 47 (May 1, 1995): 69-71.

63. U.S. Department of Health and Human Services, Public Health Service, Centers for Disease Control, *Screening Young Children for Lead Poisoning: Guidance for State and Local Public Health Officials* (Atlanta: DHHS, 1997); Nancy Tips, Henry Falk, and Richard Jackson, "CDC's Lead Screening Guidance: A Systematic Approach to More Effective Screening," *Public Health Reports* 113 (January/February 1998): 47-51.

64. Tips, Falk, and Jackson, "CDC's Lead Screening Guidance," 49.

65. Lynn Goldman and Joseph Carra, "Childhood Lead Poisoning in 1994," *JAMA* 272 (1994): 315–16.

66. Conducting investigations in Salt Lake City for her federal lead survey, Alice Hamilton was astonished when a druggist said he had never seen a case of lead poisoning from the local factories. Then the druggist explained, "Oh, maybe you are thinking of the Wops and Hunkies. I guess there's plenty among them. I thought you meant white men." Alice Hamilton, *Exploring the Dangerous Trades: The Autobiography of Alice Hamilton, M.D,* illustrations by Norah Hamilton (Boston: Little, Brown, 1943), 151–52.

67. Clair C. Patterson, "Contaminated and Natural Lead Environment of Man," *Archives of Environmental Health* 11 (September 1965): 344–60; on Clair Patterson, see Warren, *Brush with Death,* 210–19.

# CHAPTER 10

# The Cultural Politics of Institutional Responses to Resurgent Tuberculosis Epidemics

## New York City and Lima, Peru

SANDY SMITH-NONINI

Tuberculosis (TB) is an emerging health problem in two different senses. In developed countries, particularly the United States, TB had all but disappeared by the 1970s. However, by the 1980s the incidence of cases was increasing, particularly in major urban areas such as New York City. Moreover, many of the new cases that were emerging were resistant to existing antimicrobial drugs commonly used to treat TB. Internationally, in developing countries, TB had never disappeared, yet new forms of resistant TB began to emerge in the 1980s. These cases of multiple-drug-resistant tuberculosis (MDRTB) represented a new public health threat. Moreover, because of the difficulty and expense of treating MDRTB cases, they represented a major new burden on public health systems that were already chronically underfunded.

As indicated in Chapter 1, institutional responses to the emergence of MDRTB were relatively slow when compared with the response of public health institutions to epidemiologically less important illnesses such as hantavirus and Lyme disease. Anthropologist Sandy Smith-Nonini illuminates some of the factors that have contributed to this delay; in this chapter, she examines institutional responses to MDRTB in two different locations, New York City and Lima, Peru. Although the histories of institutional responses to MDRTB in these two cities reflect specific local circumstances, they are similar in that in both cases (1) cutbacks in social programs preceded the emergence of the disease in impoverished populations and (2) various sets of political and economic interests shaped the way in which local public health institutions responded to the emergence of MDRTB.

In both cases, also, this lack of response needs to be understood in terms of the marginalization of those populations most at risk and the challenges such populations face in mobilizing social movements around health issues. In this sense, those suffering from MDRTB resemble the people of the Newtown neighborhood in Gainesville, Georgia, who were exposed to industrial toxins (Chapter 7) and the residents of lead-contaminated housing in Baltimore and Philadelphia (Chapter 9). In all three cases, local mobilization was assisted by volunteer groups and nongovernmental agencies.

---

Few infectious diseases in the 1990s pose such dangers to humans and, at the same time, such a challenge to health care systems worldwide as resurgent tuberculosis. As with acquired immune deficiency syndrome (AIDS) in the 1980s, the urgency of TB was brought into the public eye when it threatened U.S. citizens. From 1990 to 1993, an epidemic of TB dominated the news in New York City and several other U.S. inner cities.

The outbreak was a shock to many who came of age in the sixties and recalled the U.S. surgeon general celebrating the medical conquest of infectious disease. But in reality the dreaded "white plague," which killed thousands during the first half of the twentieth century, has persisted all along among poor and marginalized populations (Ott 1996; Farmer 1999b). Internationally—effective drugs notwithstanding—this neglected public health menace remains the largest preventable cause of adult disability and death in the world, according to the World Health Organization (WHO).

New studies now show that the multiple-drug-resistant tuberculosis strains seen in a portion of the New York patients have caused quiet epidemics among the poor in numerous countries, particularly in settings where AIDS, warfare, or worsening poverty have reduced immunity in poor populations (Cohn, Bustreo, and Raviglione 1997; IUATLD 1997; Pablos-Mendez et al. 1998). WHO authorities have called TB—which kills 3 million people a year—"a global health emergency." The resurgence of the disease has pointed up the high cost of treating drug resistance and the potential for airborne spread of MDRTB by patients who remain contagious for months or years.

With migration, global trade, and air travel linking nations ever more tightly together, resurgent TB in poor countries rapidly becomes a First World health problem—a third of recent TB cases in the United States were diagnosed in foreign-born patients (Mckenna et al. 1995). And MDRTB rapidly jumps across socioeconomic divides. Although the TB epidemic in New York remained "hidden" among the marginalized poor for nearly a decade, once drug resistance became established, middle-class persons with AIDS, nurses, and prison guards were among those who became infected in institutional settings; and MDRTB can kill a patient with active disease in a matter of weeks.[1] So clearly, solutions to resurgent TB must transcend national borders and socioeconomic class.

On the one hand, successful control of TB requires health workers to transcend the dominant curative care paradigm of Western biomedicine—there is no "magic bullet" for curing TB. Once patients are infected, drug therapy must continue for six to twelve months, long after patients begin to feel better. Since incomplete treatment leads to drug-resistant strains of the bacillus, patient compliance remains one of the largest challenges. The best model for TB control in epidemic situations has been supervised or directly observed therapy (DOT), whereby health aides actually watch patients take their pills—an approach that calls for an unusually high level of collaboration between physicians, patients, and health workers.

But at the same time, the spread of MDRTB is demonstrating the shortcomings of international public health models that tend to overrely on standardized treatment protocols administered through hierarchical bureaucracies, pay insufficient attention to cultural difference and diversity of needs at the community level, and fail to adequately individualize treatments or provide continuity of care.

In this chapter, I discuss findings from a qualitative comparative study of two recent epidemics of resurgent TB—one in New York City and the other in Lima, Peru, one of the new "hot spots" of international MDRTB. The study is based on a series of open-ended, in-depth interviews carried out between January and June 1999 with health workers, academic observers, and current and former public health officials involved with TB control in both cities.[2] These cases highlight the ways in which political and economic interests, both locally and internationally, shaped the response of local governments to the reemergence of TB, contributing in both cases to a delay in the funding of necessary treatment and prevention programs.

In contrast to the public health system in the United States, the system in Peru is resource-poor and shaped by international health development models. After a decade of economic turmoil and war, by 1990 TB control (as well as other health programs) had failed miserably, leaving the country with the highest prevalence of TB in Latin America (Cueto 1997). Peru's

Ministry of Health, not unlike other Latin American government health programs, is highly centralized and politicized, with long traditions of leadership by urban physicians. In the aftermath of the famous cholera epidemic in 1992, the Peruvian government acknowledged its infectious disease problem and has since supported the building of a comprehensive program of TB control. Peru's program, which incorporates the WHO-approved DOT strategy, is credited with dramatically reducing the incidence of TB between 1992 and 1996 (Ministry of Health1997).

New York City's public health system, which set the standards for TB control on more than one occasion in the twentieth century, is also a highly politicized physician-dominated bureaucracy. Unlike many Third World models, however, in the United States, infectious disease strategies have long been tailored to fit into the individualized curative care agendas of private medicine. They have to be, since such a large portion of the population is served by private physicians. In New York City, after ignoring the rising incidence of TB for a decade, health authorities, aided by the Centers for Disease Control and Prevention, instituted a DOT-centered TB control program in the early 1990s that brought the outbreak under control and has become a model for urban TB control in developed countries.

The discovery in 1991 of spreading drug-resistant TB strains served as a catalyst, forcing restructuring of the health bureaucracy and drawing new funding to neglected TB programs. Patients with MDRTB were treated in hospital settings with drugs that cost twenty times as much as those used for drug-susceptible disease, at a cost to the city that ran into millions of dollars. In Peru, in contrast, the discovery of MDRTB came shortly after the National Program of Tuberculosis Control had established a reputation internationally for success in bringing down rates for drug-susceptible TB. Peruvian authorities, guided by WHO policy, treated MDRTB as a low priority. Drug resistance in Lima continues to be a problem, but since 1998 the ministry has investigated outpatient strategies for low-cost treatment of MDRTB and is now collaborating with a health nongovernmental organization (NGO) that is pioneering MDRTB treatment strategies.

I first discuss the cases of New York City and Peru in light of the prevalent socioeconomic forces that contributed to the epidemics in the 1980s and early 1990s. In the second half of the chapter I discuss the ways in which the distinctive political traditions and cultures of biomedicine and international health development shaped responses to rising TB incidences, the adoption of the DOT strategy, and strategies for treating MDRTB in New York and Peru.

## The Economic and Political Origins of Resurgent Tuberculosis

The reemergence of TB can be read in part as a biological marker for the human costs of the neoliberal movement that has championed economic growth at the cost of dismantled public programs and deepening urban poverty. The movement by Western bankers and politicians to promote economic restructuring in developing countries began in the mid-1970s and became formalized with the backing of the World Bank and the International Monetary Fund (IMF) in 1980. High interest rates in the late 1970s and the shrinkage of government encouraged by the Reagan and Bush administrations shaped both the international and national political environment within which this agenda evolved.

## The Dismantling of Tuberculosis Control in New York City, 1965–1989

Unlike with the outbreaks of "emerging" diseases like Ebola, hantavirus, or AIDS, when New York City faced an epidemic of resurgent tuberculosis in 1990–91, the famous disease detectives of the CDC faced the challenge not of identifying a new pathogen but of developing policies to correct a breakdown in institutional disease control that had been more than a decade in the making. The few remaining TB specialists in the United States had warned for years that a TB resurgence was likely due to the dismantling of the city's public health surveillance and treatment programs since the 1960s. One of those critics was Dr. Lee Reichman, executive director of the National Tuberculosis Center in Newark, N.J., who in 1992 described the upsurge in TB cases as "horrendous" to a *New York Times* reporter, remarking, "This was a 100% preventable and curable disease."[3]

Many observers have noted that TB control was in some ways a victim of its own success. There was a general sense within the medical community that the advent of new drug therapies in the fifties heralded the end of TB as a major public health problem, and physicians turned their attention to chronic problems such as cancer and heart disease that had become more common causes of death among Americans (Lerner 1993; Garrett 1995). In 1972 categorical TB project grants were phased out and replaced by general federal grants that freed states and local governments to shift funds to other purposes.[4] Over the next few years the breakdown in New York City's health infrastructure translated into the closing of thirteen of the twenty-one TB clinics that had been operating citywide in 1970 (Brudney and Dobkin 1991). During the city's 1975 fiscal crisis, health department staffing was cut by a quarter, and TB control was hit particularly hard—New York State, which had provided half of the TB budget, had terminated state support by

1979, and federal moneys for TB dropped by 80 percent (Lerner 1993). Dr. Reichman, who directed the city's Bureau of Tuberculosis Control in the early 1970s, Michael Iseman of Denver's National Jewish Medical and Research Center, and a handful of other TB experts warned of future epidemics unless funds from the cuts in defunct TB inpatient programs were diverted to strengthening outpatient programs. A series of congressional hearings were held in 1972 and periodically thereafter, but in general the TB specialists' concerns went unheeded.[5]

An indirect cause of the hidden TB epidemic was an increase in poverty, linked to economic recession and to declines in social programs. From 1969 until 1982, the incidence of poverty rose by more than half (Lerner 1993). Wallace and Wallace (1998) documented that in the Bronx, where all but one TB clinic was closed, whole neighborhoods became devastated and abandoned, with arson and the expanding drug trade claiming entire blocks, while fire and police services were cut back. Homelessness was also on the rise; by the late 1980s the city's shelters resembled refugee camps in a war zone, with wall-to-wall cots literally filling armories and warehouses.[6]

These sites became focal points of TB transmission. New York City Health Department surveys later showed that the city's TB patients were disproportionately male and nonwhite and included high percentages of homeless persons, HIV-positive patients, recent immigrants, and intravenous drug users (Frieden et al. 1995). The incidence of TB in the city began creeping upward again with small but steady increases each year after 1978, notably *before* the arrival of AIDS as a public health problem. Between 1979 and 1986 the incidence rose 83 percent (Lerner 1993).

By mid-decade health officials were preoccupied with the AIDS crisis, and when new federal monies became available, TB programs competed poorly for public health dollars. The marginalized populations affected with TB, many of whom were co-infected with HIV and needed drug rehabilitation, had few constituencies to advocate for new policies. As former New York City health commissioner Margaret Hamburg later put it, "Until 1990, TB wasn't on the health department's radar screen."[7]

The right-wing political climate had generated an atmosphere of retrenchment in the public health community. Every year from 1981 to 1987 the Reagan administration called for repeal of the federal TB program (Ryan 1992). Ironically, in 1986, the year when New York City's TB case rates suddenly rose by 20 percent over the previous year, the federally supported TB surveillance program for drug resistance was discontinued (Berkelman et al. 1994).

Dr. Dixie Snider, who headed the CDC's Division of Tuberculosis Control in 1986, became concerned about the rising TB case rates nationally and led development of a new plan for TB "elimination."[8] Despite support from the American Thoracic Society and other TB researchers, requests for the necessary funds fell on deaf ears in Washington. Snider's eradication plan was not formalized by the CDC until 1989, after which it remained unfunded until 1992.[9]

Meanwhile, the disease spread. In February 1990, after New York City health authorities had documented a dramatic rise in TB cases, budget officials at the Bush White House cut the CDC's TB control request from $36 million to $8 million.[10] In the first six months of that year the CDC received reports of nine TB outbreaks nationwide, five of which involved MDRTB strains.[11] By June 1991 the City Hospital Center in Queens, New York, had treated thirteen patients with MDRTB, and the CDC was called in to investigate. Eleven of those patients died; most had been co-infected with HIV.[12] Only after revelations of drug-resistant TB in New York and a flurry of publicity in local and national media did funding for TB control increase significantly.

## The Collapse of Public Health in Peru, 1975–1989

The tuberculosis crisis that Peru faced in 1990 must be seen in relation to the prolonged economic crisis of the late 1970s and 1980s that impoverished much of the population and decimated the public health system. In addition, the Shining Path rebellion and military repression from the government's counterinsurgency efforts resulted in massive refugee flows into urban areas and the loss of public services to large sections of the rural countryside.

At the heart of the prolonged crisis were neoliberal macroeconomic reforms that allowed the Peruvian government to restructure its external debt, conditional on currency devaluations and cutbacks in public services. Such reforms were initially put in place by a military regime following a coup d'état against a populist reform effort in 1975, but after 1978 the reforms became part of a package administered by the IMF. These policies, reinforced by the prevailing conservative bent in foreign policy of the Thatcher and Reagan administrations, unleashed high inflation, reducing real wages by more than half from 1975 to 1985. The reforms eliminated trade barriers, leading to an influx of foreign consumer goods and the collapse of export revenues.

As the gross domestic product declined and prices soared, levels of infant malnutrition increased dramatically. By 1985 food consumption had fallen by 25 percent compared with 1975 levels (Chossudovsky 1997). Popular support for the Shining Path rebellion, which originated in one of the poorest areas of the Southern Highlands, was influenced strongly by the steep drop in income and subsistence levels experienced by rural peasants (McClintock 1989).

From 1979 to 1983 the number of TB cases rose by 30 percent, making it the fifth most important cause of death in Peru (Ministry of Health 1983). In an evaluation of a twelve-month TB therapy regimen used by the Peruvian Ministry of Health in 1981, Hopewell et al (1984) reported that 41 percent of patients abandoned therapy before the tenth month. Only 47 percent were cured. Cure rates improved when the ministry adopted an eight-month regimen, according to a second survey (Hopewell et al. 1985), but the economic situation made the gains difficult to sustain. The Ministry of Health's underpaid and overworked staff was handicapped by shortages of TB drugs within the ministry[13] and by the increasing rural violence due to the Shining Path war.

A short-lived respite came in 1985 with the election of President Alan García of the reformist American Popular Revolutionary Alliance (APRA party). García promised to increase wages and control inflation by reducing the country's debt service. In retaliation the IMF, the World Bank, and international commercial banks cut off all financial support to Peru. Although real purchasing power did increase over the next year, by 1987, when local business elites also "declared war on the government," hyperinflation returned, and earning power once again plummeted (Chossudovsky 1997).

In 1986 the World Bank suspended a $40 million loan for water and sewer improvements in Lima after the government had spent less than a third of the funds. After the cholera epidemic broke out in 1991, a U.S. Agency for International Development (USAID) mission cited the condition of the municipal water and sewer system, noting that the cholera outbreak was "a disaster waiting to happen" (Harantani and Hernandez 1991, cited in Cueto 1997: 183).

Peru's Ministry of Health was a highly politicized institution, with leadership appointed by the president and a well-established system of political patronism. Davíd Tejada de Rivero, a former subdirector of WHO, took over as minister of health under García and became a strong promoter of primary health care for the poor in Peru. His campaign however, met resistance from many doctors and hospital administrators. After Alberto Fujimori's election in 1990 the ministry changed course, and primary health care became "just another program" rather than a guiding philosophy.[14] But what health workers and officials interviewed about this period most remember is the steady decline in resources. By 1991 health worker salaries had declined to

U.S. $40 to $75 per month, and workers in both the education and health sectors were on strike. That year the country's health budget spending was less than a quarter of what it had been in 1980 (Cueto 1997).

Continued civil war eventually closed down scores of rural clinics, further damaging morale. After the collapse of the public health infrastructure in the *selva* (rain forest) region, there was a resurgence of malaria, dengue, and leishmaniasis (Chossudovsky 1997). The only health program that kept functioning countrywide was vaccination against childhood diseases, owing to intervention by UNICEF and WHO, and an accord with armed rebel groups permitting vaccination in regions they occupied.

Chronic undernutrition also contributed to the rise in TB during the 1980s. More than 83 percent of the population failed to meet minimum calorie and protein requirements by 1991, and the national rate of child malnutrition rose to the second highest in Latin America (Chossudovsky 1997). Only half of TB patients received treatment in the late 1980s, and half of those abandoned treatment and were lost to follow-up. Only a fifth of the country's clinics even offered TB treatment in 1989, and the country's information system for TB was in disarray.[15]

This was the situation when Peruvians elected Fujimori on a populist platform. Once elected, after consulting with IMF advisers, Fujimori reneged on his promises of improving quality of life for the majority and instituted his famous "Fuji-shock," further compressing wages and social expenditures and laying off more public sector workers. Prices shot up overnight. Thousands of soup kitchens sprang up in *pueblos jovenes* (urban squatter communities) as families began pooling resources to make ends meet. To give just one example, the thirtyfold increase in cooking oil costs in Lima made it difficult even for the middle class to boil water or cook food, one of a combination of factors thought to have contributed to the 1991 cholera epidemic (Chossudovsky 1997). The epidemic drew international attention and overshadowed other health problems.

As cholera was being brought under control, a group of Peruvian physicians met with Fujimori about the country's TB problem and convinced him to make tuberculosis a national priority. The improvements in TB control from 1992 to 1996 paralleled the restoration of resources to public health in general following the 1992 capture of Shining Path leader Abimael Guzmán, effectively ending the civil war. Many health policy observers believe that this initiative, uncharacteristic for the neoliberal regime that had scoffed at most antipoverty programs, was influenced by the bad publicity the country received during the cholera epidemic. The risk of contagion may have been seen as a potential further threat to tourism and foreign investment at this key moment when the government sought postwar funding for reconstruction.

Despite disparate epidemiological conditions and levels of development in New York City and Peru, an accounting of the socioeconomic causes of these two TB epidemics illustrates why, in the words of one Peruvian health expert, resurgent TB is a "social disease." In both settings public health programs were dismantled along with other social services during a neoliberal period in the name of making government more efficient and diverting capital to private sector concerns that stood to gain from investments in economic growth.

The role of poverty in the epidemiology of TB is an old story. And the helplessness that many health professionals feel in the face of rising inequity has contributed to what some authors have called "public health nihilism"—or pessimism about the possibilities of controlling TB (Fairchild and Oppenheimer 1998). But as these cases show, it is not only the incidence of the disease but also the incidence of public health funds that creates epidemics. A more accurate account of the relations between poverty, public health programs, and TB should acknowledge that poverty is not just an unvarying fact of nature but is created by humans in the form of economic policies (often backed by structural violence) that sacrifice the health of the most vulnerable in order to divert funds to profit more privileged classes, a process that is as transnational as the epidemiology of TB.

## The Role of Medical and Public Health Cultures in Resurgent Tuberculosis

### "Medicalization" of TB in New York City

Social studies of medicine have documented many historical cases of "medicalization" in which diseases for which there was no easy cure suddenly became converted from "social" pathologies to individualistic "medical" problems for which specialists sought one-shot solutions. In some cases, as with the rapid dominance that male obstetricians established over childbirth (to the detriment of midwifery) in the first two decades of the twentieth century, medicalization fit the needs of an expanding medical profession, as it sought to wrest the legal rights to practice medicine away from practitioners not trained in scientifically approved schools (Starr 1982). In the cases of mental health and TB in the late 1950s, medicalization followed the development of new drug therapies that made older therapeutic regimens (such as institutionalization of schizophrenics or TB patients) obsolete. Unfortunately, the dominance of the medical model in the United States has often contributed to tossing the baby (of effective social interventions) out with the bath water (of ineffective therapies)—and such was the case with TB as effective programs of public education, outpatient care, and clinics for screening and surveillance were closed down in the same decade in which funding was discontinued for obsolete sanitariums.

Although, as we have seen, political cutbacks in funding played an important role in creating the conditions for resurgent TB, the medical community was not without complicity in this process and in many cases played a leading role in creating conditions for the return of the "white plague." Yet other physicians, working within a different paradigm within the public health sector, played heroic roles in defending public health. The clashing health care paradigms at work in this story reveal much about the cultural fissures that must be overcome to build "integrated" approaches to health care.

No better moment epitomizes medical triumphalism than U.S. surgeon general William Stewart's 1967 proclamation: "It's time that we close the book on infectious disease." In effect, Stewart was restating Louis Pasteur's 1870 prediction that someday disease-causing microbes would be eliminated across the globe (DeSalle 1999). The scores of outbreaks of emergent and resurgent infections worldwide since the 1960s point up the shortcomings of reliance on "magic bullet" solutions that ignore the role of complex systems—including the biological, ecological, economic, technological, and cultural forces—that interact in the health of populations. But in the 1960s and 1970s, such triumphalism reigned, and in the case of TB there was relatively little opposition from the medical community to the dismantling of programs for a disease that was no longer a major cause of morbidity or mortality. In 1968 the National Tuberculosis Association changed its name to the National Tuberculosis and Respiratory Disease Association (Lerner 1993). During the 1970s public health programs actually discouraged physicians from keeping newly diagnosed TB patients under their supervision (Nolan 1997).

TB had not gone away. The incidence in New York City began rising again gradually each year after 1978. But physicians had an explanation. In 1979 an influential article identified "patient compliance" as the "most serious remaining problem of TB" (Addington 1979, cited in Bayer et al. 1998: 1052). The problem was a longstanding one: people with TB begin to feel better after a few weeks of treatment and simply stop taking their drugs long before the tuberculin bacteria are eradicated. This notion that the problem of TB centered on patients' failings (with no mention of issues such as poverty, access to health care, or the actual functioning of health services) held vogue throughout the eighties in the mainstream medical literature.

And the problem was ubiquitous. New York City Health Department statistics showed increasingly poor patient compliance in TB treatment programs beginning in the early 1980s. Published articles in the medical literature warned about the dangers of this. When such patients relapsed later and sought treatment again, the likelihood of a cure would be lower. The TB strains that had multiplied in their lungs since their initial bout of therapy would be more likely to be resistant to rifampicin and isoniazid, the most commonly used antibiotics. Further, when such patients got active disease a second time, they would be a source for spread of the resistant strains.

Data existed from both international and U.S. trials showing the effectiveness of supervised therapy (or DOT) programs in such situations (Bayer and Wilkinson 1995). The CDC provided funds to New York City and two other cities for selective DOT as early as 1980—targeted at noncompliant patients. Although several hundred patients were identified as eligible for the program in the early 1980s, the number enrolled remained minuscule (Brudney and Dobkin 1991; Lerner 1993). The expense of DOT was one of the reasons officials did not seriously consider it, according to Dr. Jack Adler, a private respiratory medicine physician who in the mid-1980s directed the city's Bureau of Tuberculosis Control. Adler, who was a personal friend of the health commissioner's, had no prior experience in public health when he was appointed to the post, which at the time was only a halftime position. Like many physicians in the private sector, Adler had favored traditional approaches to TB treatment—which placed full responsibility for compliance with the patient, not health providers, and emphasized physician prerogatives in dealing with patients.[16] This stance reflected the predominant thinking about TB treatment by American physicians (Bayer and Wilkinson 1995).

The number of new research grants for TB had dropped significantly by 1979 (Ryan 1992). Compared with heart disease or cancer research, TB research lacked prestige in the medical world. Dr. Barron Lerner, who interned at Columbia Presbyterian Medical School in 1986, recalled, "We started admitting a lot of TB patients, but when physicians saw a patient with the disease it was treated like a 'curio.' TB was thought of as an interesting disease that everyone thought had disappeared. It was a case of collective lack of memory." This inattention to TB in medical schools contributed to fewer practitioners entering the field.[17]

Yet signs of a problem were mounting. In 1985 Dr. Lloyd Friedman tested welfare applicants in New York City and found that one-third tested positive for TB and 1 percent had active disease.[18] In 1986 a large hike in the city's TB cases drew the attention of local health authorities and the CDC, but little was done, in part because health staffing levels had failed to keep up with the increase in cases (Fujiwara, Larkin, and Frieden 1997). Two years later only ninety-three patients were receiving supervised treatment (Bayer and Wilkinson 1995).

Karen Brudney, who directed TB clinics for the city's Bureau of Tuberculosis Control in Manhattan and the Bronx in those years, had noticed steady increases in persons with TB and became dismayed over how many fell through the cracks in the city's underfunded and overwhelmed public health facilities.[19] In 1989 she and co-workers at Harlem Hospital undertook a study of the hospital's TB program and found an alarming rate of noncompliance, with 89 percent of patients lost to follow-up (Brudney and Dobkin 1991). Since Brudney had directed the TB program in Managua, Nicaragua, from 1984 to 1986, she had a unique perspective on New York's program. She noted that whereas case finding is regarded as the main priority in TB programs in the developing world, in the United States health workers make the false assumption that all cases are reported. One problem she identified was the policy of relying on part-time physicians, who were responsible for seeing only patients who appeared during their shifts. At that time the hospital had no organized review of lost or noncompliant patients.

In an article entitled "A Tale of Two Cities," which compared New York City's TB control (rather unfavorably) with Managua's innovative program in the mid-1980s, Brudney and Dobkin (1991) detailed the reasons that persons with TB in New York had difficulty seeking care—these included not only socioeconomic issues such as homelessness and lack of transportation but also issues related to the health system such as long waits in clinic waiting rooms and hostile treatment of patients by overworked clinic staff. Despite ample staff and resources, they wrote, New York's TB clinics' "organization and operation appear nearly calculated to alienate and frustrate patients. . . . The effect of New York's enormous staff and cascading statistics has obscured from the public as well as from the authorities the true nature and size of the problem" (269). Many of Brudney and Dobkin's observations were confirmed in a critical 1992 New York State Health Department review of the city hospitals' TB control in 1992.

Brudney and other health workers interviewed in 1999 also pointed to a crisis of leadership within the health department during the late 1980s. In their article, Brudney and Dobkin wrote that "proven approaches" such as positive incentive programs to improve compliance "had been "repeatedly suggested to the bureau's director and the health commissioner, but to date there is no such program" (270). The TB bureau's crisis of credibility began in 1987 when, following a 17 percent jump in the city's TB cases in one year, the CDC issued a highly critical review of the program, urging officials to reorganize the bureau, improve the computer system used for tracking patients, and hire additional outreach workers. It also recommended the establishment of a TB shelter for homeless patients, where infectious patients could be concentrated and their treatment supervised. This came as a slap in the face to Adler, the head of the Bureau of Tuberculosis Control, who had made recent public statements claiming that TB was *not* a problem in the shelters.

Leadership was also lacking at higher levels. Mayor David Dinkins lacked the political backing he needed to reform the city's failing social and health services. And Dr. Woodrow Meyers, the commissioner of health for the city in 1990, resigned (after a short tenure) because of a controversy over public statements he had made about the possible quarantine of AIDS patients. In his efforts to leave on good terms, Meyers failed to defend programs such as immunization, correctional health, and school health from municipal budget cutters. During the late 1980s, despite the growing increases in TB incidence, the city's share of CDC funds for TB had actually declined slightly each year because the health department, caught in a citywide hiring freeze, was not spending all the funds it received, and unspent funds were carried over to the next year.[20]

Some observers attribute the inattention to TB in New York in the mid-1980s to the distraction of AIDS, which drew funds, researchers, and public attention away from older, more common threats like TB; however, this inattention was mitigated to a degree when many AIDS researchers, aware of the risk TB posed to HIV-infected patients, began to advocate for TB control.[21] When federal funding failed to materialize for TB, the CDC obtained congressional approval to transfer a portion of HIV funds to TB control to make up some of the shortfall.[22]

Throughout the 1980s the medical leadership of New York's Bureau of Tuberculosis Control failed to respond to a significantly rising incidence in the disease the bureau was set up to control, and when confronted about the problems, it found conservative "scientific" rationales to continue ignoring them. To be sure, Adler was able to find backing for a nonaggressive response to TB in the medical literature, in which "noncompliance" was still cited as the major problem precluding a cure for TB. Yet it is no coincidence that this "medicalization" of TB took the route of least resistance by requiring little in the way of new funding. Public health in the United States has long operated within the dominant medical paradigm, and when illness is explained as a problem of individuals, the solutions are far less expensive than solutions that involve public works or social reforms (Lerner 1993; Weiss 1997). Examples of this attitude prevailing in TB care date from the early 1900s (Starr 1982).

### New York City Turns to a "Third World" Solution

At the end of 1990, when the extent of the city's TB problem was just beginning to gain public exposure, Meyers resigned, and the mayor appointed Dr. Margaret Hamburg as acting health commissioner. Hamburg, who later became the permanent commissioner, immediately lobbied (successfully) to get $20 million worth of cuts in basic health programs restored to the health department.

In 1991 Dr. Tom Frieden, a recently trained physician and Epidemic Intelligence Service officer on loan from the CDC, undertook a study of drug resistance in New York City. His research showed the presence of resistance to one or more TB drugs in 33 percent of patients diagnosed with positive cultures and resistance to two or more TB antibiotics in 19 percent of the patients (Frieden et al. 1993). Later in 1991, news reports revealed that drug-resistant TB was associated with thirteen deaths in the state prison system. A series of reports on infected hospital workers, prison guards, and persons with HIV/AIDS over the next year sparked national media coverage that drew widespread criticism of city authorities. New York City's TB funding rose sharply in the following two years. In 1992 Hamburg replaced Adler with Tom Frieden as head of the Bureau of Tuberculosis Control, now a full-time position.

With the backing of the CDC, Frieden undertook several reforms, including instituting mandatory surveillance for drug resistance, testing TB patients for drug susceptibility, a new four-drug regimen for all new patients, and a massive expansion of community-based DOT, including the offering of food supplements and incentives to low-income patients.[23] New York City's aggressive promotion of DOT, in particular, has been widely credited with a rapid drop in TB cases in the years after the policy was adopted (Frieden et al. 1995; Fujiwara, Larkin, and Frieden 1997).

Unlike the more mainstream physician leadership of the 1980s, the physicians who were involved in reforming the city's TB program had a history as public health advocates who had gone against the grain of biomedicine's individualistic model. Brudney had previously run a TB program in Nicaragua; Frieden had been active for a decade in activism for health rights in Central America and had edited *Links,* a small journal on community-based health care in the Third World. Hamburg recruited Dr. Paula Fujiwara (who later became director of the city's Bureau of Tuberculosis Control) from San

Francisco to work with Frieden precisely because she had run a successful DOT program in that city.

Frieden and his staff had to take special initiative in overcoming bureaucratic and political constraints to put the new program in place rapidly—for example, permission was sought to hire a large number of city workers in a short time. Hamburg and Frieden also worked out an arrangement with the city's Office of Management and Budget to coordinate TB-related operations across agency budgets. This coordination ensured that health professionals, as well as administrators, were involved in prioritizing TB spending. In line with this goal, an effort was made to involve nontraditional actors in meetings on TB, including representatives from corrections, education, labor, and the quasi-public agency running the city's homeless shelters. Hamburg credits Frieden with much creative initiative in dealing with bureaucratic roadblocks. For example, in order to short-circuit routine bottlenecks—like delays while all the prospective DOT outreach workers received physicals—Frieden and Fujiwara convened an "open clinic" for job applicants on a Saturday and did the physicals themselves.[24]

As New York City's reconstituted Bureau of Tuberculosis Control discovered, implementation of a labor-intensive program like DOT in a developed country is very expensive.[25] In addition, because of publicity about drug-resistant strains of TB, community outreach workers, who delivered medicines to two-thirds of the city's TB patients after 1992, were initially fearful. And in New York City these workers were unionized. Fujiwara found herself meeting frequently, and not always congenially, with health worker union representatives who took an interest in every aspect of the new policies.

The rapid expansion of DOT in New York and new CDC guidelines recommending DOT in health jurisdictions where fewer than 90 percent of TB patients were completing therapy led to debates in the medical literature over the value of supervised treatment. A conference was held on the rights of TB patients in New York, led by physicians and social scientists whose views had been shaped by recent debates over the rights of HIV/AIDS patients. Many private physicians resented the intrusion of public health authorities into the physician-patient relationship through DOT programs that required their patients to come into clinics for their medicines (and in New York City private physicians treated a significant number of patients—especially those with Medicaid or Medicare coverage). Advocates for DOT countered with data from the medical literature showing that physicians differ widely in diagnostic and prescribing practices regarding TB, contradicting the common belief among physicians that they can predict patient compliance (e.g., see Kopanoff, Snider, and Johnson 1988; Frieden et al. 1994; Liu, Shilkret, and Finelli 1998).

Hamburg blames the inattention to TB in physician education, in part, for the differences that have emerged over this issue. She and Lerner both commented on the large amount of time that Frieden devoted to responding to and educating physicians about the rationale behind the city's new policies. Frieden offered grand rounds on TB, developed diagnostic and treatment cards physicians could carry in their pockets, and distributed doctors' office charts with TB protocols.

It is worth noting, however, that close public scrutiny of public health programs is not unusual and is not necessarily a reflection of public health authorities overstepping their mandates. Weiss (1997) contrasts the controversial nature of public health, which operates in the public sector, with the low accountability enjoyed by private medicine in much of the Western world.

In the aftermath of the crisis, New York City's DOT program has been emulated by many other cities. The Web site of the Bureau of Tuberculosis Control is a model for public education on TB, and the bureau's leaders have emerged as public health heroes. The 1995 *New England Journal of Medicine* article by Frieden and his colleagues on how New York City "turn[ed] the tide" has been widely cited, and Frieden went on to work with WHO on TB in India, which has the second largest DOT program in the world. Hamburg, who later became assistant secretary for planning and evaluation at the U.S. Department of Health and Human Services, has often been called on to give public talks on how her team saved New York from contagion.

We can see several persistent themes in the course of New York City's response (and initial lack of response) to resurgent TB. The dominance of private practice and the individualized curative care medical model in the United States has balkanized health policy decision making. This shaped the cultural milieu in which the availability of new drug therapies for TB and the rise of conservative neoliberal politics became interpreted as a license to dismantle the public health infrastructure for TB. The Reagan administration's indifference to the poor population (perceived as the clientele served by public health) and the lack of funds allocated to public health further diminished the prestige of practicing or doing research in the field. Such a situation makes for weak leadership under the best of circumstances as the best and brightest seek more remunerative careers. It is also not surprising that a neglected, underfunded, and nonprestigious department would offer services as inefficient and bureaucratically stifling as those Brudney and

Dobkin described. The premise that these conditions are the predictable outcome of serving a marginalized, "noncompliant" patient population fails to explain the successful reforms of the TB health program in the mid-1990s, once political support and funding were provided to allow implementation of an evidence-driven epidemiological model of care that responded to the threat by curing MDRTB patients, screening for drug resistance, and instituting both clinic- and community-based DOT strategies.

## TB Control in Peru, 1992–1999

Ironically, the availability of more effective drugs for tuberculosis in the developed world did not translate into better disease control in poor countries. Rather, efforts to fund adequate global distribution of TB drugs languished, and the apparent success in the fight against TB in the United States and Europe created complacency in the international health community about the disease in the 1970s and 1980s. The lack of access to quality medical care (and continuity of care) in the Third World contributed to three decades of a global TB epidemic, resulting in millions of preventable deaths and the outbreaks of drug resistance that plague us today. Only the recent decision of the World Bank to focus on diseases causing disability during adults' productive years (an offshoot of the bank's 1993 introduction of economic criteria, such as productivity, into health decision making) has made TB programs, particularly DOT, an international health priority.

Physicians who had a strong commitment to international public health played a leading role in designing the reform of Peru's TB program and in promoting the program, which includes clinic-based DOT. However, classic problems of inadequate funding, professional dominance, bureaucratic intransigence, and government arrogance toward community-based approaches continue to plague Peru's TB program and interfere with a rapid response to the growing problem of drug resistance in urban areas. It is especially ironic, according to observers, that the relative success of the DOT program since 1993 may have contributed to making physician leaders complacent and slow to acknowledge the problem with MDRTB strains discovered in Lima in 1996.

Before 1990, Peru's TB situation was in crisis, with three hundred thousand persons nationally who had active disease, half going untreated. There was no national system to track the many noncompliant patients. Dr. Cesar Bonilla, a pneumonologist at Daniel Carrion Hospital in Lima, recalled that "TB wasn't really seen as an emergency in the daily work of most health workers. Many had a fatalistic view—believing tuberculosis couldn't be cured as long as poverty persists—so what's the point of trying?"[26]

Dr. Pedro Suarez, who directed Peru's National Tuberculosis Program for most of the1990s, is widely credited for building a critical mass of supporters within the medical community and successfully lobbying the Fujimori administration to back the program. Bonilla recalled that beginning in 1990 Suarez invited scores of hospital-based physicians and university- based professors to meetings on TB held all over country. The doctors were required to attend three to four meetings each year dealing with both theory and clinical issues. Eventually 200–300 health workers of all levels were attending each meeting. Daniel Carrion Hospital became the center for training. Bonilla noted that "it was good that they started with physicians because the problem [with TB control] had been the doctors, not the nurses. Many times doctors acted as obstacles to well-trained nurses. The specialists, particularly, had the attitude that nurses had to follow their orders, and even with a competent nurse, often the doctor wouldn't let her do her job."

Young doctors were especially important, according to Bonilla, because they criticized "false prophet" physicians who always talked "from [their] experience," relying on anecdotes. "We worked more from concrete investigations, and not only from studies done in the United States. We did studies of the situation in Peru. We adapted foreign models, but with a Peruvian slant. Our people became proud to work in such a program. Before this, to work in tuberculosis was to be marginalized. Today, other health programs are imitating the TB program."

Asked what he thought had convinced Fujimori to make TB a priority, Suarez responded diplomatically, calling the program "part of [the government's] struggle against poverty." He credited support and technical help from WHO, as well as the fear of contagion in the aftermath of the cholera epidemic. One Ministry of Health physician privately speculated that the Fujimori administration's support of TB control was a response to protests about hunger and health conditions in Lima's *pueblos jovenes,* since Fujimori needed "to give the impression internationally that he was very concerned about poverty."

After 1992 the state assumed almost all costs of TB medicines, and a nationwide clinic-based DOT program was put in place. The network of TB laboratories expanded from two in 1991 to more than forty-five sites. The number of cases dropped from 300,000 to around 46,000. In 1996 Peru's National Tuberculosis Program was honored by WHO as a model DOT program, and it continues to receive high marks from many public health experts in Lima for bringing down the incidence of TB in only four years. It also stands out as one of the few programs in the ministry with a reputation for efficiency.

But despite the accolades and improvements, the TB picture is far from rosy in Peru. Even with the model DOT program in place, the prevalence of TB remained among the highest in Latin America at 216 cases per 100,000 people (Farmer 1999a). Some researchers have complained that the ministry's reporting methods, which combine statistics from zones with disparate incidences of TB, tend to mask the concentration of TB in Lima's poor shantytowns (Sanghavi et al. 1998). And although the TB budget in Peru increased each year between 1992 and 1996, it stayed fixed after that, which meant a gradual decrease in real dollars allocated to the program.

Spiraling unemployment has hamstrung the poor and middle class in Peru since the 1990 "Fuji-shock," but a further economic downturn in mid-1998 added insult to injury. Neoliberal proposals to reform public health while reducing government health spending have generated widespread skepticism among public health experts—with good reason. Even several years after the restoration of health services funding, Peru's basic health indexes remain among the lowest in Latin America (Cueto 1997).

Dr. Emma Rubin de Celis, a professor of health policy at the Cayetano Heredia University in Lima, said that neoliberal reforms have pushed public health physicians to maximize numbers of consults and to process patients rapidly. Patients are urged to help finance care, "even to the extent of having family members of [poor] patients clean the hospital."[27] Meanwhile, the value of physician and nurse salaries dropped, and the Ministry of Health began to contract with many physicians, instead of hiring them for permanent positions, a practice that causes high turnover and poor continuity of care, according to Rubin de Celis.

A major premise behind neoliberal health reform in Peru has been to limit state responsibility in health to promotion and prevention, leaving most medical treatments in the private realm with the costs falling on the individual patient (Kim et al. 2000). The state guaranteed direct health care services only to the very poor, according to Dr. Julio Castro, a private physician and former legislator with the APRA party. The problem, he said, is that many studies show that poor families make food and housing their top priorities and tend to neglect health problems until they are serious.[28] The couching of public health problems in terms that blame the individual victims (e.g., for poor hygiene or poor compliance with treatment protocols) was prevalent within Fujimori's government even during the cholera epidemic, when public health messages emphasized hand washing and ignored systemic causes of the outbreak (Cueto 1997).

Professors of public health at Cayetano Heredia University like to quote a statement Castro had used to describe Peru's TB program: "The program is good but the disease is better." Asked what he meant by this, Dr. Castro said, "The government is content in having a 'good program' for control of TB, but they are measuring the program, not the size of the TB problem. And the international agencies are content with this approach also." He blamed Fujimori's neoliberal program for the continuing high rates of TB. "At the base of the problem is unemployment," said Castro, noting that since 1990 1.5 million jobs had been lost in Peru.

Dr. José Santos, a twenty-year veteran with the Ministry of Health, spent much of his time at ground zero of Peru's TB epidemic—moonlighting in NGO-run health posts in northern Lima's shantytowns, where he was working during the interview for this research. While he talked about TB policy, he examined a thin toddler with a runny nose, who had arrived at the clinic wrapped in a shawl slung over the shoulder of her mother, a small Indian woman who, like tens of thousands of Peruvians, recently immigrated to the outskirts of the capital to flee rural poverty. Santos complained that for years he had watched patients relapse after completing the ministry's TB treatment program. With their energy sapped and their families' savings depleted, they go seeking in vain for expensive second-line antibiotics they cannot afford. Santos claims he predicted that resistance to TB drugs was coming years ago. "The ministry battles the bacillus, but it's the socioeconomic conditions of the people that cause TB," he said. "The ministry ought to ask: 'If we have such a good program, then why do we have so many patients?'" Santos argues that even patients who are cured of active TB will get sick again if they continue to be malnourished. He pointed to the growth charts for the toddler and her brother, who were once dangerously malnourished but had steadily gained weight under the NGO's food supplement program. Santos has been a longtime advocate for better nutrition supplements for patients during the entire course of treatment. "You have children with TB and the Ministry has resources for pills, but not for food. What craziness is this!"[29]

Similar recommendations came from a 1998 study of the National Tuberculosis Program. The study, carried out by the Proyecto Salud y Nutrición Básica (PSNB), found that patients in Lima, where the TB epidemic is most concentrated, had received food supplements only twice in the previous year, and a third of patients interviewed had never received food (PSNB 1998). Sources in the National Tuberculosis Program at the ministry confirmed that food supplements are no longer available for most patients, a problem that was blamed on cuts in foreign aid food programs. Such drops in funding, including cuts in the Title II food program of USAID, had forced many of Lima's "popular" kitchens to close. Whereas in 1993 about 7,000 such soup kitchens were in operation, by 1999 the number was closer to

500.[30] Malnutrition had improved since 1990, but a quarter of Peruvian children under the age of five remained chronically undernourished, with anemia affecting 60 percent of children under two years of age (Cortez and Calvo 1996–97), so it is likely that inadequate nutrition continues to contribute to the incidence of TB in Peru.

The PSNB study found that, despite the ministry's efforts, TB care is not integrated with other health services, and persons who come to a clinic with other health problems often do not get referred for TB symptoms. One of the study's strongest recommendations was to strengthen the clinic-based DOT program with use of community-based promoters. To its credit, the ministry does recruit lay health promoters from communities that its clinics serve, but their dedication to these unpaid positions varies greatly, and the researchers found that promoters played little role in TB care at most urban ministry clinics. Partly as a result, home visits to follow up on patient contacts who might be infectious tend to be perfunctory. A third of patients interviewed reported they had never been seen by a social worker. The study also found that the DOT goal of supervised therapy is not met in many rural areas, where clinic staff give patients a week's supply of pills at each visit.

A community health worker with an NGO in a poor barrio of northern Lima agreed with the PSNB findings. She noted that one reason so many TB patients become noncompliant is because of the inconvenience and cost of taking public transport to a ministry clinic to receive their pills. "Many enter treatment for two months and then disappear when they feel better because they have kids and have to work. If someone has to be at work by 7 am, they can't just leave and go to the clinic when it opens at 8:30. Even if they did they'd find dozens of patients waiting to be seen and they'd have to wait," she said.

This theme came up later in a conversation with a Lima cab driver who revealed that he had an undernourished daughter in the hospital with TB-like symptoms. "Government clinics only offer bad service," he claimed. "They always tell you there are no more appointments, and that you have to come back another day." He complained bitterly about hospital costs, saying he had to go that very night to beg an administrator at the hospital to give him more time to pay a debt.

The lack of focus on community-based health work in Lima reflects, in part, the Ministry of Health's reorganization after Fujimori's election. Dr. Marcos Cueto, a public health historian at Lima's Instituto de Estudios Peruanos, said that after 1990 "the community-based folks lost ground to vertical programs. You have to remember Fujimori was trained as an engineer. He favored single factor technical interventions and didn't tend to worry about those who get left out of the plan." He noted that in recent years the Ministry of Health had tended to avoid encouraging public criticism and discussion of wider health problems (like those associated with causing the cholera or TB epidemics).

The tendency to avoid dealing with unpleasant realities that challenge the official description of a problem has peculiar consequences. A disturbing finding of the PSNB study was that some ministry clinics keep a second set of records for patients who drop out of TB treatment and then reenter the program later. Records for these "problem" patients were hidden and not reported to the central office, raising questions about the true failure rate of the National Tuberculosis Program, which is officially reported at 15 percent (PSNB 1998 ). "These patients are a headache for the nurses. They don't like to deal with them. Many are addicts or drunks, so they try to turn them away. They mess up the statistics. Everyone thinks they will fail the program again," explained Beth Yeager, a researcher on the study. Two physicians with long-term experience with Peru's National Tuberculosis Program also claimed in interviews that records for patients who abandon treatment are not included with official government statistics.

Although Suarez attended planning meetings for the PSNB study, he showed no interest in the findings, according to a researcher who worked on the project. When I asked Suarez about the PSNB recommendations in my interview, he dismissed the study as irrelevant. The PSNB recommendation to strengthen community-based health work, however, is in line with suggestions of many TB specialists who maintain that DOT is only one element of a successful TB program (see, for example, Lerner 1993; Bayer et al. 1998; Farmer and Nardell 1998; Farmer 1999a).

## MDRTB and the Limits of International Public Health Models

Another Lima-based NGO has played an important role in discovering and characterizing the city's TB drug-resistance problem. Persons with MDRTB were first identified in early 1996 by Socios en Salud (SES), a health NGO based in Carabayllo, a poor barrio of northern Lima. An earlier survey of Latin American MDRTB clusters by Laszlo and Kantor (1994) had also identified a poor neighborhood on the outskirts of Lima as having the highest level of drug resistance in the survey at 54.5 percent of isolates.[31]

Since physicians at SES's parent organization, Partners in Health (PIH) of Cambridge, Massachusetts, had experience with TB and drug resistance in other settings, PIH and SES undertook drug susceptibility testing and began to treat MDRTB patients in shantytowns of northern Lima (Farmer and Kim 1998; Farmer 1999a,b; Becerra et al. 2000).

Dr. Jaime Bayona, medical director of SES, first approached Peru's Ministry of Health in May 1996 for help to obtain expensive second-line drugs for treatment of ten MDRTB patients. The ministry declined to participate. In an interview, Bonilla, speaking for the ministry, explained that in 1996 government health authorities elected to concentrate their limited funds on drug-susceptible TB. "The way to stop MDRTB is to prevent it. This is a poor country, and there are low success rates for treating MDRTB everywhere."

The potential costs for treating MDRTB are indeed formidable. In the program that PIH/SES developed to treat MDRTB patients, the estimated costs to cure a patient with resistance to two drugs ran around $1,000, but curing patients resistant to four to six drugs might cost between $5,000 and $8,000 per patient. In contrast, the cost of curing drug-susceptible TB in Peru is only about $50 per patient.[32]

Dr. Bayona of SES and Drs. Paul Farmer and Jim Yong Kim of PIH continued to meet with Suarez and shared with the ministry their data on the rates of drug resistance that SES was encountering in northern Lima. Of 160 TB patients tested who had failed to be cured in Peru's National Tuberculosis Program, 93.8 percent had active MDRTB (Becerra et al. 2000). More than two-thirds of the drug-resistant patients SES identified were resistant to all four of the drugs used in the program (Becerra et al. 2000). Many had been treated with an incomplete regimen of the drugs used in the country's "standard" treatment protocol prior to 1992, and that is how Farmer (1999a) believes most acquired their initial drug resistance. Later patients who relapsed were reenrolled in the ministry program after the reforms and retreated with the same drugs and others. The PIH/SES researchers became convinced that WHO-approved DOT strategies of reenrolling patients who formerly dropped out of therapy and treating them with the same drugs a second time was actually amplifying drug resistance (Farmer and Kim 1998; Becerra et al. 2000). Although ministry officials continued to maintain that the MDRTB threat was minimal, they agreed to refer patients from northern Lima who abandoned TB treatment in ministry clinics to the PIH/SES program.

Significantly, it was only in 1997 that the ministry in Peru began using a four-drug retreatment program for patients who had failed the initial regimen. Before that, all patients who had failed or abandoned treatment and reentered the program were retreated by the same drugs they had initially received—a practice well known for generating drug resistance.

Marcos Espinal, head of Communicable Diseases at WHO, said the Peruvian ministry had done the right thing from a public health standpoint in declining to treat MDRTB patients. In a May 1999 phone interview Espinal said, "Most cases of TB in the world are drug-susceptible; it would be a major mistake to give the message to developing countries that MDRTB is a higher priority." He maintained that the SES model of community-based treatment was too expensive and too dependent on close physician oversight to be replicable on a larger scale.

Health policy analysts in Lima, however, note that issues of prestige may have also been a factor in the ministry decision. When confronted by the PIH/SES evidence of drug resistance, officials in the National Tuberculosis Program were basking in the approval generated by WHO's 1996 designation of Peru's TB program as a model for the Third World and tended to dismiss criticisms of the program.

Realizing that the cost of second-line drugs would rapidly become prohibitive for their NGO undertaking, Farmer and Kim began lobbying WHO, USAID, and other international health institutions to draw attention to the need for affordable second-line drugs and international funding for MDRTB. PIH hosted several small conferences on drug resistance and pulled together a multi-institutional working group on MDRTB to advise international agencies.[33]

Their efforts were boosted when a new survey of drug resistance in the world commissioned by WHO and the International Union Against Tuberculosis and Lung Disease was published in 1998. The survey, conducted from 1994 to 1997 in thirty-five settings, found drug resistance in every country and region surveyed and identified MDRTB "hot zones" in Russia, Estonia, Latvia, the Dominican Republic, and Argentina (Pablos-Mendez et al. 1998). Peru also ranked high on the list. WHO estimates that 50 million people may be infected with drug-resistant TB strains already, and the problem is growing rapidly—with drug resistance developing in 10 percent of the 8 million new TB infections occurring each year.

In fall 1998 the Peruvian government, with WHO approval, began its own pilot program, using a standardized drug regimen to cure drug resistance, which was designed to cost no more than $1,500 per patient. "We became convinced we had to do something because each of these patients was a focus for transmission," recalled Bonilla.

There were big differences between the ministry's model and the PIH/SES model for treating MDRTB. The medical team from the two NGOs found that it had to tailor regimens for no fewer than thirty-four drug-resistance profiles in the cohort of patients (Becerra et al. 2000). To manage this, SES health workers met daily, with follow-up communication by e-mail with Boston-based PIH to monitor each patient's progress and adjust dosages and drugs as indicated.[34]

Although potential patients for the ministry trial were given drug susceptibility tests to determine if they had MDRTB,[35] in initial trials the ministry did not use the test results to tailor treatments to individual resistance patterns, as did PIH/SES. Instead the ministry sought to test a more affordable and replicable "one-size-fits-all" approach to therapy. A health worker in the ministry's TB division attributed the decision not to individualize treatment in part to the difficulties of training doctors and nurses in a system in which so many worked on contract and often moved from one post to another. In interviews, several TB experts who had experience dealing with drug resistance expressed concern that, given the variety of MDRTB drug susceptibility profiles (86% of the patients in the ministry trial were resistant to three or more drugs),[36] the ministry's one-size-fits-all drug treatment plan would not be effective against MDRTB.

By mid-1999 it was clear that the PIH/SES treatments were curing drug-resistant TB. At that time, PIH/SES reported that 85 percent of the eighty drug-resistant patients who had completed individualized eighteen- to twenty-four- month treatment protocols had remained smear- and culture-negative.[37] A few months later, as the Peruvian health authorities analyzed the ministry's trial data, it appeared that some of the fears the NGO physicians expressed about the government trial were valid. Less than a third of the first cohort completing the ministry's trial were cured.[38] Since then the ministry has joined forces with the PIH/SES team to treat MDRTB in Lima, an effort that has recently been greatly aided by new funding from the Gates Foundation. In late 2002, PIH reported that the rates of "likely cures" for the program's first 367 drug-resistant patients ranged from 68 to 86 percent, depending on patients' levels of drug resistance.[39]

The findings that a standardized one-size-fits-all regimen results in such low cure rates for MDRTB patients present a special challenge to WHO's communicable disease program. Another major challenge to international protocols was illustrated by the differences in outpatient care in the two trials on Peruvian patients. In the treatment of MDRTB, patients must take highly potent drugs for up to eighteen to twenty-four months; these drugs cause myriad side effects from upset stomachs to dizziness, loss of appetite, and generalized aches and pains. Although both the PIH/SES and the initial ministry trials for treating MDRTB were outpatient programs, the levels of community-based services differed greatly. To deal with the serious side effects of the drugs, the PIH/SES program put great emphasis on training full-time, salaried health promoters who spend most of each day visiting patients' homes, where they observe them take their drugs, talk with them about symptoms, and advise them on how to deal with problems. Promoters also worked with family members to establish a support system for the sick person.

In contrast, in the ministry trial, MDRTB patients had to travel to the closest clinic each morning to get their drugs, and support services were provided at the clinic. Some ministry clinics had volunteer health promoters who went to look for patients who missed treatments, but, as the PSNB study showed, the effectiveness and coverage of the ministry's health promoter network were spotty. Dr. Bayona of SES worried that the lack of community-based outreach and support in the ministry program would lead to a high dropout rate that could amplify existing resistance, writing a death sentence for such patients and putting their families and friends at risk. He noted that some patients simply would not be able to visit the clinic daily. "If you have MDRTB are you going to sit in your house or are you going to go to work? You have to support yourself. People with MDRTB are working as waiters, mechanics, bus drivers, etc. and they are spreading [drug-resistant] disease."[40]

Dr. Robert Gilman, a U.S. physician who does research on TB with PRISMA, a Lima-based health NGO, concurred with Bayona about the long-term risks of MDRTB patients going untreated. He cited the WHO estimate that a drug-resistant patient who remains infectious continues to infect about ten other people each year, one of whom, on average, will come down with active MDRTB. Gilman, who teaches at the Johns Hopkins School of Hygiene and Public Health in Baltimore, Maryland, then made a chilling observation about this frequently quoted statistic: "You have to remember that *the other nine who are infected with MDRTB, but who don't get the disease right away may get it later if their immune systems become weakened by malnutrition or other illnesses*" (emphasis added). In a country with the malnutrition and poverty rate of Peru, that prospect did indeed seem daunting.

The dilemma that MDRTB poses to international health experts is a Faustian one. If the international spread of MDRTB is to be checked, TB specialists may have to turn to more community-based and education-intensive approaches that are more costly and complicated to implement. As Dr. Snider of the CDC points out, from an economic standpoint, international agencies have not yet made a full commitment to curing drug-susceptible TB. It was not until the early 1990s that the resurgent TB epidemics in the United States and changes in health priorities at the World Bank drew more funds to international TB, boosting the promotion of DOT programs in poor countries. In 1999 only 16 percent of TB patients globally had access to such programs.[41] Yet without additional funds and a drastic drop in prices for the necessary antibiotics, most experts found it hard to imagine how MDRTB could be dealt with on a global scale.

The rub, according to Snider, is that "where there's a large pool of MDRTB and drug-resistant patients enter existing TB treatment programs without being cured, even good TB programs lose effectiveness. It hurts their reputation with the public." For this reason, the physicians at PIH elected to undertake a rare advocacy effort with governments, agencies, drug companies, and private foundations, seeking increased funding to treat MDRTB and less expensive prices for capromycin and other second-line TB drugs.

Regrettably, the United States, one of the countries best able to fund new international health efforts, has shown relatively little leadership on the global TB problem. In 1993, just as the CDC and WHO were gearing up to confront resurgent TB, the USAID cut its entire overseas budget for TB control (Garrett 1995). New policies adopted by the Clinton White House in 1996 to deal with emerging disease threats focused heavily on surveillance for TB at the country's borders and education of travelers but failed to address the core problem of financing responses to the global epidemic (Lederberg 1996).

Although PIH physicians Farmer and Kim emphasize the humanitarian argument for curing MDRTB, they found that they made more headway with international health authorities once they began emphasizing that in "hot spots" for drug resistance, DOT programs for drug-susceptible TB may perversely amplify resistance if patients drop out and then reenter the program. Clearly new strategies are needed to expand DOT programs for drug-susceptible TB while also preventing the spread of drug-resistant tuberculosis.

## Toward an "Integrated" Public Health

The silent epidemics of tuberculosis were slowly building in New York City and Lima during the periods of cutbacks in public health programs. In both case studies the building of effective DOT programs followed changes in health administrations (accompanied in Peru by a change in national government).

In New York City, where the average standard of living is far higher than in Peru, the populations most affected by the poorly run TB program were ghettoized and "invisible" from a public policy perspective, and public health authorities found it convenient during the retrenchment years of the1980s to ignore epidemiological indicators of rising TB incidence among the marginalized poor. It took a belated intervention by the CDC, several deaths, adverse media coverage, and the fears of middle-class citizens at risk for drug-resistant disease (articulated through the media) before local or federal authorities backed changes in policy and renewed funding.

In Peru the population affected by TB and other diseases of poverty is far larger, but advocacy for public health reforms was undermined, initially by military intervention in the 1970s, then by economic intervention by international lenders (and national business elites) during the 1980s. A strong program for TB in Peru grew out of the efforts of a group of reformist physicians, but funding and political support for the program was tied to Fujimori's efforts to restore Peru's image in the eyes of tourists and foreign investors after the war and the embarrassing cholera epidemic.

Whether in Lima or New York, one thing public health programs have in common is financial instability over time. The tendency of U.S. policymakers to discontinue disease-specific programs after the problem is contained has been dubbed the "U-shaped curve of concern" (Reichman 1997). Likewise, Cueto (1997) has noted a longstanding pattern in Latin America of health services being funded in spurts, in response to an immediate epidemic threat, and then declining.

The New York case study illustrates how the CDC, which lacks regulatory authority, must rely on its reputation for competence and its ability to nurture relationships and fund collaborative projects at the state and local level when confronted with an emerging or resurgent disease outbreak. As Foreman (1994: 27) has noted, in the United States, "the response to emergent public health hazards may be federalized, but it is not centralized."

On the international scene, WHO, lacking a large budget or any regulatory authority, relies heavily on the reputation of its professional medical staff and a limited capacity to fund specific programs (e.g., diarrhea control, AIDS prevention, TB control, essential drugs) for its influence and leverage in member countries. WHO's capacity to monitor countries' DOT programs for TB is limited, as the organization has played a relatively minor role in assisting ministries of health with strategic planning or management of health services.[42]

Despite the substantial improvements in drug-susceptible TB in Peru, many in the public health community worry that gains in this decade, including the reformed National Tuberculosis Program, will not be sustainable, given the new spread of MDRTB, the persistent high poverty and unemployment, and continued neoliberal pressures to reduce government commitments in health. Central to sustainability in any health program must be strong advocacy on the part of public health authorities to maintain support for prevention and screening infrastructure and for rapid responses to rising rates of disease. Notably, in both New York City and Lima in the early

1990s, constructive criticism and leadership for reforms in TB programs often came from unconventional players and depended on strong political advocacy both within the health community and in the larger political arena. Recent advocacy from PIH is likewise promising to be one of the most important factors in persuading WHO and Peruvian government authorities to confront the problem of drug resistance. In all these situations advocacy was not a substitute for epidemiological data but served as an impetus to conducting the necessary studies and acting on the findings.

But in situations with low advocacy (e.g., New York City's Bureau of Tuberculosis Control in the 1980s) or with little pressure for accountability (e.g., Peru's National Tuberculosis Program after receiving WHO recognition in 1996), conservative interpretations of epidemiological risks often become substitutes for evidence-based decision making. PSNB's critical study of Peru's clinic-based DOT program such a short time after the WHO accolades also raises questions about the criteria used by WHO to evaluate DOT initiatives in developing countries. Are the shortcomings of clinic-based DOT programs glossed over in the name of maintaining goodwill with national leaders?

The complacency that Peruvian health workers complain about in the administration of the National Tuberculosis Program may well reflect fear that, given the recent neoliberal zeal to cut social programs, critical debate about TB services will upset the applecart of a "good program"—even though, given Peru's poverty and MDRTB problem, the program may not be good enough.

Interestingly, in congressional hearings held on the New York TB crisis in December 1991, Congressman Ted Weiss (D-N.Y.) charged that federal, state, and municipal agencies had all been complacent about tuberculosis funding, even after signs of rising TB incidence. He chided the public health authorities in the room for failing to advocate strongly enough when funds for programs like TB control were cut back. Alan Hinman, who at the time headed the CDC's National Center for Prevention Services, noted that funding for prevention often depended on public attention being drawn to a disease threat. But Weiss countered that it was the job of public health authorities, not Congress or the public, to assess threats to public health and advocate for new programs when they are needed.

Although Weiss was certainly grandstanding in this opportunity to impugn Republican budget cutters, it is true that studies of the history of U.S. public health point to a legacy of weak advocacy and lax enforcement of regulations, in part because of the beleaguered position of public health vis-à-vis private medicine and other interest groups.[43] Some TB experts interviewed at the CDC for this study expressed discomfort about advocacy in the public arena. Snider, who pushed for the CDC's TB eradication program in the 1980s, pointed out that advocacy often involves overstating one's case, using hyperbole, and confronting authority—positions that seem at odds with the scientific stance of objectivity. In recent years, in response to the many "emerging" or "resurgent" infectious disease outbreaks, the CDC has begun to develop educational materials on how to better advocate about health concerns with politicians, the press, and the public. But in a world in which TB causes a quarter of all preventable deaths[44] (with prospects for tens of millions more as MDRTB spreads and works its deadly synergy with AIDS),[45] these efforts are only the beginning of what is needed.

Advocacy goes hand in hand with public education. In an ideal world, said Snider, "public health ought to be pro-active—in regular contact with radio stations, with the Hispanic community, with gay men." As neglected as it is, especially when funds are short, public education about health risks may also be central to developing a strong constituency for sustaining health programs. The public education efforts on TB developed in New York City in the mid-1990s serve as a model for urban programs: the Web site of the Bureau of Tuberculosis Control is a font of clearly written information presented in a question-and-answer format. Outreach workers took witty comic books on TB into the communities. Unfortunately, in the United States, despite the advent of the so-called information age and clear evidence that the public is concerned about health issues, public education about preventive health has remained a low priority.

This is also true in Peru, where National Tuberculosis Program workers lament the lack of resources for promotion of public information on TB. Other than banners over a couple of highways, there is little to indicate government concern. Dr. Pablo Campos, who studies infectious disease at Cayetano Heredia University in Lima, noted that this lack of promotion of public health information is especially risky in a country where so many people are illiterate. "There is no popular concept of TB or AIDS in Peru," he said, but he noted that he was hopeful about a new program to incorporate health education into the schools.[46] At Daniel Carrion Hospital, Bonilla worried about what effect the lack of public awareness has on TB case finding. "The extreme poor don't go to clinics. And the middle class has a very low awareness of TB. Yet as the program progresses, we see more and more atypical patients. It's not just 'pobrecitos.' who come into our clinics with TB nowadays."

One of the best approaches to public education may be to follow the PSNB recommendations (and the SES example) and put resources into community-based health outreach work. Dr. Campos agreed: "We need to redefine the roles of health promoters. They should be more involved in promoting the rights of the patient, instead of working as medical assistants."

Interestingly, many of the ingredients of academic proposals for a more "integrated" public health dovetail with the goals of community-based DOT, in that there is a heavy emphasis on education (of health workers and the public) and on improving communication and participation both within the health system and with patients. Such programs require investment in people and an embrace of longer-term futures and a more complex social and epidemiological big picture than is usually acknowledged in individualized, curative care models or in neoliberal political ideologies.

Tuberculosis is a social disease. It is spread by poverty, but by striking men and women in their prime, it also *creates* poverty. It is convenient for the developed world that most of the global TB epidemic is elsewhere. But, in the words of Farmer and Kim (1998: 673), "outbreaks of tuberculosis are only briefly local." MDRTB recently turned up in thirty-four out of thirty-five countries surveyed (Pablos-Mendez et al. 1998), and there are now several documented cases of TB transmission on airliners.[47] Bifani, Plikaytis, and Kapur (1996) and Farmer (1999a) cite multiple cases of MDRTB strains making their way across borders and between U.S. cities. Clearly, in the era of globalization and "free trade," we are all in this together, and closing the borders simply is not a practical (or humanitarian) solution.

We have had the means to cure tuberculosis for fifty years. But just as magic-bullet approaches fail to see the social and epidemiological forest for the individual trees, now, with MDRTB, we are seeing that international public health models also have to change—and bring the trees (and quality of care!) back into their concept of the forest. But structural barriers to care and to funding of TB control are usually at the root of bad programs, however they are designed. If we are going to cure TB, the welfare of *people,* in all their aggregate and individual complexities, will have to be at the forefront of health planning, funding, and advocacy. The success of New York City's DOT program, and the encouraging outcomes to date from the SES program of community-based treatment for MDRTB, demonstrate that this can be achieved in the Third World as well as in the First.

### NOTES

This research was made possible by a Mellon-Sawyer Post-Doctoral Fellowship at the Center for the Study of Health, Culture and Society, Rollins School of Public Health, Emory University, during the Spring Semester, 1999, and by a Ford Foundation Peru Academic Exchange Fellowship, administered through the Duke University–University of North Carolina Program in Latin American Studies in June–July 1999. I would like to express special thanks to the staffs of Partners in Health and Socios en Salud and to the many other public health workers, scholars, and persons with TB who shared their knowledge and experience with me during interviews in Atlanta, Boston, New York City, and Lima, Peru.

1. See Farmer 1999a,b for excellent discussions of how TB "hides" among the poor.

2. I also interviewed TB specialists at the Centers for Disease Control and Prevention in Atlanta and at WHO in Geneva. Most interviews were done in person, but some were conducted by telephone, with follow-up by e-mail. While in Lima, I visited and observed the TB programs at a government-run TB clinic and in a nongovernmental organization that specialized in community-based TB care. Interview data were supplemented with a Lexis-Nexis search of media coverage of the New York epidemic and a review of the TB literature in medical, social science, and health policy journals.

3. Sandra Friedland, New Jersey Q & A: Dr. Lee B. Reichman, in "Waging a War on Drug-Resistant TB," *New York Times,* May 10, 1992, sec. 13NJ, p. 3.

4. See American Lung Association 1996; Sbarbaro 1996; and Reichman 1997.

5. See Ryan 1992: 390 and Reichman 1997.

6. Dr. Karen Brudney, interview with author, New York City, February 1999.

7. Margaret Hamburg, phone interview with author, April 1999. All quotations by Hamburg in this chapter are from this interview.

8. After Snider's plan became accepted at the CDC, the division was renamed Division of Tuberculosis Elimination.

9. Dr. Dixie Snider, interview with author, Atlanta, Ga., April 1999; all quotations by Snider in this chapter are from this interview. See also CDC 1989 and CDC 1996. In 1987 CDC officials estimated that it would cost $36 million a year to eradicate TB by 2010, but after the severity of the disease in the inner cities was recognized in 1992, the agency revised its estimate radically upward, asking for $540 million a year. In contrast, the amount the Bush administration requested for TB was only $12.3 million in 1992 and $35 million in 1993. See Michael Specter, "Tuberculosis: A Killer Returns," *New York Times,* October 11, 1992, 1.

10. Philip J. Hilts, "Victory over TB Seen as Thwarted by Budget Unit," *New York Times,* February 28, 1990, sec. 1, p. 24.

11. Elisabeth Rosenthal, "The Return of TB: A Special Report—Tuberculosis Germs Resurging as Risk to Public Health," *New York Times,* July 15, 1990, sec. 1, p. 1.

12. Mireya Navarro, "New York Asks U.S. for Help in Tracking New TB Cases," *New York Times,* January 24, 1992, sec. B, p. 6.

13. Ministry of Health 1982.

14. Dr. Marcos Cueto, interview with author, Lima, Peru, June 1999. All quotations by Cueto in this chapter are from this interview.

15. Dr. Pedro Suarez, director, National Program of Tuberculosis Control, Ministry of Health, Lima, interview with author, Lima, Peru, June 1999. All quotations by Suarez in this chapter are from this interview.

16. Dr. Jack Adler, interview with author, New York City, February 1999.

17. Dr. Barron Lerner, phone interview with author, April 1999.

18. Rosenthal, "Return of TB."

19. Brudney interview.

20. Weiss 1991.

21. Dr. Jim Curran, dean of the Rollins School of Public Health, Emory University, and former director of AIDS research at the CDC, interview with author, Atlanta, February 2000.

22. In 1991 the CDC was using about $10 million in HIV funds for TB activities (testimony by Alan Hinman before the House of Representatives, Human Resources and Intergovernmental Relations Subcommittee of the Committee on Government Relations, December 18, 1991).

23. Two-thirds of New York City's TB cases were managed through community-based DOT under health department supervision, and one-third of the cases were managed in clinics and private practices, in collaboration with municipal health authorities. Most of these received their drugs in a clinic-based DOT program, using multidisciplinary health teams overseen by the health department (Fujiwara, Larkin, and Frieden 1997).

24. Dr. Paula Fujiwara, interview with author, New York City, February 1999.

25. For example, in the 1993 fiscal year, at the height of the city's response to the epidemic, New York City's health department spent more than $100 million on TB control.

26. Dr. Cesar Bonilla, interview with author, Lima, Peru, June 1999. All quotations by Bonilla in this chapter are from this interview.

27. Dr. Rubin de Celis, interview with author, Lima, Peru, June 1999.

28. Dr. Julio Castro, interview with author, Lima, Peru, June 1999. All quotations by Castro in this chapter are from this interview.

29. Dr. José Santos, interview with author, Lima, Peru, June 1999.

30. Beth Yeager, researcher on the Proyeto Salud y Nutrición Básica, interview with author, Lima, Peru, June 1999. All quotations by Yeager in this chapter are from this interview.

31. The extent of the drug-resistant TB problem in Lima remains unclear because so much of the information is based on the patients who come to the clinics or hospitals. As of mid-1999, Dr. Robert Gilman, a U.S.-trained TB specialist at PRISMA (a Lima-based NGO), reported that pediatric TB rates have not risen, which is a good sign. In addition, unlike in U.S. urban centers and Africa, relatively few TB patients in Peru are co-infected with HIV. On the other hand, in the Social Security Hospital where he works, Gilman reported that about 50% of AIDS patients are getting active TB, and about 50% of those get MDRTB. Dr. Robert Gilman, interview with author, Lima, Peru, June 1999. All quotations by Gilman in this chapter are from this interview.

32. Interviews with PIH researchers and ministry officials.

33. Interviews and background materials provided by Drs. Paul Farmer and Jim Yong Kim, Boston, February 1999.

34. I conducted interviews with patients, health professionals, and community health workers during a two-day visit in June 1999 to Socios en Salud, Carabayllo, Peru, where I accompanied a health promoter on his daily rounds to patients' homes. I also visited an SES primary care clinic in a poor barrio.

35. MDRTB is defined as resistance to at least two drugs: rifampicin and isoniazid, the two antibiotics in first-line TB treatment protocols.

36. Dr. Robert Canales of the ministry's Technical Unit for TB, interview with author, Lima, Peru, June 1999.

37. Interviews with SES administrators in Lima, and Dr. Paul Farmer 1999a and interview with author, Atlanta, Ga., February 2000.

38. Canales interview and Farmer interview.

39. The cure rates were 68% for those who had previously undergone more than two TB drug regimens, 79% for those who had undergone exactly two previous treatment regimens, and 86% for those who had undergone only one or less previous treatment regimen, according to Sarah Van Norden, "MDR TB Treatment in Peru: An Update," *PIH Bulletin,* Winter 2003.

40. Dr. Jaime Bayona, interview with author, Lima, Peru, June 1999.

41. Dr. Marcos Espinal, Communicable Diseases Division, World Health Organization, Geneva, Switzerland, phone interview with author, May 1999.

42. Walt 1994 and De Cock 1999.

43. Weiss 1997.

44. Alan Hinman, cited in Hilts,"Victory over TB Seen as Thwarted," 24.

45. A person who is HIV positive and is exposed to (drug-susceptible or drug-resistant) TB has a far higher chance of becoming infected, and of coming down with active disease, compared with those who are HIV negative. Once a person who is HIV positive tests positive for TB on a skin test, that individual runs a 10% chance of coming down with active TB within a year (Curran interview).

46. Dr. Pablo Campos, interview with author, Lima, Peru, June 1999. All quotations by Campos in this chapter are from this interview.

47. See *MMWR* 1995 for a CDC review of six such cases.

REFERENCES

Addington, W. 1979. Patient compliance: The most serious remaining problem in the control of tuberculosis in the United States. *Chest* 76 (Suppl.): 6.

American Lung Association. 1996. Maintaining momentum: America's TB challenge. A public policy brief of the American Lung Association. American Lung Association—Washington Office, 1726 M. St., NW, Suite 901, Washington, D.C. 20036.

Bayer, Ronald, and David Wilkinson. 1995. Directly observed therapy for tuberculosis: History of an idea. *Lancet* 345:1545–1548.

Bayer, Ronald, et al. 1998. Directly observed therapy and treatment completion in the United States: Is universal supervised therapy necessary? *American Journal of Public Health* 88:1052–1058.

Becerra, M. C., J. Bayona, J. Freeman, P. E. Farmer, and J. Y. Kim. 2000. Redefining MDR-

TB transmission "hot spots." *International Journal of Tuberculosis and Lung Disease* 4(5): 387–394.

Berkelman, Ruth, et al. 1994. Infectious disease surveillance: A crumbling foundation. *Science* 264: 368–370.

Bifani, P. J., B. B. Plikaytis, and V. Kapur. 1996. Origin and interstate spread of a New York City multidrug resistant m. Tuberculosis clone family. *JAMA* 275:452–457.

Brudney, Karen, and Jay Dobkin. 1991. A tale of two cities: Tuberculosis control in Nicaragua and New York City. *Seminars in Infectious Disease* 6(4): 261–272.

CDC. *See* Centers for Disease Control and Prevention.

Centers for Disease Control and Prevention. 1989. A strategic plan for the elimination of tuberculosis in the United States. *Morbidity and Mortality Weekly Report,* Supplement, 38(S-3): 1–25.

———. 1996. Tuberculosis morbidity—United States, 1995. *Morbidity and Mortality Weekly Report* 45(18).

Chossudovsky, Michel. 1997. IMF shock treatment in Peru. In *The Globalisation of Poverty: Impacts of IMF and World Bank Reforms,* 191–213. London: Zed Books.

Cohn, David, F. Bustreo, and Mario Raviglione. 1997. Drug resistant tuberculosis: Review of the worldwide situation and the WHO/IUATLD Global Surveillance Project. *Clinical Infectious Diseases* 24(Suppl. 1): S121–130.

Cortez, Rafael, and Cesar Calvo. 1996–97. La nutricion infantil en el Peru. *Punto de Equilibrio,* no. 46, ano 6 (December/January): 29–33.

Cueto, Marcos. 1997. *El regreso de las epidemias: Salud y sociedad en el Peru del siglo XX.* Lima: Instituto de Estudios Peruanos.

De Cock, Kevin. 1999. International responses to HIV/AIDS. Presentation in Sawyer Seminar on Emerging Illnesses and Institutional Responses, Rollins School of Public Health, Emory University, April 2.

DeSalle, Rob. 1999. *Epidemic! The World of Infectious Disease.* New York: New Press.

Fairchild, A. L., and G. M. Oppenheimer. 1998. Public health nihilism vs. pragmatism: History, politics and the control of tuberculosis. *American Journal of Public Health* 88:1105–1117.

Farmer, Paul. 1999a. Hidden epidemics of tuberculosis. Working Paper No. 239, Latin American Program, Woodrow Wilson International Center for Scholars, Washington, D.C.

———. 1999b. *Infections and Inequalities: The Modern Plagues* (Berkeley: University of California Press.

Farmer, Paul, and Jim Yong Kim. 1998. Community-based approaches to the control of multi-drug resistant tuberculosis: Introducing DOTS-Plus. *British Medical Journal* 317:671–674.

Farmer, Paul, and Ed Nardell. 1998. Editorial: Nihilism and pragmatism in tuberculosis control. *American Journal of Public Health* 88:1014–1015.

Foreman, Christopher H., Jr. 1994. *Plagues, Products and Politics: Emergent Public Health Hazards and National Policymaking.* Washington, D.C.: Brookings Institution.

Frieden, Thomas, et al. 1993. The emergence of drug-resistant tuberculosis in New York City. *New England Journal of Medicine* 328(8): 521–526.

Frieden, Thomas, et al. 1994. Tuberculosis clinics. *American Journal of Respiratory and Critical Care Medicine* 150: 893–894.

Frieden, Thomas, et al. 1995. Tuberculosis in New York City—Turning the tide. *New England Journal of Medicine* 333(4): 229–233.

Fujiwara, Paula, Christina Larkin, and Thomas Frieden. 1997. Directly observed therapy in New York City: History, implementation, results, and challenges. *Clinics in Chest Medicine* 18(1): 135–148.

Garrett, Laurie. 1995. *The Coming Plague: Newly Emerging Diseases in a World Out of Balance.* New York: Penguin Books.

Harantani, Joseph, and Donald Hernandez. 1991. *Cholera in Peru: A Rapid Assessment of the Country's Water and Sanitation Infrastructure.* Washington, D.C.: USAID Mission to Peru.

Hopewell, Philip, et al. 1984. Operational Evaluation of treatment for tuberculosis: Results of a "standard" 12-month regimen in Peru. *Amererican Review of Respiratory Disease* 129:439–443.

Hopewell, Philip, et al. 1985. Operational Evaluation of treatment for tuberculosis: Results of 8- and 12-month regimens in Peru. *American Review of Respiratory Disease* 132:737–741.

International Union Against Tuberculosis and Lung Diseases. 1997. *Anti-Tuberculosis Resistance in the World.* Report. Geneva: International Union Against Tuberculosis and Lung Diseases and the World Health Organization.

IUATLD. *See* International Union Against Tuberculosis and Lung Diseases.

Kim, Jim Yong, et al. 2000. Sickness amidst recovery: Public debt and private suffering in Peru. In *Dying for Growth: Global Inequality and the Health of the Poor,* edited by Jim Yong Kim, Joyce V. Millen, Alec Irwin, and John Gershman, 127–154. Monroe, Maine: Common Courage Press.

Kopanoff, Donald, Dixie Snider Jr., and Martha Johnson. 1988. Recurrent tuberculosis: Why do patients develop disease again? A United States public health service cooperative survey. *American Journal of Public Health* 78:30–33.

Laszlo, A., and I. N. Kantor. 1994. A random sample survey of initial drug resistance among tuberculosis cases in Latin America. *Bulletin of the World Health Organization* 72(4):603–610.

Lederberg, Joshua. 1996. Infectious disease—a threat to global health and security. Editorial in *JAMA* 276:417–419.

Lerner, Barron H. 1993. New York City's tuberculosis control efforts: The historical limitations of the "war on consumption." *American Journal of Public Health* 83:5, 758–766.

Liu, Z., K. L. Shilkret, and L. Finelli. 1998. Initial drug regimens for the treatment of tuberculosis: Evaluation of physician prescribing practices in New Jersey, 1994–1995. *Chest* 113(6):1446–1451.

McClintock, Cynthia. 1989. Peru's Sendero Luminoso Rebellion: Origins and trajectory. In *Power and Popular Protest: Latin American Social Movements*, edited by Susan Eckstein, 61–101. Berkeley: University of California Press.

McKenna, Matthew T., Eugene McCray, and Ida Onorato. The epidemiology of tuberculosis among foreign-born persons in the United States, 1986 to 1994. *New England Journal of Medicine* 332:1071–1076.

Ministry of Health [Peru]. 1982. Situación actual de la enfermedades transmisibles en el Pais—1982. Unpublished internal document.

———. 1983. Cuadro no. 5—Mortalidad y morbilidad por tuberculosis, informe estadistico anual de enfermedades transmisibles, 1963–1983, OGIE.

———. 1997. *Tuberculosis in Peru: Informe 1997*. Evaluación del Programa Nacional de Control de la Tuberculosis en el Peru—Año 1997. Lima, Peru.

Nolan, Charles. 1997. Topics for our times: The increasing demand for tuberculosis services—a new encumbrance on tuberculosis control programs. *American Journal of Public Health* 87:551–553.

Ott, Katherine. 1996. *Fevered Lives: Tuberculosis in American Culture since 1870*. Cambridge: Harvard University Press.

Pablos-Mendez, Ariel, et al. 1998. Global surveillance for antituberculosis-drug resistance, 1994–1997. *New England Journal of Medicine* 338(23): 1641–1649.

Proyecto Salud y Nutrición Básica. 1998. *Estudio sociomedico sobre la tuberculosis: Lima*. Informes de Investigacion 12, Lima, Peru: Proyecto Salud y Nutrición Básica.

PSNB. *See* Proyecto Salud y Nutrición Básica.

Reichman, Lee. 1997. Defending the public's health against tuberculosis. *JAMA* 278: 865–867.

Ryan, Frank. 1992. *The Forgotten Plague*. Boston: Little, Brown.

Sanghavi, Darshak M., et al. 1998. Hyperendemic pulmonary tuberculosis in a Peruvian shantytown. *American Journal of Epidemiology* 148(4): 384–389.

Sbarbaro, John A. 1996. Commentary on tuberculosis surveillance. *Public Health Reports* 111 (January–February): 32–33.

Starr, Paul. 1982. *The Social Transformation of American Medicine*. New York: Basic Books.

Wallace, Rodrick, and Deborah Wallace. 1998. *A Plague on Your Houses: How New York Was Burned Down and Public Health Crumbled*. New York: Verso.

Walt, Gill. 1994. *Health Policy: An Introduction to Process and Power*. London: Zed Books.

Weiss, Lawrence. 1997. *Private Medicine and Public Health: Profit, Politics, and Prejudice in the American Health Care Enterprise*. Boulder, Colo.: Westview Press.

Weiss, Ted. 1991. Testimony in "Tuberculosis in New York City: An Epidemic Returns." Proceedings of a hearing before the Human Resources and Intergovernmental Relations Subcommittee of the Committee on Governmental Operations, U.S. House of Representatives, December 18.

# CHAPTER 11

# Institutional Responses to the Emergence of Lyme Disease and Its Companion Infections in North America

## A Public Health Perspective

ANDREW SPIELMAN, PETER J. KRAUSE,

AND SAM R. TELFORD III

The emergence of Lyme disease as a distinct biomedical entity, linked to infection by a specific tick-borne pathogen, the spirochete *Borrelia burgdorferi*, dates from the mid-1970s. Lyme disease, as indicated in Chapter 1, is an example of an illness that rapidly gained public health recognition, despite initially being of limited epidemiological importance and lacking a test that could definitively indicate the presence or absence of the infection. The institutional response to Lyme disease was accelerated in part because the initial population at risk included affluent communities with connections to the biomedical and public health establishment. Lyme also gained attention because researchers and patient activist groups constructed it as a "new threat" from a new organism.

The complex constellation of social and biomedical forces that shaped the early emergence of Lyme disease as a "new" illness, despite its evident relation to other recognized disease conditions, has been described earlier by Robert Aronowitz.[1] In the present chapter, Andrew Spielman and Sam R. Telford, public health entomologists, and Peter J. Krause, an infectious diseases physician, describe the ecological transformations that produced Lyme disease and draw our attention to the roles played by an extraordinary array of individual and institutional actors in the emergence of Lyme disease. These included local physicians, conservation foundations, residents' associations, academic institutions, private industries, patient advocacy groups, state and local governments, and the courts. By exploring the role played by each of these groups, the authors illustrate how the movement of an illness along the pathways described in our "Processual Model of Social Emergence" (Chapter 1) can involve multiple institutions. These institutions or organizations may have conflicting interests in shaping the process of emergence. The chapter warns us to question more simplistic models of disease emergence that focus only on the role of public health institutions and communities of suffering.

---

The novel guild of tick-borne pathogens that emerged in the northern United States during the closing decades of the twentieth century has assumed broad public health importance, and much of Europe and parts of Asia have suffered similarly. Unique policy issues arise because these conditions disproportionately afflict the affluent, at least in the northeastern United States and Europe. The growing reluctance of people to walk through vegetation in the affected regions may inhibit economic growth in many of the most desirable residential and recreational locations. One of the resulting infections, Lyme disease, has gained particular notoriety as a source of severely debilitating and chronic symptoms, and its lesser-known companion diseases, human babesiosis and human granulocytic ehrlichiosis (HGE), may be fatal. These tick-borne infections continue to demand attention. We have studied the epidemiology and epizootiology of these infections since the guild first began to emerge in North America and have helped implement various public health responses. The discussion that follows, accordingly, provides our personal view of the manner in which the various societal institutions that operate in the affected parts of North America have responded to the emergence of these deer tick–transmitted infections.

## Environmental Conditions Leading to Outbreaks

The tick-borne pathogens associated with Lyme disease include diverse, phylogenetically unrelated microbes that share common reservoir as well as vector hosts: the Lyme disease spirochete itself (*Borrelia burgdorferi*), a sporozoan (*Babesia microti*), a rickettsia (*Anaplasma phagocytophilum*), and a flavivirus related to Powassan virus (Spielman et al. 1979; Bakken et al. 1994; Telford, Dawson, and Halupka 1997; Solberg et al. 2003). The *Ixodes ricinus*–like ticks that transmit these infections include the wood tick of Europe (*I. ricinus*), the taiga tick of Eurasia (*I. persulcatus*), and the deer tick of northeastern North America (*I. dammini* or *I. scapularis*). We differentiate *I. dammini* from its nonvector sibling species, *I. scapularis*, a tick that occurs well south of the range of these zoonotic infections. Although these agents of human disease appear to have originated prior to the ice ages (Marshall et al. 1994), they remained little noticed until recently because they perpetuated in limited foci where vector ticks were scarce or were perpetuated by other ticks that feed solely on rodents. Another *B. burgdorferi*–like spirochete, *Bo. andersoni*, for example, is maintained by *I. dentatus*, a rabbit-specific tick (Telford and Spielman 1989) that rarely attaches to human hosts. Other spirochetes are maintained similarly in *I. spinipalpis* ticks, a tick that feeds on wood rats in Colorado and California but not on people (Maupin et al. 1994). Although this guild of tick-borne pathogens may be present in a site, human risk is not necessarily implied.

The earliest direct evidence of the presence of this guild of pathogens in eastern North America was derived in the 1930s and involved *Ba. microti* and *A. phagocytophilum*, which parasitized voles and mice on the island of Martha's Vineyard in Massachusetts (Tyzzer 1938). The absence of human infections at that time is attributable to the feeding habits of the presumed vector tick, *I. muris*, a host-specific, nest-associated parasite of small rodents that has since become locally extinct. Human infections could not have occurred until a less fastidious tick became established that could serve as a "bridge between mouse and man."

People become infected by the various pathogens borne by deer ticks solely where infestations of *I. ricinus*–like ticks have become established and where the deer that are their main definitive hosts are abundant. This health burden, however, does not solely depend upon dense deer populations (Schulze et al. 1984; Glass et al. 1994) because other various environmental factors appear to be critical to the life cycles of these vector ticks. These zoonotic pathogens emerge where (1) deer have proliferated, (2) the old-field brush habitat has developed as a result of the abandonment of farmland, and (3) numerous people reside in such brush and forested sites or visit there for recreational purposes. Lyme disease and the deer tick pathogen guild in general, therefore, should be considered as anthropogenic, environmental infections.

Given the anthropogenic nature of the infections carried by deer ticks, locally applied environmental policies greatly affect risk of human infection. Suburban residential developments, in particular, promote white-tailed deer density and that of white-footed mice. Natural predators have largely been eliminated and dogs leashed. Deer may be fed and their presence encouraged in other ways. Although recreational hunting would limit deer abundance, the use of firearms is banned around houses. In general, firearms cannot be discharged within 500 feet of a residence or within 150 feet of a hard-surfaced road. Suburban affluence thereby promotes contact between people, deer, deer ticks, mice, and the guild of diverse pathogens that proliferate in this interface. Risk of human infection by deer-associated pathogens has increased and is amplified in the northeastern United States as a result of a convergence of anthropogenic environmental events that developed silently during the twentieth century.[2]

## Community-Based Responses

The person who became the index case for human babesiosis in North America was infected on Nantucket Island in Massachusetts; she continued to live on Nantucket after she recovered and was well known to one of us (AS). Her firsthand experience with the disease helped her play a key role in diagnosing this infection in the second case, four years later, in a neighbor and close friend. Together, these women encouraged Harvard personnel to study Nantucket fever and generously provided their own personal funds in support of this fieldwork. Another Nantucket resident, the late Dr. Gustave Dammin, vacationed near their homes. His work as chief of pathology at Boston's Brigham Hospital prepared him to take an active role in these investigations. He helped recruit funds that supported the work of other investigators and organized the residents of the island. The late Dr. Alexander Langmuir, the retired founder of the Epidemic Intelligence Service of the Centers for Disease Control, performed a similar function on Martha's Vineyard. A biological research station that was maintained on the island by the University of Massachusetts became the venue for much of this work. The late botanist director of that station, Dr. Wesley Tiffney, worked for more than two decades in support of these research efforts. Together, these and other individual residents of these well-known vacation islands played a crucial role in the early investigations of tick-borne infections.

Recognition of the first cluster of Lyme disease cases owes much to the efforts of two lay residents of Old Lyme, Connecticut, whose children suffered in 1975 from symptoms that had been misattributed to juvenile rheumatoid arthritis. Polly Murray and Judith Mensch independently noted that neighbor children suffered similar arthritic symptoms and realized that such a clustered distribution is inconsistent with this diagnosis. They described the anomaly that they discovered to a group of research workers at the Yale School of Medicine that included Drs. Allen Steere and Stephen

Malawista. This event launched the original investigation that resulted in the recognition of Lyme disease as a North American clinical entity. Like the residents of Nantucket and Martha's Vineyard, those of Old Lyme are exceptionally affluent and accomplished. The late Roger Tory Peterson, a much honored naturalist, lived there. These tick-borne infections differentially afflict the people who are best equipped to deal with such a threat.

Formally organized residents' associations are present on various other affected islands in the region and often serve to initiate intervention measures. Extended family associations facilitated our research access to Naushon Island and Great Island in Massachusetts. The residents of Great Island, particularly Dr. Wilson (Roly) Nolen, former vice president of the Becton Dickinson Company, have been particularly generous in their financial as well as physical support of this work. They permitted their blood to be drawn annually in support of a longitudinal epidemiological study conducted by Harvard School of Public Health (HSPH) investigators. Dr. Norman Dahl, president at that time of the Block Island Residents' Association (BIRA) and a former director of the Rockefeller Foundation India development program, sought out and actively recruited the participation of research workers from HSPH and the University of Connecticut. BIRA members donated operating funds and have permitted their blood to be sampled annually for more than a decade. The residents of Gibson Island in Maryland, including many retired physicians and Johns Hopkins faculty, similarly sought out and supported the work of HSPH investigators. The residents of certain other affected sites, however, have been far less encouraging. Even some exceptionally affluent and well-educated people appear to dislike the notoriety associated with ownership of land where health is at risk. Local organizations that are organized to serve general social interests can contribute to the public health.

## Local Physicians

The response of physicians practicing in sites that are burdened by these deer-related zoonoses has been diverse. Lyme disease was first diagnosed in North America in 1969 by a physician practicing in a small town in northwestern Wisconsin (Scrimenti 1970). The case report describing this seminal event was remarkably complete and included references to a potential spirochetal etiology and penicillin therapy. This observation in the *Archives of Dermatology* escaped notice by American physicians until retrospective searches of the literature were undertaken during the late 1970s. More recently, Johan Bakken of the Duluth Clinic established the first American diagnosis of HGE. Other practicing physicians have enthusiastically contributed to research efforts, including our collaborators on Block Island and Nantucket and in southeastern Connecticut.

Although numerous Lyme disease publications have appeared (1,392 references cited in Medline before 1990 and more than 3,518 thereafter), not all local practitioners have kept up with the literature. As recently as 1990, certain physicians on Cape Cod remained skeptical when their patients presented with the symptoms of Lyme disease or babesiosis, even when these physicians were prompted by their patients. In contrast, other physicians have incorrectly diagnosed cases of Lyme disease or human babesiosis in their patients on the basis of spurious laboratory tests or subjective criteria. Some of these physicians also recommend excessively long parenteral treatment regimens. Although the vast majority of physicians diagnose, treat, and report cases appropriately and participate enthusiastically in research efforts, some complicate treatment and surveillance efforts.

### Conservation Foundations

Various nonprofit conservation organizations have come to terms with the emergence of these deer-associated infections. In Massachusetts, these agencies include the Trustees of Reservations, the Audubon Society, and the Nantucket Conservation Foundation (NCF) and are dedicated to the preservation of large parcels of forested land. On Nantucket Island, for example, hunters had removed 250 or so deer each year during the one-week legal shotgun hunting season, keeping constant the estimated population of 1,500 animals on an island that is roughly fifty square miles in area. Increases in the prevalence of infections due to the deer tick pathogen guild stimulated community leaders, particularly James Lentowski, director of the NCF, to action. Lentowski himself had suffered attacks of chronic Lyme arthritis, babesiosis, and ehrlichiosis over a four-year span and led the community drive that culminated in the petition to the Massachusetts Division of Fisheries and Wildlife that extended the shotgun deer season on Nantucket. Largely owing to his leadership, the NCF has been proactive in alerting its members and those who enjoy their properties to "check for ticks." The NCF widely distributes without charge a pamphlet on ticks and risk reduction written in collaboration with the HSPH lab.

### Residents' Associations

Local residents' associations are present in many sites in which these tick-borne infections are zoonotic. They form cooperative groups dedicated to removing brush, reducing deer density, and applying insecticide. The Block Island Residents Association engaged Harvard and the University of Connecticut to determine the extent of the tick-borne disease problem on that island, funding the first two years of work there with a generous research grant. A prospective epidemiological study began in 1990 (with federal funding) and continues to this day. Virtually all of the one thousand residents respond regularly to questionnaires and provided baseline serum samples. Interestingly, although the original intent of BIRA was to gather robust evidence to be used in obtaining full community support for comprehensive interventions directed against these tick-borne infections, including possible deer eradication, no such action has yet taken place. BIRA funding, however, provided the initial stimulus that recruited National Institutes of Health funding of the epidemiological work. This has resulted in numerous publications that have considerably enhanced our knowledge of the burden of infection due to tick-borne pathogens, particularly that due to concurrent infection (Krause et al. 1996, 1998, 2003).

Gibson Island, a peninsula located southeast of Baltimore, on the Chesapeake Bay, provides another example of how communities initiated efforts to reduce the risk of tick-borne infection. This gated community of six hundred relatively affluent residents includes retired physicians as well as university faculty. The "tick committee" of the residents' association completed a questionnaire-based survey under the direction of a prominent Johns Hopkins epidemiologist who resided there. An annual incidence of 12 percent for Lyme disease was estimated, qualified with the notation that some reporting bias was inherent. Fear of acquiring infection had become so prevalent on Gibson Island that residents avoided contact with vegetation and even their own backyards. Their "tick committee" provided funds to the Harvard laboratory over a period of three years to initiate studies and make recommendations for risk reduction. Basic epidemiological studies indicated that Lyme disease was occurring only infrequently on Gibson Island, although the agent was enzootic there. Instead, a Lyme disease mimic, manifesting as erythema migrans and associated with Lone Star tick bites, was prevalent there, as it was elsewhere in the southern and central United States (Armstrong et al. 1996, 2001). The etiology of Masters disease, the name we have coined for this syndrome (Telford, Dawson, and Halupka 1997), remains obscure. The aggressive Lone star tick had been misidentified by Gibson Island residents as the deer tick, providing additional anecdotal support for the perception that Lyme disease was epidemic there.

## Academic Institutions

Entomologists from the Harvard School of Public Health who had previously focused their efforts on mosquito-borne infection invested research effort in tick-borne disease after the second case of human babesiosis was recorded in the region. Funds were raised locally and later by contract with the CDC. Soon thereafter, rheumatologists at the Yale School of Medicine documented an anomalous clustering of juvenile rheumatoid arthritis diagnoses in south-central Connecticut (Steere et al. 1976). Dr. Allen Steere led this work at Yale and continues to provide a "court of last resort" at the Massachusetts General Hospital for people suffering from Lyme disease. Many academic institutions have engaged in work on these infections.

Although few infections carried by deer ticks occur in the southeastern United States, university investigators in that region have sought to join in this body of work. Dr. James Oliver at Georgia Southern University, for example, lamented the "general attitude among physicians and veterinarians . . . that Lyme disease is not a problem in that area" and proposed "to demonstrate that *I. dammini* is not distinct from *I. scapularis*" (Oliver et al. 1993:55). This objective was satisfied when no differences could be detected between two isolates of these ticks that had been maintained in the same laboratory for many years. Contamination seems likely because these isolates share identical mitochondrial characteristics (Norris et al. 1996). Synonymization satisfies the CDC surveillance criterion for endemic Lyme disease, which requires that a recognized vector of Lyme disease must be present. The resulting taxonomic controversy continues to confuse the epidemiology of these infections (Telford 1998).

Agricultural research stations have participated in work on the deer-related infections. The Connecticut State Agricultural Experiment Station, under the direction of Dr. John Anderson, has registered a series of important observations on these infections. Investigators at the federal Agricultural Research Service in Kerrville, Texas, have developed a device for applying acaricide to the bodies of deer. Although this "four-poster" device appears useful for reducing tick abundance in intensely zoonotic sites in the northeastern states, the threat of chronic wasting disease in deer may obviate any measures that involve feeding stations for these animals.

## Wildlife Organizations

The deer that had virtually been eliminated from the eastern United States began to return after the turn of the twentieth century (Spielman et al. 1985), particularly after 1937, when the Pittman-Robertson Act began to organize federal and state wildlife management efforts throughout the country. The goal was to manage deer density in a manner that would provide hunters with game and would serve the conservationist community.

State wildlife organizations also seek to protect the public health, although these contrasting roles can be contradictory.

An operational effort to eliminate Lyme disease was launched in a collaboration linking personnel of the Massachusetts Division of Fisheries and Wildlife with colleagues working at the HSPH. This "Great Island Experiment" began in 1981 on a three-hundred-hectare tombolo attached to Cape Cod in Massachusetts (Wilson et al. 1988). Some thirty-five deer appeared to be present on this one-square-mile site. Sharpshooters selectively eliminated virtually all these deer during 1983 through 1984. Fewer than six remained. Only about a tenth as many ticks infested mice after the deer were shot than before, and the cases of human disease that annually occurred among the two hundred residents of the site declined to nil. Only two cases of Lyme disease and one of babesiosis have subsequently been recognized. This collaboration between university personnel and a state wildlife organization resulted in a long-term improvement in the health of a recreational community and serves as the model for future efforts.

Deer density was reduced in at least two subsequent campaigns that were designed to protect the public health. The effort launched in northeastern Massachusetts, in Ipswich, was conducted by the Trustees of Reservations, a nonprofit organization dedicated to land conservation, in collaboration with personnel from the HSPH. Deer on an isolated landmass were destroyed by a sharpshooter and by a controlled hunt (Wilson and Deblinger 1993) and reduced to a density that restored the health of the deer herd and reduced tick burdens on mice. Another such effort served to eliminate the sixty or so deer that infested isolated Monhegan Island, Maine. Although the evaluation of this experiment is ongoing, tick densities have severely declined (Dr. Robert P. Smith Jr., pers. comm.). Other deer-targeted interventions include an effort to exclude deer from a site in Westchester County, New York, by means of a high fence (Daniels, Fish, and Schwartz 1993).

State wildlife managers and policymakers in Massachusetts worked closely with HSPH personnel in a cooperative relationship that fostered a body of knowledge that helped change the way deer are managed. This wildlife management policy is based on a concept termed "wildlife cultural carrying capacity" (Ellingwood and Spignesi 1986), which takes into account human tolerance for various kinds of wildlife and how that tolerance may wane if, as in the case of deer, human health is threatened. Even though deer do not transmit Lyme disease directly to people, the public will associate the disease with deer if research has shown a connection between the two. (The same can be said for beaver and giardiasis.) As a result, Connecticut and Massachusetts have reduced their coastal deer population goals based on their understanding of cultural carrying capacity. In Massachusetts, for example, goals for deer density are 8 deer per square mile in the eastern part of the state and 15 farther west. Because of biological carrying capacity, however, densities in the eastern part of the state tend to be higher than those in the western part, owing mainly to mild winters in eastern Massachusetts. Habitats there can support more deer per square mile than those in western Massachusetts. For this reason, the Crane Reservation in the northeastern corner of Massachusetts supported 120 deer per square mile and Nantucket Island as many as 80, whereas in western Massachusetts deer density never exceeded 50-60 per square mile. Such densities greatly exceed optimal levels. With the objective of starting to reduce the size of the deer herd slowly so that their grandchildren might be less afflicted, Nantucket citizens petitioned the state wildlife agency to expand the regular season for shotgun hunting by one week. The number of deer taken each year for the three years following the expanded hunting season exceeded four hundred, a doubling of the harvest. Although we cannot yet evaluate its effect on deer or tick density, this approach may serve as a model for gradual management of deer herds by increased hunting pressure. Unfortunately, such an intensification in the removal of deer by extending the hunting season by a week may now be negated by the concomitant efforts to promote tourism on Nantucket during the late fall. The popular "Christmas Stroll" season has been extended, beginning before Thanksgiving and ending in early January, and as a result hotel owners no longer reduce their rates in time for the hunting season; virtually all off-island hunters who would normally have benefited from bargain hotel rates now seek to hunt elsewhere.

### Industry

A veterinary vaccine designed to protect dogs against Lyme disease began to be marketed by the Fort Dodge Company soon after a method for culturing *Bo. burgdorferi* in BSK medium was developed. The product received regulatory approval based on its apparent safety, but long before efficacy ultimately was demonstrated by Levy, Lissmann, and Ficke (1993). This preparation of killed cultured spirochetes continues to be marketed aggressively by suburban veterinarians. Instead of informing potential customers that this disease tends to be mild and self-limiting in dogs (Greene 1992), the company's advertisements state that "Lyme disease is a devastating bacterial disease which can cause permanent painful disability in both humans and animals . . . and is . . . often fatal." In fact, only a few human fatalities have conclusively been attributed to this infection. Although an infected dog poses little threat to the health of its owners, various users of this product have informed us that they vaccinated their dog because they thought that this purchase protected the health of their children as well as the lives of their dogs. By stating that Lyme disease "has been reported in 47 states," proponents of canine vaccination imply that properties located virtually anywhere within the continental United States convey risk. Risk, of course, is exceedingly local. We note that this canine vaccine continues to be distributed broadly, including in many sites in which risk of infection is virtually nil.

A vaccine designed to protect people against Lyme disease was developed by SmithKline Beecham. Marketed under the trade name LYMErix, this product was approved by the Food and Drug Administration based on a series of safety and efficacy trials (Steere et al. 1998). The immunogen is a recombinant outer surface protein of the spirochete (designated as OspA) that is expressed mainly by spirochetes residing in the guts of nonfeeding ticks and by spirochetes that have been produced in artificial culture. This novel vaccine causes spirochetes to be destroyed in the gut of the infected tick before they can migrate to the salivary glands and before the OspA epitope can be presented to the vaccinated host (Telford and Fikrig 1995). Efficacy appears to be about 70–80 percent. Immunity, of course, is specific for certain Lyme disease spirochetes and does not include the agents of ehrlichiosis or babesiosis. Various "genospecies" of the Lyme disease agents may similarly escape vaccine protection. Advertising for this product, which included children exposed happily in wooded sites, may have encouraged people to relax their efforts to protect themselves against ticks in the mistaken impression that they are fully protected against Lyme disease as well as all other deer-tick borne infections. The availability of LYMErix appears to have reduced motivation, in at least two coastal New England communities, for efforts to reduce deer density. Despite these problems and the withdrawal of the vaccine by the manufacturer in response to disappointing sales, the development of this vaccine was an important public health milestone. A vaccine directed against tick-borne infection conveys unique public health implications.

Various acaricides designed to protect against infections transmitted by deer ticks have been marketed aggressively. Lawn care and pest control companies contract to apply acaricidal sprays to residential properties, and this seems to reduce risk locally, although at some cost to the environment. Permethrin sprays are marketed under the trade name Permanone for application to the clothing for people walking through tick-infested brush; the US. Armed Forces rely on permethrin-impregnated uniforms to reduce the risk of bites by hematophagous arthropods, and this approach seems to reduce the risk of acquiring a tick bite (Schreck, Snoddy, and Spielman 1986). Although permethrin is harmless for people, anecdotal reports of neurotoxicity due to overapplication of diethyltoluamide (DEET, the only effective mosquito repellant) by mothers concerned that their children may be bitten by ticks suggests the potential for chemical hazards associated with risk reduction. Cost-benefit analyses of the environmental and health implications of the various chemically based measures remain to be performed, but these chemicals continue to be used with uncertain degrees of efficacy.

A method for delivering acaricide to the rodent hosts of these deer-related infections was developed and patented by personnel of the HSPH in 1988. Tubes containing permethrin-impregnated cotton were distributed across sites that were perceived to convey risk of infection. A company known as EcoHealth was established (by one of us, AS) to market this product. Although efficacy was demonstrated in a series of studies conducted in Massachusetts (Mather, Ribeiro, and Spielman 1987), this conclusion was challenged in a series of studies conducted in New York (Daniels, Fish, and Falco 1991) and Connecticut. The resulting controversy has limited the distribution of the product.

Serological diagnosis for Lyme disease appears to have rendered various small businesses profitable. A hospital or commercial clinical laboratory may charge as much as $75 for a single serological test, and at least fifty different commercial diagnostic kits have become available on the market. Unfortunately, failure to recommend standards and lack of a national program for proficiency testing have promoted the perception that serodiagnosis for Lyme disease is inaccurate or unreliable. Concordance was poor when various of these diagnostic kits were used to assay identical serum samples (Bakken, Case, and Callister 1992). More recently, standardization of reagents and approaches has allowed greater specificity of serological assays for the Lyme disease spirochete. Then too, the publication of recommended standards by the CDC (Anonymous 1995) has promoted a greater sense of confidence in interpreting serological results for Lyme disease.

### Advocacy Organizations and Lay Individuals

Numerous advocacy groups have lobbied state governments as well as the federal government and seek to disseminate information about this group of infections. Many such groups have provoked scientific controversy, for example, by promoting the opinions that congenital Lyme disease is underrecognized, that chronic infections may persist indefinitely, that a prolonged course of antibiotic therapy for Lyme disease frequently is required, and that risk of infection is ubiquitous and intense. Unfortunately, a politically charged climate has resulted from the dialogue (or lack thereof) between various of these patient advocacy groups and many scientific research workers and from a failure of the lay public to distinguish anecdote from peer-reviewed data. The financial support that the NIH and CDC have provided for research on the deer-related zoonoses, of course, owes much to the efforts of these advocacy groups.

Numerous private individuals and advocacy groups have prepared and distributed newsletters that advance one or another opinion about Lyme disease and established home pages on the Internet. Prominent among these is a newsletter and Web site named "Lymetruth," which described an advocacy action as follows: "Protesters outside the school denounced Yale's claim that this bacterial infection is 'over-diagnosed, over-treated, and simple to cure.' They spoke of chronic Lyme disease with crippling arthritis, paralysis, vision problems, heart problems, dementias, and clinical depression, leading to suicide" (Dodge 1998). In contrast to these adversarial activities, the substantial energies of these articulate and passionate people might more usefully be directed toward efforts to promote community awareness of tick-borne infection and expanding dialogue with scientists.

The Connecticut-based Friends of Animals has stridently advocated another point of view. This organization has opposed efforts to reduce the density of deer, and its members have demonstrated publicly to disseminate this point of view. The organization launched a vigorous protest demonstration in 1989 in response to the effort by the Trustees of Reservations to reduce deer density on the Crane Reservation. Its efforts included the formation of a human chain that blocked road access to the reservation and invasion of the site by people dressed as deer. This well-publicized and televised event may have inhibited subsequent efforts to reduce deer density.

Following a suggestion contained in a letter in the *New England Journal of Medicine,* certain physician groups have recommended "malariatherapy" for the treatment of Lyme disease. This treatment was inspired by the knowledge that malarial infection was once induced in syphilitics to produce a febrile condition that reduces the symptoms of this spirochetal disease and by the notion that other spirochetoses should respond similarly. Indeed, a source in the northeastern United States has distributed human blood containing viable *Plasmodium vivax* organisms with the recommendation that they be injected therapeutically (Rawlings et al. 1991). At least one episode of self-induced malaria was reported. It may be that an autochthonous case of *P. vivax* in a New Jersey boy in 1993 was the result of local anophelines feeding on such a nearby malariotherapy patient.

Individual residents of enzootic sites in the northeastern United States that are affected by Lyme disease played a major role in the earliest responses to this threat to the public health. These contributions included activities initiated by the affected residents themselves as well as by residents' associations. Numerous physicians, public administrators, industrialists, and philanthropists are included among these exceptionally sophisticated people. At least four U.S. presidents, for example, are among them, and one, Bill Clinton, has visited Martha's Vineyard repeatedly. He visited Nantucket Island in 2000. These prominent people possess the attitudes and abilities necessary to recruit whatever resources may be necessary to protect their health and that of their neighbors.

## The Courts

Lyme disease is responsible for a major change in tort law that permits suits to be filed in the face of this and other vector-borne infections. If property owners are aware that a hazard is associated with their property, they are expected to warn their visitors or employees and to take due care to protect their health. But because arthropods tend to be regarded as "hidden hazards" and "parts of nature," no vector-borne infection could be anticipated, and any preventive measures would be unnatural. Negligence could not be demonstrated. Track workers employed by a railroad operating in a zoonotic part of Long Island, however, argued successfully in 1989 that they should have been warned about Lyme disease and advised to protect themselves against ticks. The case was facilitated because federal law obliges employers to provide a safe workplace and warn their employees in the event of a perceived hazard. Businesspeople are governed by less rigorous court precedent and are required only to follow the "industry standard" in preventing exposure while applying reasonable measures to reduce risk to patrons and business visitors. Homeowners need take only the "ordinary care" that is practiced in the neighborhood. The 1989 precedent has been followed by other cases that were successfully pursued. Various states have recently been stratifying risk of acquiring Lyme disease and have demonstrated that risk is great in certain sites. This creates an obligation to "warn," and such warning signs have begun to proliferate in certain neighborhoods. The obligation to "take care" has been rendered more critical by the recent availability of various protective measures. Numerous public agencies have recommended measures for reducing risk of infection. Proof of "negligence" in the case of Lyme disease is facilitated because risk has been recognized as place-specific and risk of injury can be reduced. Ticks are no longer considered to be ubiquitous and a part of nature.

The commercial availability of a vaccine designed to protect against Lyme disease or other tick-borne infection might increase the frequency of litigation because such a product would set a new standard of care. Property owners or employers would acknowledge this standard if they delivered this vaccine to some users of their properties and may be under some obligation to extend this protection to others. By purchasing such a vaccine and delivering it generally to the members of a household, the property owner establishes a standard of care for that household that might be extended to visitors or employees. Vaccination for Lyme disease or other deer-related infections carries a special burden because such products serve uniquely place-related diseases.

The direct correlation between deer density and risk of the infections transmitted by deer ticks in enzootic regions may come to affect the policies of agencies concerned with conservation or wildlife management. Risk is likely to correlate with proximity to properties on which these activities take place. These deer-related zoonoses carry special burdens.

**Local Government**

The Board of Selectmen of the Town of Nantucket began an inquiry soon after four summer residents and one permanent resident of the island acquired babesial infection in 1974. This first multiple-case transmission season resulted in an aggressive distribution of notices at the local airport and the ferry landing. Notification soon became far more passive, perhaps because of a fear that widespread hysteria might inhibit tourism. One of us, posing as a tourist, periodically visited the Nantucket information bureau from 1987 to 1997 to inquire about the risk of tick-borne diseases. Until the mid-1990s, the standard response was that "ticks were no more common on Nantucket than elsewhere"; informational brochures about Lyme disease were stored out of sight behind the counter and were available only when more information was requested specifically. Since 1995, such information has been displayed more prominently. The local newspapers covered significant developments in the ongoing investigations but subsequently limited their reporting to isolated informational news articles that were presented during the height of "tick season," perhaps as a result of commercial influence. The *Block Island Times,* in contrast, routinely placed a one-sixteenth-page community service advertisement in each issue that warned of possible tick-related hazards and recommended preventive measures. Other local newspapers are tending to follow suit, and even local chamber of commerce directories now contain "tick advice."

## State Government

The state of New Jersey has led the various affected states in responding to the emergence of Lyme disease. Following two decades of discussion, its legislature passed a bill in 1997 that authorized the various county authorities to designate agencies to "provide surveillance, education, training, and recommendations on integrated pest management for the management of Lyme disease or other tick-borne disease vectors" (Chomsky 1999). Connecticut's exemplary tick-borne disease surveillance program, in collaboration with the CDC, was largely stimulated by citizen advocacy operating through state representatives. Various other states have entered into cooperative agreements with the CDC to describe the distribution of these infections within their borders.

The attorney general of the state of Connecticut held a public hearing to resolve a dispute relating to insurance coverage for the diagnosis and treatment of Lyme disease (Feder 2000). One "Lyme Camp" included "academic-based physicians who use[d] limited antibiotic therapy to treat patients with Lyme disease"; this camp was pitted against a second camp, which included "community-based physicians and patient advocacy groups who believe[d] in open-ended antibiotic therapy" (Feder 2000: 855). The first camp argued for an objective set of case criteria and a treatment regimen based on scientific evidence, whereas the second favored a subjective system of diagnosis and a "therapeutic alliance between physician and patient" (855). Extensive arguments were presented. Several months later, the Connecticut legislature passed Public Act 99-2, which denied the objective arguments and now requires health insurance companies to pay for the prolonged courses of treatment specified by the activist Lyme camp. State legislatures are vulnerable to arguments that are not scientifically based.

## Federal Government

The first of these infections carried by deer ticks to be recognized in a resident of North America was discovered as a result of a parasitological referral to the CDC. This 1968 case of human *Ba. microti* infection was acquired on Nantucket Island in Massachusetts (Western et al. 1970). The case originally was referred to a university teaching hospital in New Jersey, where it was misdiagnosed and reported to the CDC as an autochthonous *Plasmodium falciparum* infection, a reportable disease. An accompanying microscopic preparation was transmitted to Dr. George Healy of the CDC Parasitology Branch in Atlanta. His correct diagnosis laid the basis for an understanding of the various key elements in the epizootiology of the entire guild of rodent-borne zoonoses. An Epidemiological Intelligence Service (EIS) officer, Dr. Karl Western, was assigned to describe the case, an event that was regarded as a medical oddity. No further response seemed necessary.

The CDC later assigned another EIS officer, Dr. Trenton Ruebush, to investigate the epidemiology of "Nantucket fever" after the series of additional cases that were diagnosed during the mid-1970s. At the same time, a research contract was issued to the HSPH, where one of us (AS) was investigating the epizootiology of this infection. The resulting collaboration produced a series of publications that were coauthored by CDC and Harvard University personnel.

Lyme disease eventually served to stimulate the formation of a branch of the CDC's Division of Vector-Borne Infectious Disease devoted to surveillance, epidemiology, diagnosis, and intervention research on this infection and on plague. Dr. David Dennis served as the first chief of the Bacterial Zoonosis Branch (BZB), and an HSPH doctoral graduate who had undertaken his thesis on Nantucket fever, Dr. Joseph Piesman, serves as Lyme disease vector branch chief. This unit established a dialogue that arrived at an ever evolving case definition and has posted recommendations for vaccine distribution. The CDC records Lyme disease incidence and geographical distribution and has issued a series of cooperative agreements that are designed to monitor the geographical distribution of Lyme disease. Some $4.3 million in contract support are distributed annually to regional health authorities and universities for Lyme disease–related work (in 1999). The BZB operates on an internal budget of just over $2 million. It seems curious that CDC personnel working on the infections transmitted by deer ticks are assigned to different laboratories. Those focusing on human babesiosis or on HGE continue to work in Atlanta, Georgia, where the early investigators on this subject were based. Those focusing on Lyme disease or the various virus infections work in Fort Collins, Colorado.

Personnel of the NIH first became involved in studies related to tick-borne disease when a malariologist, Dr. Louis Miller, now head of the Laboratory of Parasitic Disease, undertook a series of studies of Nantucket fever during the late 1970s. A fundamental discovery concerning Lyme disease resulted when a scientist at the Rocky Mountain Laboratories (RML) of the NIH, Dr. Willy Burgdorfer, received a shipment of ticks that were sampled from Shelter Island in New York, a site in which both human babesiosis and Lyme disease were frequent (Burgdorfer 1984). The etiology of Lyme disease remained entirely unknown until his serendipitous discovery of spirochetes within the guts of these deer ticks that he was analyzing for evidence of spotted fever rickettsiae. A colleague at the RML, Dr. Alan Barbour, collaborated in this work and developed the basic method for culturing these organisms. The RML soon dedicated a laboratory to studies relating to Lyme disease, where Dr. Thomas Schwan has defined crucial elements in the vector-pathogen relationship. NIH personnel have participated directly in the study of Lyme disease and its co-transmitted infections.

The NIH began to provide investigator-initiated grant support for external research on infections related to deer ticks during the 1970s, soon after this complex of infections began to emerge as a threat to the public health. An office for coordinating Lyme disease grants was established during the late 1980s. This program of research support enjoys a gratifying level of success. Indeed, the NIH distributed $16 million in 1999 to more than forty external laboratories for research related to Lyme disease.

## Conclusions

The emergence of Lyme disease and its associated infections in the northeastern United States stimulated a broad variety of societal responses during the past several decades. The affected populations mobilized themselves and recruited assets from many levels of government and stimulated the efforts of university research workers. The extraordinary affluence, educational status, and societal prominence of these people resulted in a huge research investment in this disease, involving about as many federal research dollars as are invested in malaria, a disease that annually kills millions. The quality of the resulting countermeasures has been mixed, largely because conflicting interests may have motivated certain individuals, advocacy groups, industrial operations, and university laboratories. Although incidence of these tick-borne infections continues to increase, meritorious programs have been launched by various local, state, and federal governmental agencies as well as many of those in nongovernmental sectors. This extraordinary diversity of effort provides a model for what is possible when a novel disease emerges among an exceptionally motivated group of people. Where extensive resources are available, diverse institutions will respond to the challenge posed by an emerging infection.

### NOTES

Research for this chapter was supported in part by grants from the National Institutes of Health (AI 42402, AI 39002 and AI 19693) and a contract from the Centers for Disease Control and Prevention.

1. Aronowitz 1998: 57–83.

2. The landscape of the eastern United States changed during the several centuries that followed the arrival of European settlers in North America (Cronon 1983). The

first such visitors were greeted by vast open meadows and brush-free groves of stately trees. Verrazano in 1583, Gosnold in 1604, and the Pilgrims in 1619 were entranced by the vistas that greeted them. The land was soon converted into fields and the forests into cordwood, charcoal, and masts for British warships. Few large mammals survived. Henry David Thoreau wrote in the 1830s that he knew an elderly woman who, in her youth, knew someone who had once seen a deer. But the rise of agriculture in the Great Plains destroyed the agricultural economy of the East, and much land was abandoned. Although 75% of the land in the region was devoted to farming in 1880, brush came to cover more than half of the countryside by the 1920s. The resulting old-field habitat created an ideal environment for white-tailed deer, which feed by browsing rather than grazing on grass. Brush-free forests, of course, provide little sustenance for such animals, and the new brush-choked, predator-free landscape that developed during the twentieth century provided a fruitful habitat for deer and their acarine ectoparasites.

Landscape changes of similar magnitude occurred in the upper Midwest. The mature forest that characterized the region during the 1850s harbored some five deer per square kilometer, but as many as thirty-five by the 1940s (Alverson, Waller, and Solheim 1988). The dairy industry cleared large expanses of meadowland, which was surrounded by a brush-choked ecotone. In addition, wildlife management practices sought to increase the deer herd by creating "wildlife openings" in the forest, which created increased edge. These anthropogenic changes in the landscape, both in the upper Midwest and in the Northeast, promoted the density of deer.

Although deer ticks are not present wherever deer are abundant, the relationship between deer density and that of these ticks is direct. They may infest deer in all their developmental stages and feed mainly on these host animals as adults (Piesman et al. 1979). All but 5% of adults feed on deer, with the remainder parasitizing such alternative hosts as raccoons, opossums, or feral cats (Wilson et al. 1990b). Islands that are free of deer sustain no deer ticks (Anderson et al. 1987), and the density of their scat correlates strongly with the density of these ticks (Wilson, Adler, and Spielman 1985). Indeed, deer ticks are most abundant where deer are most dense (Wilson et al. 1990a). It follows that tick density on mice increases after deer density is increased by baiting with acorns (Jones et al. 1998). Risk of human infection correlates directly with the density of deer around houses (Lastavica et al. 1989). Taken together, these observations demonstrate that risk of tick-borne infection relates largely to deer density.

REFERENCES

Alverson WS, Waller DM, Solheim SL. 1988. Forests to deer: Edge effects in northern Wisconsin. Conserv. Biol. 2:348–358.
Anderson JF et al. 1987. Prevalence of *Borrelia burgdorferi* and *Babesia microti* in mice on islands inhabited by white-tailed deer. Appl. Environ. Microbiol. 53:892–894.
Anonymous. 1995. Recommendations for test performance and interpretation from the Second National Conference on Serologic Diagnosis of Lyme disease. MMWR 44:590–591.

Armstrong P et al. 1996. A new *Borrelia* infecting Lone Star ticks. Lancet 347:67–68.

Aronowitz R. 1998. Making Sense of Illness. Cambridge: Cambridge University Press.

Armstrong PM, Rosa Brunet L, Spielman A, Telford SR III. 2001. Risk of Lyme disease: Perceptions of residents of a Lone Star tick-infested community. Bull WHO 79:916–925.

Bakken JS et al. 1994. Human granulocytic ehrlichiosis in the upper Midwest United States. A new species emerging? JAMA 272:212–218.

Bakken LL, Case KL, Callister SM. 1992. Performance of 45 laboratories participating in a proficiency testing program for Lyme disease serology. JAMA 268:891–896.

Burgdorfer W. 1984. Discovery of the Lyme disease spirochete and its relation to tick vectors. Yale J. Biol. Med. 57:515–520.

Chomsky MS. 1999. Tick-Tock-Doe tick management legislation for New Jersey. Wing Beats 10:18–34.

Cronon W. 1983. Changes in the Land: Indians, Colonists, and the Ecology of New England. New York: Hill and Wang.

Daniels TJ, Fish D, Falco RC. 1991. Evaluation of host-targeted acaricide for reducing the risk of Lyme disease in southern New York state. J. Med. Entomol. 28:537–543.

Daniels TJ, Fish D, Schwartz I. 1993. Reduced abundance of *Ixodes scapularis* (Acari: Ixodidae) and Lyme disease risk by deer exclusion. J. Med. Entomol. 30:1043–1049.

Dodge D. 1998. Lymetruth, Special Yale Issue Guilford, Conn., no.5 (June).

Ellingwood MR, Spignesi JV. 1986. Management of an urban deer herd and the concept of cultural carrying capacity. Trans. Northeast Deer Technical Committee 20:6–7.

Feder HM Jr. 2000. Differences are voiced by two Lyme camps at a Connecticut public health hearing on insurance coverage of Lyme disease. Pediatrics 105:855–857.

Glass GE et al. 1994. Predicting *Ixodes scapularis* abundance on white-tailed deer using geographic information systems. Am. J. Trop. Med. Hyg. 51: 538–544.

Greene RT. 1992. Questions "push" for vaccination against *Borrelia burgdorferi* infection [letter]. J. Am. Vet. Med. Assoc. 201:1491–1493.

Jones CG et al. 1998. Chain reactions linking acorns to gypsy moth outbreaks and Lyme disease risk. Science 279:1023–1026.

Krause PJ et al. 1996. Concurrent Lyme disease and babesiosis: Evidence for increased severity and duration of illness. JAMA 275: 1657–1660.

Krause PJ et al. 1998. Persistant parasitemia after acute babesiosis. N. Engl. J. Med. 339:160–165.

Krause PJ et al. 2003. Increasing health burden of human babesiosis in endemic sites. Am. J. Trop. Med. Hyg. 68:431–436.

Lastavica CC et al. 1989. Rapid emergence of a focal epidemic of Lyme disease in coastal Massachusetts. N. Engl. J. Med. 320:133–137.

Levy SA, Lissmann BA, Ficke CM. 1993. Performance of a *Borrelia burgdorferi* bacterin in borreliosis-endemic areas. J. Am. Vet. Med. Assoc. 202:1834–1838.

Marshall WF et al. 1994. Detection of *Borrelia burgdorferi* DNA in museum specimens of *Peromyscus leucopus*. J. Infect. Dis. 170:1027–1032.

Mather TN, Ribeiro JMC, Spielman A. 1987. Lyme disease and babesiosis: Acaricide focused on potentially infected ticks. Am. J. Trop. Med. Hyg. 36:609–614.

Maupin GO et al. 1994. Discovery of an enzootic cycle of *Borrelia burgdorferi* in *Neotoma mexicana* and *Ixodes spinipalpis* from northern Colorado, an area where Lyme disease is nonendemic. J. Infect. Dis. 170: 636–643.

Norris DE, Klompen JSH, Keirans JE, Black WC. 1996. Population genetics of *Ixodes scapularis* (Acari: Ixodidae) based on mitochiondrial 16S and 12S genes. J. Med. Entomol. 33:78–89.

Oliver JH et al. 1993. Conspecificity of the ticks *Ixodes scapularis* and *I. dammini* (Acari: Ixodidae). J. Med. Entomol. 30:54–63.

Piesman J, Spielman A, Etkind P, Ruebush TK, Juranek DD. 1979. Role of deer in the epizootiology of *Babesia micoti* in Massachusetts USA. J. Med. Entomol. 15:537–540.

Rawlings J, Perdue J, Perrota D, Simpson D. 1991. Update: Self-induced malaria associated with malariotherapy for Lyme disease. MMWR 40:665–666.

Schreck CE, Snoddy EL, Spielman A. 1986. Pressurized sprays of permethrin or deet on military clothing for protection against Ixodes dammini. J. Med. Entomol. 23:396–399.

Schulze TL et al. 1984. *Ixodes dammini* (Acari: Ixodidae) and other ixodid ticks collected from white-tailed deer in New Jersey, USA. J. Med. Entomol. 21:741–749.

Scrimenti RJ. 1970. Erythema chronicum migrans. Arch. Dermatol. 102:104–105.

Solberg VB, Miller JA, Hadfield T, Burge R, Schech JM, Pound JM. 2003. Control of *Ixodes scapularis* (Acari: Ixodidae) with topical self-application of permethrin by white-tailed deer inhabiting NASA, Beltsville, Maryland. J. Vector Ecol. 28:117–134.

Spielman A et al. 1979. Human babesiosis on Nantucket Island, USA: Description of the vector, *Ixodes (Ixodes) dammini*, n.sp. (Acarina: Ixodidae). J. Med. Entomol. 15: 218–234.

Spielman A et al. 1985. Ecology of *Ixodes dammini* borne human babesiosis and Lyme disease. Annu. Rev. Entomol. 30:439–460.

Steere AC et al. 1976. Lyme arthritis: An epidemic of oligoarticular arthritis in children and adults in three Connecticut communities. Arthritis and Rheumatism 20:7–17.

Steere AC et al. 1998. Vaccination against Lyme disease with recombinant *Borrelia burgdorferi* outer surface lipoprotein A with adjuvant. N. Engl. J. Med. 339:209–215.

Telford SR III. 1998. The name *Ixodes dammini* epidemiologically justified. Emerg. Infect. Dis. 4:132–133.

Telford SR III, Dawson JE, Halupka KC. 1997. Emergence of tickborne diseases. Sci. Med. 4:24–33.

Telford SR III, Fikrig E. 1995. Prospects towards a vaccine for Lyme disease. Clinical Immunotherapeutics 4:49–60.

Telford SR III, Spielman A. 1989. Enzootic transmission of the agent of Lyme disease in rabbits. Am. J. Trop. Med. Hyg. 41:482–490.

Tyzzer EE. 1938. *Cytoecetes microti*, n.g., n.sp., a parasite developing in granulocytes and infective for small rodents. Parasitol. 30:242–257.

Western KA et al. 1970. Babesiosis in a Massachusetts resident. N. Engl. J. Med. 283:854–856.

Wilson ML, Adler GH, Spielman A. 1985. Correlation between abundance of deer and that of the deer tick, *Ixodes dammini* (Acari:Ixodidae). Ann. Entomol. Soc. Am. 78:172–176.

Wilson ML, Deblinger RD. 1993. Vector management to reduce the risk of Lyme disease. Pp. 126–156 in Ecology and Environmental Management of Lyme Disease, ed. Ginsberg HS. New Brunswick, N.J.: Rutgers University Press.

Wilson ML, Telford SR III, Piesman J, Spielman A. 1988. Reduced abundance of immature *Ixodes dammini* (Acari: Ixodidae) following elimination of deer. J. Med. Entomol. 25:224–228.

Wilson ML et al. 1990a. Microgeographic distribution of immature *Ixodes dammini* (Acari: Ixodidae) correlated with that of deer. Med. Vet. Entomol. 4:151–160.

Wilson ML et al. 1990b. Host dependent differences in feeding and reproduction of *Ixodes dammini* (Acari:Ixodidae). J. Med. Entomol. 27:945–954.

# CHAPTER 12

# The Politics of Institutional Responses

## CDC and the Controversy over Maternal and Newborn HIV Testing

LYDIA OGDEN

The emergence of the HIV/AIDS epidemic posed an unprecedented challenge for public health institutions in the United States and abroad. The Centers for Disease Control and Prevention, charged with investigating infectious disease outbreaks in the United States, played a central role in the early identification and investigation of AIDS. From the outset, AIDS posed a significant challenge to the CDC for three reasons.

First, AIDS emerged at a historical moment when many in the public health community had concluded that infectious disease threats were a thing of the past. As Ruth Berkelman and Phyllis Freeman note in their essay on emerging infections and the CDC (Chapter 13), by the mid-eighties the CDC's mission had shifted away from infectious diseases and toward the study of chronic health problems. With the emergence of AIDS, resources had to be redirected to cope with this new emerging infection.

Second, as indicated in chapter 1, the nature and dimension of the AIDS epidemic did not coincide with the CDC's corporate culture. AIDS was a different kind of infectious threat, one that required a different level of commitment, community cooperation, and sustained response than the CDC was initially prepared to deliver.

Third, the AIDS epidemic rapidly became politicized and a target of massive media attention. As Lydia Ogden describes in the following analysis of the CDC's HIV Survey of Childbearing Women (SCBW), the institutional culture of the CDC was not prepared to deal effectively with either the politics or the media attention generated by the AIDS epidemic. Ogden's study is important for what it tells us about the internal workings of the nation's front-line defense against emerging illnesses. Her conclusions concerning the agency's handling of the controversy that developed around the SCBW serve as a backdrop to explain the CDC's evolving agility and capability in serving as the government's lead organization in informing the public about more recent health threats, including anthrax, smallpox, West Nile virus, severe acute respiratory syndrome (SARS), and other emerging health threats.

---

Over the past two decades the Centers for Disease Control and Prevention has demonstrated remarkable leadership in guiding the nation's war against HIV/AIDS. At the same time the history of the CDC's institutional response to the HIV/AIDS epidemic has been marked by a series of controversies. These have included debates over the CDC's initial identification of AIDS as gay-related, its early definition of "at-risk populations," and whether early case definitions of the disease excluded many infected women. The present case study examines the controversy that developed over the CDC's Survey of Childbearing Women. The survey was initiated in 1988 as a means of tracking the course of the AIDS epidemic in the United States. The agency suspended the survey in 1995 after a long and costly public battle waged in national and local media and in federal and state legislatures. The CDC's handling of the survey and the controversy that developed around the survey eventually resulted in a loss of credibility and political capital for the CDC. The analysis that follows traces the history of the SCBW controversy. By doing so, it attempts to demonstrate the ways in which changes in science and politics influenced the CDC's institutional responses to AIDS.

## The Survey of Childbearing Women

The CDC began the survey in 1988 as part of a national serosurveillance system designed to track the epidemic of HIV infection in the United States by monitoring blood samples from certain groups in the broader population. The effort to construct a system to monitor asymptomatic HIV infection was driven by epidemiological and health care planning needs and gained steam under the Reagan administration, which encouraged the CDC to develop reliable estimates of the numbers of Americans infected. One of the suggestions was nationwide household blood specimen collection; however, a pilot door-to-door survey in Washington, D.C., planned by the National Center for Health Statistics (now part of the CDC) was abruptly canceled following protests.[1] After a similar pilot in Dallas, researchers determined that nonresponse bias would render any household serosurveillance unreliable. In contrast, blinded surveillance—testing blood specimens without links to individuals' identity—did not present the same hurdles and was more likely to render an accurate picture of at least a sample of the U.S. population. In this context, blinded surveillance was deemed effective and expedient, and the CDC's "family of surveys" was developed.

In 1987, state and local health departments were conducting a variety of serosurveys to track HIV prevalence and incidence.[2] With specially appropriated funding, the CDC set about the task of systematizing the patchwork of current efforts. Over the course of a year, protocols were drafted and approved, funds awarded, and surveys started from scratch or greatly expanded. As of May 1, 1989, the family of surveys included blinded surveillance in sexually transmitted disease (STD) clinics, drug treatment centers, women's health clinics, tuberculosis clinics, sentinel hospitals (those in large metropolitan areas that provide leading indicators of trends), the sentinel clinical practice network, sentinel laboratories, and the SCBW, which was then conducted in forty-one states, Washington, D.C., and Puerto Rico. In addition to the surveys it conducted, the CDC collaborated with other agencies that conducted either anonymous or identified testing, including HIV testing of college students and migrant and seasonal farmworkers. Blood collection agencies tested donors, the military tested all civilian applicants for military service, and the Job Corps tested all entrants, but not anonymously. Various prison systems tested prisoners, but the protocols differed: some states required incarcerated populations to be tested anonymously, for surveillance purposes only, and some were tested individually.[3]

*Blinded surveillance* is conducted by testing leftover blood specimens that have been permanently and irrevocably stripped of personal identifiers. It is not, therefore, conducted on "individuals." It is not the same as *anonymous testing*, which is conducted on individuals whose identity is delinked from the blood specimen and can be restored. Anonymous testing can be conducted either for surveillance purposes or for individual testing purposes, or both. If anonymous testing is conducted for individual testing purposes, results are provided to the person being tested, and tracking is conducted through an anonymous, unique identifier system, such as a combination of numbers and letters. Blinded surveillance and anonymous testing are therefore separate and distinct tools. This distinction was typically elided in the controversy over the SCBW.

By 1995, when the survey was suspended, the CDC was funding the SCBW in forty-five states, the District of Columbia, Puerto Rico, and the Virgin Islands. The five states not conducting the survey were Idaho, Nebraska, North Dakota, South Dakota, and Vermont, all low-prevalence areas.[4] The SCBW supplied estimates of the total number of HIV-positive women delivering children in a particular place and time and monitored prevalence over time. Results from the survey were used to estimate the need for HIV services for women and children, to stimulate the development of HIV prevention programs, to target resources, and, combined with other data, to estimate HIV prevalence in all women and the U.S. population. It provided warning signs that HIV was moving into the heterosexual population and stimulated the development of prevention programs as well as treatment and care programs for women and children in an environment in which the majority of funding was still dedicated to gay men and injection drug users. Indeed, when the survey was canceled, AIDS activists, particularly advocates for women, denounced the CDC. Eileen Hansen, president of the Women's AIDS Network, said, "Suspending the survey makes absolutely no public health sense. . . . It caves in to political pressure from those who would mandate testing and ignores the importance of the survey. No one disputes the fact that this is one of the best tools for planning the fight against HIV infection in women. Why cut it now, when women are one of the fastest growing populations of people with HIV?"[5]

The survey was based on HIV testing of leftover blood specimens collected on filter paper for routine newborn metabolic screening, an existing public health program in all areas conducting the survey. Leftover blood taken from newborns to test for phenylketonuria (PKU) and other treatable conditions was stripped of personal identifiers and tested for HIV antibodies. Some areas tested year-round; others tested only a sample of specimens. Because maternal antibodies to HIV cross the placenta during pregnancy and are exhibited by the newborn at birth, a baby's positive antibody test result reveals infection in the mother. It reveals HIV exposure, but not necessarily infection, in the infant. Without treatment, approximately 75 percent of infants born to HIV-infected mothers are not themselves infected; however, all will have a positive antibody test, owing to the mother's antibodies. Uninfected children lose maternal antibodies to HIV by the time they are twelve to eighteen months old—therefore, a positive antibody test after that point reveals infection in the child. At the time the SCBW was developed, antibody testing was the only practical method to assess infection; later, testing methods for the virus itself were refined, so that infection could be directly assessed, rather than simply antibodies to infection.

To preserve anonymity, only limited demographic data, including the month and county of birth, were abstracted from the metabolic screening form; in some states, the mother's age group and race/ethnicity were also collected. State health departments maintained a database of these results and periodically transferred the information to the CDC. It is important to note that no blood was taken specifically for the survey and that the samples were not identifiable. Within the scientific community, and under federal ethics guidelines defining human subjects research, those provisions meant that no extra blood was required and no identifiable individuals were being tested, two conditions that met federal ethical and legal criteria for determining that the survey did not constitute "human subjects research" at the time the SCBW was developed and implemented[6,7] and later revised.[8,9]

In December 1987, the CDC's Institutional Review Board (IRB) reviewed and approved a series of protocols for use in the family of serosurveys, including the SCBW. Although the internal IRB approved the surveys, the CDC requested that the Office for Protection from Research Risks (OPRR) at the National Institutes of Health (which initially funded some of the surveys) also review the protocols. The OPRR is designated as the Department of Health and Human Services' oversight office to ensure that participants in research conducted by DHHS agencies are protected.

In March 1988, the OPRR convened an ad hoc Regulation Interpretation Advisory Group to review the surveys. The OPRR determined that the regulatory definition of "human subject" was not applicable, since no extra blood was taken and the tests were anonymous, and therefore concluded that the HIV tests were not conducted on "human subjects," that is, on individuals who could be identified. Based on this determination, the CDC concluded that the SCBW did not constitute human subjects research under DHHS regulations and that the survey did not need either IRB review or local participating sites' Assurance of Compliance (required by the same DHHS regulations). Even so, the CDC's review board continued to review the survey on an annual basis;[10] state-level IRBs also reviewed and accepted the survey.

The National Academy of Sciences is a congressionally chartered, private organization that advises the federal government on matters of science; the National Research Council draws its members from three affiliate organizations: the National Academy of Sciences, the National Academy of Engineering, and the Institute of Medicine. In February 1989, a committee formed by the National Research Council, part of the National Academy of Sciences, recommended that blinded HIV surveillance be conducted on blood samples from all newborn infants, not just a sample. The CDC rejected the suggestion as too costly and unnecessary, asserting that the sample populations captured by the SCBW were sufficient to monitor infection trends adequately.[11] A letter to the editor of the *Los Angeles Times* following news reports of the council's report signaled public sentiment: the recommendation "seems guaranteed to offend just about everybody. . . . Were I the parent of a newborn child, I would demand, as an absolute right, to know of any condition which represented a potential health hazard to my child." Furthermore, the letter writer stated, "This 'recommendation' makes me wonder just how much the council members, for all their medical and scientific expertise, know about how real flesh-and-blood human beings feel and operate."[12]

A year later, the Institute of Medicine convened another panel to examine prenatal and newborn screening for HIV infection. Its report, issued in 1991, endorsed blinded newborn surveillance but stipulated that testing and screening policies "must be responsive to advances in diagnostic technology, scientific understanding of the disease and medical therapy."[13] The panel stated that universally screening pregnant women "for the purpose of early diagnosis and treatment is both an achievable and compelling objective." It recommended against formulating HIV screening policy through legislative or regulatory routes because, it said, that would "not permit the flexibility and latitude required to respond to new developments" in diagnosis, treatment, and scientific understanding of HIV infection. The committee asserted that HIV test results should be confidential but that mothers should be strongly encouraged to share their results with their children's pediatricians. It counseled that, as treatment options improved and testing methods refined, testing should be expanded to identify those in need of prophylaxis or treatment as soon as possible.

The 1991 report reinforced the CDC's separate policies of blinded surveillance of leftover newborn blood specimens to monitor the prevalence of HIV infection in childbearing women and voluntary testing for diagnostic purposes of pregnant women and newborns, initially advised for those deemed to be at high risk for infection and later for *all* pregnant women, regardless of identifiable risk or lack thereof. The CDC had recommended counseling and voluntary testing for women at high risk for HIV since 1985. But it did not issue recommendations for routine counseling and voluntary testing for all pregnant women or their newborns until 1995, a full year after its recommendations for using azidothymidine (AZT) during pregnancy to interrupt mother-to-child HIV transmission.

## The Changing Environment: The Science of AIDS

The CDC's policy regarding family surveys was formulated on the basis of existing knowledge of AIDS and available testing technologies and treatment options. Over the period of time in which the survey policy was implemented, however, these factors changed. These changes should have raised questions within the CDC concerning the appropriateness and process of conducting blinded serosurveillance. When the SCBW was begun, no approved treatments existed for asymptomatic HIV infection, for either adults or children. Shortly afterward, however, the CDC issued recommendations for preventing *Pneumocystis carinii* pneumonia (PCP), a common opportunistic infection in both adults and children with AIDS.[14] The guidelines stressed the importance of early identification of those with HIV infection in order to begin prophylactic treatment before symptoms appeared. Guidelines for PCP prophylaxis for children, disseminated in the CDC's weekly *Morbidity and Mortality Weekly Report* (*MMWR*) in 1991, reiterated the importance of early diagnosis and treatment.[15]

In its revised SCBW protocol issued in 1992, the CDC noted that "recommendations for prophylactic and antiretroviral therapy have made early diagnosis of HIV infection important. . . . Although HIV-infected infants cannot be identified at birth using readily available laboratory tests, asymptomatic infants clearly benefit from early clinical intervention and therapy."[16] At this point, CD4+ cell counts were used as a marker for HIV virus infection among children. Specifically, the CDC guidelines for PCP prophylaxis recommended that infants born to HIV-infected mothers (i.e., children clearly exposed to HIV and possibly infected with the virus) be monitored closely after birth for any evidence of immune dysfunction, as evidenced by a decline in age-normal CD4+ levels.

But many children weren't being identified as being HIV-infected before developing PCP. Writing in the *Journal of the American Medical Association* in 1993, CDC staff reported a decade's worth of surveillance showing that more than half the cases of PCP among perinatally infected children occurred between three and six months of age. In 44 percent of reported cases, the child had not been evaluated for HIV infection before PCP struck. They concluded, "Effective efforts to prevent PCP in this population will require identification as early as possible of children who may be infected with HIV."[17]

The CDC revised the guidelines for PCP prevention for children in 1995, just before the SCBW was suspended. "Effective prevention of PCP among HIV-infected infants requires that exposure to HIV be identified either before or immediately following birth so that prophylaxis can be initiated before 2 months of age," the CDC's *MMWR* stated.[18] A study of children with PCP was cited, showing that "59% of the children who had not received prophylaxis had not been identified as being at risk for HIV infection soon enough for prophylaxis to be initiated." Announcing the new guidelines in the *New England Journal of Medicine,* CDC staff noted that PCP had not substantially decreased among infants infected in the womb because the children were still not being identified as being HIV exposed or infected before contracting PCP. "These cases represent a failure of current strategies for identifying exposure to HIV."[19] The evolving CDC policy concerning the importance of early detection of HIV in infants was potentially in conflict with the policy of blinded studies of newborn exposure to HIV in order to monitor the epidemiology of the epidemic, a conflict the institution did not recognize, even after it was described by politicians and the press.

During this same period, in addition to staving off opportunistic infections among children already infected with HIV, researchers were working on ways to prevent mother-to-child HIV transmission (also known as vertical transmission). Breast-feeding was relatively quickly identified as a mode of transmission, and as early as 1985 the CDC recommended that HIV-infected women not breast-feed.[20] In 1992, the World Health Organization issued a bulletin recommending that in settings where infectious diseases are not the primary cause of infant death (such as the United States), pregnant women known to be infected with HIV or whose HIV status is unknown should be advised not to breast-feed. In order for this policy to be put into effect, mothers needed to know their HIV status. WHO further recommended that "voluntary and confidential HIV testing should be made available to women along with pre-and post-test counseling, and they should be advised to seek such testing before delivery."[21] The CDC's continued support for blinded surveys, without widepsread implementation of voluntary prenatal HIV counseling and testing, was bound to raise questions in face of this recommendation.

The viability of the survey was further challenged by the development of an effective means of preventing vertical transmission of HIV. In early 1994, the NIH announced that it was stopping a trial (known by its code, ACTG 076) designed to prevent perinatal transmission of HIV because the preliminary results were so dramatic. ACTG 076 demonstrated that the drug AZT (also known as zidovudine [ZDV]) could interrupt vertical HIV transmission, reducing by more than two-thirds the infant's chances of infection. For the first time in the history of the epidemic, researchers had found a way to prevent HIV transmission with a drug. AZT had already been used to suppress AIDS symptoms; this was the drug's first successful use as a preventive agent.

The treatment required that pregnant women take AZT five times daily throughout their pregnancy, beginning as early as the fourteenth week of gestation, and receive AZT intravenously during labor and delivery and that their infants receive liquid AZT orally for the first six weeks after birth. The final results of ACTG 076 were published in November 1994.[22] Earlier, in August 1994, the CDC's *MMWR* carried guidelines for using AZT to reduce perinatal transmission.[23] A year later, in July 1995, the CDC issued guidelines for routine HIV counseling and voluntary testing of all pregnant women or their infants, if the mothers weren't tested during pregnancy, to facilitate identification of women and newborns who might benefit from the 076 protocol.[24]

Finally, at the time the CDC developed the family of surveys, individual AIDS antibody testing was a two-step process: an initial enzyme-linked immunosorbent assay (ELISA or EIA) was confirmed by a second test, the Western blot. This process could take several weeks for batching and results notification. Moreover, the tests merely revealed antibodies to the virus; they did not test for the presence of the virus itself—a point often lost in the later debate about unblinding the SCBW. The technology used for the SCBW could not be used to individually diagnose infants as being HIV-infected or HIV-uninfected, only HIV-exposed. The CDC never intended the SCBW to be used as a diagnostic tool for individual infants, and the timing and procedures of testing were not designed to provide individual-level information. The agency clearly stated that "these surveys are not to be used in lieu of prenatal/neonatal screening and testing and counseling programs."[25] At the time the SCBW began, CDC guidelines recommended that all women of childbearing age with identifiable risks for HIV infection should be routinely counseled and tested for HIV antibody during family planning visits, as part of prenatal care, or at the time of delivery.[26]

Concurrent with changing understandings of the importance of early testing for mothers and infants and the development of new treatments were infant testing developments. These too challenged the assumptions underlying the CDC's blinded survey policy. In June 1989, the CDC reported that polymerase chain reaction (PCR) testing accurately identified infants infected with HIV.[27] This meant that HIV infection could be diagnosed within a few months after birth, instead of at twelve to eighteen months, when maternal antibodies faded to nondetectable levels, or at the time of an opportunistic infection such as PCP. News reports included quotes from the

CDC lead scientist on perinatal transmission issues, who praised the results, calling it "a major step," and said, "We have been very hindered by the fact that we don't have diagnostic tests in infants. . . . We need that critically because people would like to start treatment with AZT very early on."[28,29]

In 1989, 1991, and 1993, news reports about tests for infants facilitating early treatment led the AIDS topics covered by the media. During this period of rapid advances in perinatal prevention, diagnosis, and treatment, the CDC maintained its support for a bifurcated testing strategy: blinded newborn testing for public health monitoring and voluntary maternal and newborn testing for prevention, diagnostic, and treatment purposes. This approach reaffirmed the CDC's commitment to voluntary HIV testing and its opposition to "unblinding" the SCBW—because that would represent a form of mandatory testing, as all women giving birth would have been tested for HIV, with or without their consent. However, to the press, politicians, and the public, the distinction between surveillance and diagnostic testing was not clear, and the SCBW was seen as a simple way to extend screening, diagnosis, and treatment.

### The Changing Environment: The Politics of AIDS

Advances in the science of AIDS prevention and treatment were not the only changes that occurred over the course of the family of surveys' history. The politics of AIDS was changing as well, and these changes further challenged the viability of the Survey of Childbearing Women. By the time the SCBW and CDC's other blinded surveys were launched in 1988, America had lived with the AIDS epidemic for seven years. In that time, political and activist "party lines" had developed, intensified, and hardened. In the early years of the epidemic, amid calls for tattooing or quarantining people with AIDS, closing bathhouses, and criminalizing "homosexual activity," the gay community had developed a deep distrust for mechanisms perceived as curtailing individual liberties masquerading as means for controlling the epidemic. In this environment, HIV testing was anathema. Gay leaders expressed concerns about losing insurance coverage, losing jobs and housing, and other discriminatory actions. Public health officials who recommended widespread antibody testing were denounced as "fascists."

In 1986, when Surgeon General C. Everett Koop issued his report on AIDS, he concluded that "mandatory testing would do little more than frighten away from the public health establishment the people most at risk for AIDS, the people who most needed to be tested. . . . A push for more testing should be accompanied by guarantees of confidentiality and nondiscrimination."[30] Conservatives "retaliated with a call for AIDS testing, lots of it." There were demands for "massive, even compulsory, AIDS testing" and assertions that public health officials who opposed testing were "patsies for homosexual militants."[31] Conservatives latched on to the testing issue, implicitly (and in some cases explicitly) arguing that homosexuals would willfully endanger the public health for their own interests and should be forced to submit to testing, to protect the uninfected.

Public opinion polls revealed that most adult Americans favored widespread AIDS testing. In 1987, 58 percent of likely voters responding to a poll sponsored by the Democratic Governors' Association and Democrats for the 80s agreed that AIDS tests should be mandatory "for every citizen, with the government guaranteeing the confidentiality of each person's test results."[32] That same year, 74 percent of those responding to a *U.S. News & World Report* / CNN poll said that patients entering hospitals should be required to have an AIDS test, and 77 percent thought a test should be required for a marriage license.[33]

Summer polling leading up to the 1988 presidential election showed that 74 percent of all adults and 75 percent of registered voters supported or strongly supported mandatory AIDS tests for people getting married,[34] a possible plank in the Republican platform. In a national poll of adults by the *Los Angeles Times* in 1989, 57 percent of respondents asserted that AIDS testing should be mandatory for people at high risk, even if there was a possibility that the test results would not be confidential.[35] That same year, a poll for the Associated Press showed that 72 percent of adults nationwide believed that doctors should be required to notify the spouse or sexual partner of a person who tested positive, and 67 percent believed that the state or local health department should be notified.[36] The percentage of Americans favoring mandatory testing did not change significantly over the next five years. In 1993, 77 percent of adults responding to a national telephone survey sponsored by the American Association of Blood Banks said that the names of people who test positive for HIV should be reported to health officials.[37]

### The Debate over AIDS Testing in New York

In this politically charged environment, beginning in 1989, New York activists and politicians would clash over testing newborns and pregnant women for HIV. Those engaged in the debate represented a range of interest groups with diverse agendas. Their protracted battle's effects were felt nationwide. The CDC was absent from the public debate in New York because the agency has historically adopted a position that it should not become overtly involved in state debates. It was little heard in the national discussion, instead relying on partners (AIDS organizations, state and local health departments) to carry the public health message, which focused on the blinded survey's utility. In the media, where the battle was fought, AIDS activists framed the debate as a mother's rights versus baby's rights issue and asserted that unblinding the survey was the first step on a slippery slope leading to mandatory testing for other groups and perhaps all Americans, while health departments were essentially silent in the face of the activists' arguments and a furious backlash from politicians.

New York City is a leading indicator in the American AIDS epidemic. As Larry Altman of the *New York Times* wrote in 1998, "Since the early days of the AIDS epidemic, New York City has been one of the hardest hit of American cities. Although it has only 3 percent of the nation's population, it has 16 percent of the AIDS cases. As a result, New York City has been a bellwether for AIDS trends nationwide."[38] And not just for statistical trends: among public health professionals, New York is widely perceived as a testing ground for community reaction to various AIDS interventions, including testing, treatment, and prevention programming.

By 1988, in New York, CDC scientists reported, "HIV/AIDS was the first and second leading cause of death in Hispanic and black children 1 through 4 years of age, accounting for 15% and 16%, respectively, of all deaths in these age-race groups."[39] In January 1988, New York public health practitioners, armed with information from blinded newborn surveillance, foresaw dire consequences: "Health officials predicted Wednesday that the number of infants exposed to the AIDS virus will increase by 40% within the next two years in New York state, taxing pediatric facilities in New York City and straining the social fabric of some of its minority neighborhoods."[40]

In this context, in March 1988, New York's *Newsday* reported that "a heel-prick blood test routinely given to newborns when they are two or three days old has been enlisted in the battle against AIDS."[41] In a page 1 story in the *Washington Post*, the CDC's lead scientist on blinded surveillance stated, "This is an enormously powerful public health management tool to target prevention programs. But it is not meant for individual diagnosis."[42] The American Civil Liberties Union (ACLU), signaling the position it would hew to throughout the debate, stated that when AIDS testing is "completely blind, we have no problems with it." The article noted that the ACLU opposed mandatory testing, "saying it can lead to discrimination and personal turmoil." A leading ethicist, LeRoy Walters, director of the Center for Bioethics at the Kennedy Institute for Ethics, Georgetown University, endorsed blinded surveillance, but with a caveat: "If effective treatment becomes available, it would be necessary to reconsider the ethics of blind testing."

In July 1988, the New York State Department of Health first released data from blinded surveillance. The testing, the media reported, "ordered by Gov. Mario M. Cuomo and David Axelrod, state health commissioner, provides an invaluable tool for measuring the scope of the epidemic, detecting infection trends and focusing services and prevention services on specific groups."[43] At this point, no one in New York was speaking against blinded surveillance; in fact, politicians and political appointees were taking credit for it.

About a year later, in June 1989, New York health officials floated the idea of allowing mothers the option of obtaining the results of their babies' HIV tests "as part of a push for early treatment." *Newsday* reported, "Public health officials believe that promising treatments for AIDS-related illnesses make it increasingly advantageous for people to know whether they are infected with the virus. State Health Commissioner David Axelrod is expected to announce the new programs this week in conjunction with recommendations by the federal Centers for Disease Control for preventive treatment of *Pneumocystis carinii* pneumonia, leading killer of people with acquired immune deficiency syndrome."[44] The *New York Times* reported that the state would "permit women whose babies have been anonymously screened to learn the results of the tests."[45] The perception was thus created that the state—and the survey's national sponsor, the CDC—had the power to unblind the survey and give the results to anxious mothers. This perception was false. Not every infant's blood was tested. Those tests that were done took weeks, or even months in some cases, to obtain results, reducing their utility as a diagnostic tool. By the time the testing took place, all personal identifiers had been removed from the specimens; and the CDC's IRB had ruled that the survey could not be unblinded or else it would require informed consent. The perception was also created that the survey was the only mechanism for women and infants to be tested, although it was not. Mothers could always be tested as part of prenatal care, and they could request that their babies be tested.

Resistance to unblinding the survey quickly surfaced: activists decried the move as an end run to mandatory testing, and because of ethical and legal problems with unconsented testing, the CDC threatened to pull the state's funding for the survey if the tests were no longer blinded. The CDC contributed a fourth of the survey's cost in New York, and the state paid the remainder; the state retreated, and the issue stalled.

Two years later, on April 21, 1991, the first New York editorial urging that the SCBW be unblinded ran in *Newsday*.[46] It demonstrated a good command of the public health issues, noting that a mandatory test went against the grain because "medicine should not be adversarial" and doctors and mothers should work together (a common theme in the CDC's much later public statements). But, it added, nearly twenty thousand women gave birth without prenatal care in New York in 1990. "And some mothers—often those most at risk—resist testing because they don't want to face up to the ghastly possibility of AIDS."

Clinically, the newspaper was also on target, correctly pointing out that a positive result for the baby does not necessarily indicate infection but does reveal "newborns most at risk" and that doctors could then closely monitor their CD4+ counts to know when to begin treatment to ward off PCP. "At the moment, health officials in Albany are rethinking their position on routine mandatory HIV testing," the editorial noted. "It should be easy enough to implement, since the state routinely does blind HIV tests now." It concluded that, if the state tests babies, "it must be certain it can offer them decent care. . . . What a cruel trick it would be to offer mothers hope, then condemn them and their babies to suffer in a helpless, exhausted health-care system."

This conclusion embodied the horns of the dilemma for politicians. Calling for a test was simple—everybody wants to save babies. But honest debate about the costs of putting in place the health care and social support systems to provide treatment and services for HIV-positive mothers and babies was virtually nonexistent in New York over the next several years, as well as in the national debate over proposed federal legislation to mandate newborn testing. In the year preceding the suspension of the survey, the CDC's sound bite was "A test isn't treatment." But a review of press coverage shows that the CDC did not actively advocate for services for pregnant women or infants. Typically, health service issues are seen as the province of other federal agencies, notably the Health Resources and Services Administration (HRSA) and the Health Care Financing Administration (HCFA). News reports show no evidence that the CDC strongly articulated the ethical problems involved with forced testing absent case management (including health care and social and support services) for those who were found to be HIV-positive.

Two years after the first *Newsday* editorial, in 1993, Nettie Mayersohn, a New York State assemblywoman from Queens and a Democrat, proposed her first legislation to mandate disclosure of the SCBW test results. The bill failed in committee after a narrow 10-9 vote, owing to strong opposition from AIDS activists. The committee asked a panel of physicians, ethicists, patients' advocates, and others to examine the issue and make a recommendation for the next legislative session.

In an article entitled "Testing Newborns for AIDS Virus Raises Issue of

Mothers' Privacy," the *New York Times* reported on the debate and the schisms it had caused among usual allies because of the "struggle over how to reconcile the health needs of children with the right to privacy of their mothers."[47] It cited state health commissioner Mark Chassin's concerns about mandatory testing, "including the potential for great harm if pregnant women were in any way driven away from health care for fear of the test." Some women refuse to take the test, the article stated, because they fear losing their children or suffering discrimination if they are positive. Sounding a general theme, the article reported that, according to counselors, a common reason for refusing the test "is that many women at high risk do not want to know the results." Four days later, in "AIDS Babies Pay the Price," the paper's editorial page exhorted lawmakers to unblind the tests, concluding, "It seems cruel and misguided to protect parental privacy when the welfare of tiny babies is at stake."[48] Thus the lines were drawn that would be recapitulated throughout the debate at the state and federal levels.

In December 1993, the twenty-three-member New York panel, a subcommittee of the state AIDS Advisory Council, drafted its recommendation, which became final in February 1994, that hospitals be required to "strongly encourage" AIDS tests for pregnant women rather than mandate tests for either mothers or infants. Four physicians, including prominent pediatrician Louis Z. Cooper, director of pediatrics at St. Luke's–Roosevelt Hospital Center, dissented. Cooper stated that "reliance on counseling that encourages voluntary testing ignores the unacceptably high failure rate of such an approach" and that it "siphons off resources which could be focused more effectively for needed care."[49] Cooper had been quoted earlier as saying, "We recognize that the only way to [test] at the moment does intrude on the mother's choices. But we hope it also reminds women that having a child is not a license to do anything you want, that society does have expectations of its parents."[50] Commenting on the final report, and deploring what has become known as "AIDS exceptionalism," Cooper condemned the counseling recommendation: "It is insufficient to offer the protection which every infant deserves, protection which has been guaranteed newborn infants in New York state for other serious diseases."[51] Throughout the state and federal debate, the medical community (as distinct from public health practitioners) was divided over the notion of testing, with pediatricians generally supporting it and obstetricians generally opposing. The lack of congruence among physicians contributed to confusion and frustration among the press, politicians, and the public as they tried to understand and respond to this complicated issue.

Advocates applauded the decision, saying that mothers and children would be subjected to stigmatization and discrimination if the confidentiality of test results was breached. *Newsday* decried it, reminding readers, "All babies born in New York are already tested for the AIDS virus. But until now, State Health Commissioner Mark Chassin has kept the results anonymous, using them only as a statistical tool."[52] The paper pointed out that 13 percent of New York City women "show up at a hospital just before delivery, so there's little opportunity to counsel or coax a patient into taking an AIDS test," and it cited state statistics showing that some hospitals had a 90 percent success rate in getting pregnant women to test whereas others had only a 5 percent success rate. It asked rhetorically whether a mother's privacy rights should "eclipse a newborn's right to have a complete bill of health as early as possible" and asserted that an AIDS test "should be as routine as checking a newborn for sickle cell or any of the eight genetic diseases." It closed by praising Mayersohn's bill and added, "If Chassin won't move aggressively in this area, lawmakers should." The editorial reasoned that if public health officials could not be trusted to do the right thing for children medically and the right thing for public health, lawmakers were the next best resort. This theme also would be repeated throughout the New York and national debates.

In early 1994, amid clamoring editorials, letters to the editor by AIDS activists and experts as well as politicians on both sides of the issue, columns demanding newborn testing, and news articles reporting the plight of infants whose first diagnosis of HIV infection was when they were hospitalized with PCP, Mayersohn brought her bill again, paired on the Senate side with a Republican from the Bronx, Guy Velella. It again stalled in committee.

In May 1994, U.S. representative Gary Ackerman, a Democrat from New York, inspired by the debate over Mayersohn's bill back home, introduced federal legislation to unblind the survey. The proposed Newborn Infant HIV Notification Act required states conducting newborn HIV testing to "promptly disclose" the results to the biological mother, any state agency acting as the legal guardian, prospective foster and adoptive parents, or other legal guardians. Ackerman's bill remained stalled in committee in the 103rd Congress. However, it catalyzed AIDS organizations nationwide, which had been watching the fight in New York but generally staying out of the fray. Washington-based groups moved to block Ackerman's bill, fanning out across Capitol Hill and encouraging constituents to write or call their elected representatives.

Mayersohn's bill remained stuck in New York, short of the votes needed to bring it to the General Assembly. On July 4, the assembly passed a bill requiring doctors to counsel pregnant women and new mothers about HIV testing. After an angry late-night session, the state Senate failed to vote. Governor Mario Cuomo and Health Commissioner Mark Chassin refused to take a public stand on testing throughout the debate.

Also on July 4, *Time* magazine took the story national, with a report on advances in treatment, including AZT during pregnancy, the CDC's blinded testing, and the New York impasse. Commenting on blinded surveillance, the article said: "The strategy made sense at first, but advances in treatment have changed the ethical equation."[53] In August, the CDC issued recommendations for voluntarily treating HIV-positive pregnant women and newborns with AZT to interrupt perinatal HIV transmission.[54]

In November 1994, when the final results from the ACTG 076 study were published in the *New England Journal of Medicine,* the report again heightened interest in testing.[55] "Until now, proposals for mandatory testing were quashed by those who said that forced testing violated people's civil rights," wrote Gina Kolata in the *New York Times.*[56] But, quoting the chief of pediatrics at the National Cancer Institute, part of the NIH, which sponsored ACTG 076: "These data do change the agenda" because of the possibility of preventing infection in infants by offering AZT to infected pregnant women. For that reason, he supported mandatory testing for pregnant women. CDC scientists backed voluntary testing, the article reported, even though they acknowledged that "testing was still not offered routinely to pregnant women, even in inner city clinics whose patients are most likely to be infected," despite the fact that the medical community had known about the 076 results for more than half a year.

The article gave the last word to an ethicist. The real debate, said Arthur Caplan, director of the Center for Bioethics at the University of Pennsylvania, was not over testing but was "a fight to make sure testing remains confidential." The solution was to address confidentiality concerns. He concluded, "To give a kid a preventable case of AIDS in the name of civil rights seems wrong."

In 1995, the battle continued in New York to move Mayersohn's legislation and, in Washington, to get Ackerman's out of committee. In March, Ackerman reintroduced the Newborn Infant HIV Notification Act with 217 original cosponsors: 155 Democrats and 62 Republicans. The bill was again furiously opposed by Washington-based AIDS groups, who by this time had organized a working group within the National Organizations Responding to AIDS (NORA), an umbrella advocacy group comprising Washington-based lobbying groups and the ten largest AIDS service organizations in the nation, such as AIDS Project Los Angeles, the Gay Men's Health Crisis in New York, and AID Atlanta.

Mayersohn's bill was similarly under attack, this time with help from organizations outside New York. Meanwhile, news articles, editorials, and columns continued to push for unblinding newborn testing; in the wake of the 076 results, some even called for mandatory prenatal testing, most notably the *Washington Post.* The CDC attempted to work with NORA and other groups, encouraging them to move away from a civil rights–based argument, pitting mother's rights against child's rights, and to an argument based on effective medical intervention during pregnancy based on routine counseling and voluntary testing and treatment to interrupt perinatal HIV transmission. But the agency was not visible on Capitol Hill because of the prohibition against lobbying under which federal agencies are legally bound to operate, and others carried the message.

### Suspending the SCBW

In April 1995, the New York Senate passed the Mayersohn bill, although it was still stalled in the Assembly. In May 1995, at the federal level, Ackerman succeeded in getting the Health and Environment Subcommittee of the House Commerce Committee, where his bill was stuck, to hold hearings. On May 11, after Ackerman made an impassioned plea to protect babies, the committee heard from witnesses who supported or opposed mandatory newborn testing. Supporters of the Survey of Childbearing Women repeated many of the arguments they had used in New York for several years: ACTG 076 was only one study and should not be heavily relied upon; testing newborns was too late; the focus should be on preventing HIV infection in women, so that children weren't exposed in utero.

CDC officials Drs. Helene Gayle and Jim Curran (both now former directors of the agency's AIDS programs) testified that the best way to prevent infection in infants was to promote routine HIV counseling, voluntary HIV testing, and AZT treatment for pregnant women, as recommended in CDC guidelines. They noted that the guidelines called for voluntary newborn testing if the mothers had not been tested during pregnancy or at delivery. They then reported that the SCBW was being "suspended" in order for the agency to consult with external partners about the best way to implement the guidelines and use scarce resources wisely.[57] The letter informing states of the suspension was signed by Dr. Philip R. Lee, assistant secretary for health, but the CDC was responsible for dealing with the political, public, and press reaction.

The decision to suspend the survey was condemned on almost all fronts and characterized as "political" by almost all observers, no matter their vantage point. AIDS activists were furious that the sole population-based mechanism to track infection in women had been lost and publicly thrashed the CDC for a lack of leadership. "This is one more misstep in a series of missteps," said Terry McGovern of the HIV Law Project in New York.[58]

State health departments were likewise infuriated because they faced a loss of funding to conduct the survey and, therefore, the possibility of shutting down their programs to conduct the testing, including laying off staff. But although they were active on Capitol Hill through organizations like the Association of State and Territorial Health Officials (ASTHO) and the National Alliance of State and Territorial AIDS Directors (NASTAD), states were essentially absent from the national public debate, judging from media coverage, both before and after the survey was ended, leaving the CDC alone to answer criticisms of the survey and the decision to end it.

Politicians decried the CDC's decision as a craven action to avert unblinding the survey and avoid mandatory testing for newborns. An Associated Press story quoted Representative Ackerman: "Nobody wants to play politics with AIDS and we didn't expect the CDC to. All we want to do is identify children in time to help them. They have a right to health care."[59] He vowed to introduce legislation to force the CDC to resume testing. Mayersohn denounced the CDC for treating the AIDS epidemic differently than any other disease. "I can't believe that this is the guardian of public health in this country," she said.[60] DHHS secretary Donna Shalala, speaking to the *New York Times,* said, "There really are profound ethical questions involved. Sometimes you do something for one purpose and it has unintended consequences. . . . We thought there was a serious question about collecting information about which we didn't inform parents."[61]

Media attention, both reporting and editorializing, was uniformly negative. The *New York Daily News* said that the survey, because it was used only to gather statistics, was "virtually useless in combating AIDS," but it denounced the decision to suspend testing that might have led to treatment, attributing it to "politics and special interests, pure and simple."[62] The *Los Angeles Times* headlined its editorial "How Politics Undermines Good Science" and called for expanded prenatal testing, pointing out that the survey was "a victim of well-meaning but misplaced efforts to halt AIDS in newborns."[63] "Feds Cave on AIDS Testing" was *Newsday*'s headline on an op-ed that asserted the survey was stopped for "raw political reasons."[64] Media coverage shows that these allegations essentially went unanswered by public health officials.

At the end of the month, First Lady Hillary Rodham Clinton opened the Pediatric AIDS Foundation's campaign to promote prenatal testing in an East Room ceremony. "Mothers should be tested for HIV so babies can be saved," she said, giving ammunition to those who supported mandatory testing for pregnant women and again conflating a test with treatment.[65] What would "save" babies would be providing AZT during pregnancy, labor, and delivery and postpartum, not a test alone. In July, the CDC issued its guidelines for routine counseling and voluntary testing for all pregnant women.[66] Also in July, the CDC held a consultation with states, other federal agencies, and activists to discuss methods to implement the guidelines and, equally important, means to evaluate the success of prenatal intervention. Meeting participants stressed the need for continuing the SCBW to monitor trends in HIV infection in women.[67] Despite this review by individuals outside the agency, the survey was not resumed.

Throughout 1995, Mayersohn's bill failed to gain the votes needed to be reported out of committee and go to the floor for a vote. In mid-March, however, the Association to Benefit Children, a New York children's advocacy group, filed suit against the Pataki administration to force the state to release the results of the newborn HIV tests. (At this time, New York was conducting the survey with its own funds, using CDC moneys for other pediatric prevention activities, so its SCBW continued even after the CDC suspended the survey nationwide.) The suit was settled in October, when the state health department issued a regulation requiring doctors to counsel mothers about prenatal testing and, after birth, to notify mothers about the HIV tests on their babies and to inform them that they could learn the results if they chose. Mothers were required to sign a consent form stating they did or did not want the results. Children deemed to be at risk for HIV infection could be tested without parental consent.

In July 1995, at the federal level, three months after the House Commerce Committee hearing, the Senate voted to renew the Ryan White Care Act after rejecting an amendment that would have required mandatory HIV testing not only for newborns but for pregnant women as well. Ryan White funds are used to provide services for people with AIDS. In September 1995, Representative Michael Bilirakis of Florida introduced the House version. Section 204 authorized grants to states for HIV counseling and testing pregnant women and newborns; made states eligible for Ryan White funding only if they required HIV counseling, testing, and disclosure; mandated that states determine whether counseling and testing are routine obstetrical practice; prohibited states from receiving Ryan White Title II funds unless they were able to prove by the year 2000 that the HIV status of 95 percent of all newborns was known; and required states to determine the rates of perinatal HIV transmission and the possible causes. AIDS advocates fought the bill tooth and nail until it became apparent that Ryan White funding would be held up indefinitely because of disagreements over Section 204. That was unacceptable. The amended bill passed both House and Senate on May 8, 1996, and was signed by President Clinton on May 20, becoming PL 104-146 and opening the possibility of nationwide mandatory newborn testing, exactly what the CDC had fought all along. The federal legislation gave Mayersohn's bill the push it needed, and it passed in early June. The agreement was announced June 5, 1996, and Governor George Pataki signed the legislation June 26, making New York the first state in the nation to require newborn HIV testing.

## Improving Response to Emerging Illness: Assessing Institutional Deficiencies

How did the CDC arrive at a point in 1995 where its survey, providing much-needed information about the extent and trends of HIV among women, was canceled? How could the agency not have seen the gathering storm, given the sea changes in science, politics, and media coverage? According to Jim Curran, former associate director for HIV/AIDS, "We just didn't come to grips with the problem soon enough," and the CDC especially did not address the issue of care for women and children.[68]

An influential report from the Institute of Medicine called *The Future of Public Health* cites inadequate policy development as one of the primary impediments to sound public health practice. Public health policy development at all levels of government "is often ad hoc, responding to the issue of the moment," resulting in "inadequate consideration of options, unintended side effects, long-term results, or effective allocation of resources based on impact on health status."[69] The report stipulates that legitimate public policy decisions "reflect a full examination of the public interest" and "sound analysis of problems and interventions" and "consider the interests of all affected parties, especially the general public" or their democratically elected representatives, who are "accountable to the public in ways executive branch agency personnel are not." Good science conducted by publicly funded institutions like the CDC must consider the will of the public. The IOM report notes that, when it is used judiciously, the knowledge base of public health can mediate partisan politics and inform policy decisions. "Technical knowledge will have the best effect, however, when used in the context of a positive appreciation for the democratic political process, by professionals who are politically as well as technically astute."[70]

Internal systems shortcomings and external communications deficiencies contributed to the CDC's lack of an adequate understanding of the political landscape, to its being blindsided and unable to advocate successfully for the SCBW or for voluntary testing for pregnant women and newborns as a gateway to care and services. These deficiencies are not unique to the CDC and can serve as a checklist for all institutions in assessing their capacity to respond to emerging illnesses or changing scientific situations and to other policy matters that involve the public, the press, and the political system.

## Internal Systems Shortcomings

The CDC's internal shortcomings included the following:

• *Inadequate Environmental Scanning—Media:* Environmental scanning—ongoing monitoring of media coverage, listserves, and other external communications about relevant topics—is the lifeblood of any external communications activity; the vast majority of people get their information about government from the media. But the CDC clip file on the SCBW provided to the author began in 1994 with six clips; the vast majority were from 1995 and covered the suspension of the survey. By 1994, the public debate on blinded testing of newborns was well under way—and far out of the CDC's control. Because of the inadequacy of its media tracking before and during the New York battle over the survey, the CDC was deaf to much of the media-reported debate. The upshot was that many nonroutine, subject-specific media activities—including tracking, response, and proactive relations—fell through the cracks because no systems were in place and insufficient resources were allocated to communications activities. Even when the agency became aware of the New York debate, it refrained from joining it, not fully appreciating that New York was likely a bellwether for the rest of the nation. "We avoid state issues—sometimes to a fault—and this was seen for too long as a New York issue," Curran believes. "We treated it like a state issue, not a national issue. Then Ackerman nationalized it."[71] He adds that the CDC perceived the SCBW solely as a surveillance issue (the agency's assigned role) and not a care issue (which was not the agency's role), which precluded addressing the larger issues with which the press, public, and politicians were concerned.

• *Inadequate Environmental Scanning—Statutory and Case Law:* Although the Equal Protection Clause of the Fourteenth Amendment of the U.S. Constitution may not be sufficient to enforce HIV testing or treatment for children, some legal scholars believe that there "is sufficient legal precedent to justify mandatory testing of newborns. Justification can be found in legal provisions that permit testing of newborns for STDs, that allow for mandatory vaccination of children, and that support the general right of the state to intervene to protect the health and safety of children. . . . Finally, evidence indicating that early intervention on behalf of newborns is critical to their survival provides states a compelling justification for intervention."[72] There is no indication from internal documents provided to the author or from external assertions reported in the media that these legal issues were analyzed as they related to the SCBW.

Nor was the HIV/AIDS staff able to analyze the actual probability of adverse consequences of mandatory testing for women and children because the nation had limited experience with mandatory HIV testing and the legal ramifications of forced testing. Activists raised the specter of women being charged with prenatal neglect if they were not tested or if they were tested and found positive but elected not to take AZT and their children were HIV-positive at birth. There were also concerns that, if children were positive and their mothers elected not to put them on AZT or other medications, the state would remove the children from their mothers. These fears are not groundless. A number of states include prenatal exposure to drugs or alcohol in their definition of child abuse and neglect; even absent state law, case law supports charging mothers with abuse or neglect or to find the child neglected if prenatal exposure is found.[73] But without hard information, evaluating the likelihood that women would be charged and found guilty and assessing how that probability should factor into the policy calculus were beyond the program's ability.

Likewise, the staff was unable to assess accurately the possible impact of "Baby Doe" regulations (orders issued in the early 1980s by the Reagan administration requiring treatment for multiply handicapped newborns and requiring medical personnel to report instances of withheld treatment), the U.S. attorney general's September 1988 expansion of the 1973 Rehabilitation Act (which protects handicapped persons) to include "a person with AIDS, AIDS-related conditions, or someone perceived as having AIDS,"[74] or the Americans with Disabilities Act, passed in July 1990, which defined HIV infection as a handicapping condition. In this instance, the handicapping condition would have been revealed through a linked HIV test. The legal position of the survey was therefore an important question to address.

Internal records do not include or mention any analysis of the legal environment, which would have facilitated policy review as statutory and case law changed over the course of the survey. With such information—for example, an analysis of case law involving mother's versus children's rights and the intervention of the state in instances in which children were deemed to be at risk in utero—the program might have been more forceful in steering external partners away from arguments based on mothers' civil rights.

In addition to being unfamiliar with existing statutory and case law affecting newborn testing, the CDC HIV staff was also not well informed about proposed state-level legislation, again because of nonexistent tracking and reporting systems. Some information was sporadically provided to program staff from external groups such as the American Academy of Pediatrics[75] and the AIDS Policy Center at the George Washington University,[76] but no system existed to ensure that this information was shared with the communications or policy staff and management.

• *Unwieldy Clearance System:* The CDC's system for clearing "controlled correspondence"—that is, letters responding to official requests for information from federal, state, or local lawmakers or executives; fact sheets; editorials; letters to the editor; and other external communication—impeded quick responses or preemptive strikes. Press coverage, particularly critical coverage, tends to speed up the policymaking process and to move decisions to higher bureaucratic levels.[77] "The advantages to policymakers of being able to frame the issue, rather than having it framed for them in the press, are enormous,"[78] but sluggish federal clearance policies precluded those advantages in the fight for the SCBW and contributed to the suspension of the survey. Because press coverage affects the agency's capacity to see policies successfully adopted and implemented, it affects not only policy process but also policy content.[79] Negative coverage, as almost all the reporting on the SCBW eventually was, decreases the number of viable policy options and the chances that government agencies can accomplish their goals.

• *Inadequate Media Training for Scientific Spokespersons:* The CDC staff was not prepared to respond to criticism or, more important, to shape the debate on testing in the media. Scientific staff generally view the media with suspicion; they do not see the media as having a legitimate role to play in public health policy. These attitudes are apparent to journalists. Nat Hentoff, the author of several op-eds about the survey, including one that compared it to the infamous Tuskegee syphilis study, said that the CDC scientists he spoke with "were indignant that [he] would question the survey's value to society."[80] He added that the scientists seem to "drift above the plain, messy business of human existence." Jim Dwyer, another op-ed writer, who wrote extensively on the survey and infants and children with AIDS, described the agency's scientists as "arrogant."[81] Those perceptions may have negatively affected the CDC's communications about the SCBW.

As the debate grew more heated and the stakes for the CDC higher, news coverage more and more frequently quoted politicians (and even columnists) instead of scientists, public health officials, or the CDC, demonstrating public health's inability to frame the debate on its own terms.

Public health officials relinquished the bully pulpit, allowing politicians to set the public health agenda. And as politicians came to be accepted as legitimate sources of information on health issues, the credibility and perceived expertise of public health officials and others were diminished. According to Shilts, this trend had begun early in the AIDS epidemic, when "public health officials hadn't helped by framing issues politically themselves. The public was used to hearing health officials sound like politicians, so it didn't sound jarring when politicians starting talking like they were health officials."[82] Additionally, communication too often focused on barriers, problems, or negative consequences, such as women fleeing the health care system. For example, a CDC scientist was quoted in the *New York Times* in 1994 complaining about the fact that, although the medical community had known about the 076 results for more than six months, "H.I.V. testing was still not offered routinely to pregnant women, even in inner city clinics whose patients are most likely to be infected."[83] For many outside public health, this reinforced the notion that testing should be mandatory because they felt that women could not be trusted to get tested of their own volition and doctors were not pursuing voluntary testing avidly enough.

· *Inadequate Ethical Assessment:* In March 1988, Rod Hoff, one of the developers of blinded newborn testing in Massachusetts, predicted, "Once there is an effective treatment for infected infants, we will very quickly convert to a case-detection system."[84] In August 1988, leading ethicist LeRoy Walters said that once effective treatment became available, "it would be necessary to reconsider the ethics of blinded testing."[85] And in 1990, Ron Bayer, who later defended the SCBW, said that "the ethical issue surrounding blind testing without informing those who are positive of the results has muddied" and that blind tests "shouldn't be done."[86] But because of the long lead time before HIV tests were actually performed, the existing system for blinded surveillance of newborns was not seen as suitable for a diagnostic tool, even if the test could be unblinded after informed consent was obtained, and changes to the system (to speed up testing, for example) were not believed feasible.

The CDC's IRB reviewed the survey, but because it was staffed predominantly by scientists, the review board was insulated from external realities, such as changes in public and political opinion. A 1995 OPRR review of the IRB, undertaken in response to a complaint about the survey (and another concerning CDC's measles vaccine research), found "a number of weaknesses" in the CDC's system for protecting human research subjects.[87] These were technicalities, but the OPRR also determined that the IRB did not meet statutory requirements for diversity, "including consideration of race, gender, and cultural backgrounds and sensitivity to such issues as community attitudes.[88] Additionally, the decision from the OPRR and the CDC's IRB that the survey did not test human beings was a legally based opinion that confounded the press, politicians, and the public, who perceived the agency as testing babies and withholding the results. The CDC's explanations, grounded in legalistic OPRR/IRB language, did not help to ameliorate the situation, even though they were accurate.

The human dimension dominated the public discussions of the SCBW, but the CDC virtually ignored it in public statements. Public health practitioners must balance principles such as beneficence and utility against the principles of autonomy and justice.[89] The majority of CDC scientific staff, who may have had medical school ethics training geared toward clinical practice or no formal ethics training at all, are generally inadequately prepared to assess their planned or ongoing activities within an ethical framework appropriate to public health.[90]

• *Organizational Paradigms and Biases:* Bureaucracies excel at standardizing processes; unfortunately, they all too often routinize thinking, including biased thinking. At the CDC this tendency contributed to the agency's delay in reacting to the changing scientific and political environment within which maternal testing was occurring. According to Curran, "it's easier to avoid stopping something than to do something," especially when that something (e.g., health care for women and children) is beyond your agency's control.[91] Policy biases and pressures inherent in a large and variegated organization like the CDC are caused by asymmetry in information and sources of information, disproportionate attention to preferred information sources, prior intellectual commitment on the part of those developing policy, selectivity in organizational recruitment, and bureaucratic pressures, such as budgets, deadlines, and political concerns.[92] All these biases affected policy discussions surrounding the SCBW. Lindblom notes that policymakers simplify complicated analyses by two main means: "concentrating on policies that differ only incrementally" and "ignoring important possible consequences of possible policies, as well as the values attached to the neglected consequences.[93]

As early as 1989, an internal document provided to the author by survey director Marta Gwinn asserted that the SCBW needed "to be re-evaluated in light of several developments during the past year.[94] Developments included changes in infant testing and treatment; greater advocacy for more widespread HIV testing and counseling for women of childbearing age; the IOM report recommending anonymous screening for all newborns nationwide; and state-led policy deliberations, including New York's move to conduct voluntary HIV testing using newborn screening specimens and New Jersey's move to require physicians to offer HIV counseling and testing during prenatal visits. The survey was not significantly altered to address those developments. Five years later, a July 1994 e-mail said that the survey needed to be "reconsidered, recast and repackaged for the post-076 era," which was accomplished by emphasizing the survey's ability to assess the effectiveness of routine prenatal HIV counseling, voluntary testing, and voluntary AZT treatment for pregnant women and newborns.[95] The CDC chiefly discounted other policy options, such as mandatory testing for newborns, and many staff did not appreciate the bigger picture. "Many at CDC never understood that the debate was not about surveillance, it was about prevention and care," Curran said.[96] The agency generally avoids treatment and care issues, which are seen as the responsibility of other federal agencies. In the federal public health bureaucracy, turf is jealously guarded, and the CDC did not have support for taking on the broader issue. This piecemeal federal approach to public health caused the CDC to not be involved in the testing debate soon enough.

Hesitance to address treatment and care issues resulted in cognitive dissonance regarding the SCBW—scientific staff rarely connected these issues to their calls for earlier identification of infants exposed to HIV in order to begin PCP prophylaxis. But the CDC's thinking trailed the changes in treatment, reported so thoroughly by the media, as well as public and political opinion changes. At least as early as 1992, according to internal e-mail communication, Jim Curran was pushing to tie "eligibility for funding for the SCBW to progress in promoting testing for women/infants."[97] Despite this, the two efforts were not integrated. According to Curran, all the blinded surveys, including the SCBW, were "on autopilot."[98]

CDC policy decisions seemed to conform to so-called AIDS Orthodoxy, the ACLU/activist-led rhetoric that focused on civil rights, privacy, confidentiality, and warning of women fleeing the health care system. CDC staff credited those assertions, even absent strong evidence. During the public debate on the SCBW, both CDC scientists and activists often cited the Illinois experience with HIV screening for marriage licenses as proof that women would opt out of medical care.[99,100] But that analogy is imperfect—the state began requiring HIV testing for marriage licenses on January 1, 1988, before significant advances in treatment, and, more important, obtaining a marriage license is not comparable to carrying a pregnancy to term, giving birth, and assuming responsibility for a child.

Finally, the agency lacked a routine system for resolving scientific disagreement and resulting policy differences, whether they arose between staffs, such as the scientific staff and the communications staff, or between staff and management (who are typically scientists promoted through the organization). Scientific historian Thomas Kuhn has noted that most scientists faced with facts that are inconsistent with their paradigms fight to defend and then fix up the paradigms, rather than accept new ones. "They do not renounce the paradigm that has led them into crisis"[101]—in this case, the value of blinded surveillance and the data it generates. In contrast, nonscientists at the agency had little or no deep-seated attachment to the survey or the scientific schema that created it. They placed great value on a pragmatic assessment of what was optimal, given external realities. Without a system to resolve differences about the survey and its continuation, the survey remained at the center of an internal tug-of-war of conflicting ideologies.

 • CEO Acting as Policy Manager: As observed by the author throughout the SCBW controversy, the head of the Office of HIV/AIDS (later the Division of HIV/AIDS in the National Center for HIV, STD, and TB Prevention [NCHSTP]) acted as the policy manager for discussions on the Survey of Childbearing Women. His attention to the gathering storm over the SCBW was diverted by an internal reorganization, as well as by other thorny policy issues, such as syringe-exchange programs as a means to prevent HIV infection and the growing call for HIV infection reporting to supplement the existing system of AIDS case reporting. Accordingly, significant policymaking often occurred ad hoc, at the program level, without his full involvement. This lack of oversight further hampered the agency's response to the emerging controversy over maternal testing.

Policy management was run on the collegial model as described by George,[102] but without clear roles and lines of authority, responsibility, and accountability and matrixed across several centers and offices. The director oversaw a freewheeling "'debate team' that consider[ed] information and policy options from the multiple, conflicting perspectives of the group members in an effort to obtain cross-fertilization and creative problem-solving." Although the collegial system can have significant benefits, such as providing the decision maker with unfiltered information, it has serious costs. It places substantial demands on the decision maker's time, requires unusual interpersonal skills to mediate differences and maintain teamwork, and risks degenerating into a closed system of mutual support. These costs were all apparent in the development of policy related to the SCBW.

The collegial system employed by the CDC during discussion of the survey was an incomplete version of the multiple advocacy system described by George and Porter, among others.[103,104] Policymaking around the SCBW lacked structured, balanced debate centrally coordinated by management through an "honest broker" responsible for shepherding the policy process.[105] It suffered from maldistributions of power, weight, influence, competence, information, analytic resources, and bargaining and persuasion skills among the various staff involved.[106] It was not "grounded in the concept that a competition of ideas and viewpoints is the best method of developing policy—not unregulated entrepreneurial advocacy but orderly, systematic, and balanced competition."[107] Genuine multiple advocacy has several benefits: it ensures that all points of view will be aired and examined, mitigates against the suppression of conflicting views, mirrors external opinions and interests, enhances staff morale, improves the quality of policy options and the arguments to support them, bridges the gap between policy development and implementation by ensuring that those who will be responsible for policy implementation are part of the formulation process, and lessens the likelihood that dissenters will undermine the policy because their views have been heard and discussed.[108]

· *Undefined "Public" to Whom the CDC Should Be Responsible and Responsive:* In *The Future of Public Health,* the Institute of Medicine identifies the lack of consensus on mission and content of public health as one of the chief barriers to effective problem solving. "Progress on public health problems in a democratic society requires agreement about the mission and content of public health sufficient to serve as the basis for public action. There is no clear agreement among public decision-makers, public health workers, private sector health organizations and personnel, and opinion leaders [or the public] about the translation of a broad view of mission into specific activities."[109] The CDC's scientific staff frequently dismissed external concerns or opposition to the SCBW as "just politics."

The IOM report cites "poor interaction among the technical and political aspects of decisions" as a significant problem,[110] explaining, "Technical experts may not understand or appreciate the appropriate and fundamental role for the political process in public policy-making, especially as it expresses society's values as criteria for selecting among options that have been defined with appropriate technical competence."[111] In *Creating Public Value,* Moore asserts flatly that "it is fundamentally wrong and dangerous for [government] managers to become so arrogant that they no longer trouble to check their ideas about what is publicly valuable with citizens and those who represent the public."[112] Aside from the pragmatic necessity of expanding support and authorization for public health activities, there is a broader philosophical and ethical question that the CDC—as any public agency—must grapple with: How responsive should it be to what the public wants? (E.g., polling data previously cited show that, over time, the majority

of the public wanted mandatory testing.) Who is "the public"? How responsive should it be to constituent groups, such as AIDS activists? Or to those
most at risk compared with everyone else? To states, within a federalist system? Or to legislators, who are democratically elected and accountable to
the public in ways CDC staff are not? The agency lacked a mechanism in the
policy development, implementation, and evaluation process to ask and
explicitly answer these questions.

### External Communications Deficiencies

The internal shortcomings just described resulted in inadequate external
communications with the press (and thus the public), political leaders, and
public health partners. Key insufficiencies included the following:

  • *Inadequate reactive media communications and extremely limited proactive
communications:* The "media are one of the few available levers for building
and sustaining political support inside and outside the government," and "a
receptive public and a favorable political environment are essential for the
performance of day-to-day tasks."[113] But internal confusion about authority
lines for communications, the unwieldy clearance system, and the inability
to resolve internal debates about the viability of the survey and messages
about it meant that the CDC's responses to columns, editorials, letters to the
editor, and articles were unacceptably delayed or often never sent.

  • *Spokespeople who were not credible:* The public's belief in experts has
diminished markedly over the past three decades, as has its confidence in
federal—not state—government.[114] The CDC has not been hit as hard by this
as have other areas of national government; favorable reporting on disease
investigations, such as SARS and West Nile virus, and first-run movies like
*Outbreak* have bolstered the CDC's position. But the agency lives with the
legacy of Tuskegee and is still perceived with skepticism, simply because it is
part of the federal government. AIDS activists, the bulk of the partners the
CDC relied upon to carry the SCBW message, are generally viewed as stri-
dent complainers,[115] and media analysis of the SCBW debate underscores
that point, although it cannot register the intensity of statements, only their
content. Other partners, such as scientists, clinicians, and national organizations representing state-level public health officials, were not effective for
other reasons: their messages were not consistent and unified, they were perceived as being self-interested, or they were not engaged in the issue. In
states where HIV prevalence and incidence were (and are) low, the survey
was never a hot issue, as it was in New York. Accordingly, many of the CDC's
traditional partners did not become involved in the debate until the survey
was ended and the funding lost, but by then it was too late.

• *Need for improved congressional relations:* There was no systematic communication from the CDC as a whole or from the HIV/AIDS program specifically to Capitol Hill; instead, the CDC's messages were delivered by the administration and by advocacy groups, and the agency's messages were frequently confused with theirs. Compounding the problem was that formal communication, in the form of controlled correspondence to politicians, was delayed, often for months. For example, a letter on the SCBW from David Bonior, at that time the majority whip, dated July 13, 1992, was prepared by program staff on July 31 but did not leave the CDC until September 16.[116]

• *Inadequate communication with the DHHS and the administration:* As the debate on the SCBW grew more intense, so did interest and concern at the DHHS and the Office of National AIDS Policy at the White House. But clear channels of communication did not exist to either the department or the White House, and communication was based more on personal relationships than on lines of authority or responsibility. Among the problems that resulted were incomplete communication up and down the chain, mixed public messages, a confused policymaking environment, and redundant or unused work products such as question-and-answer fact sheets.

### The Chief Institutional Failing

Ultimately, however, the chief institutional failing was a failure of leadership. The Institute of Medicine report lists four components of national leadership for public health: identifying and speaking out on specific health problems, allocating funds to accomplish national public health objectives, building constituencies to support appropriate actions, and supporting development of public health knowledge and data bases.[117] With regard to the Survey of Childbearing Women, the CDC succeeded only in the last sphere. According to Jim Curran, the CDC "was unable to demonstrate leadership."[118] The agency was not part of the public debate, it failed to accomplish its public health objectives, and it failed to build supportive and coherent constituencies to support its efforts. The CDC failed to establish that the real debate was not about a surveillance tool but about treatment and care for HIV-infected women and their children. Funding for the survey was not redirected to programs to increase routine maternal counseling and voluntary HIV testing; nor was the CDC able to secure additional funding so that the survey could be continued while individual testing was also ramped up (the preferable solution). Finally, the CDC failed to mobilize states to defend the need for the survey while at the same time dedicating other resources to increasing maternal testing.

According to Curran, the major barriers to effective action were the compartmentalized responsibilities in the federal public health system for surveillance, prevention, and care; federalism, that is, the tension between federal and state roles, responsibilities, and autonomy; and the CDC's inability or unwillingness to engage in the New York debate. He also cited the inability to act freely in government because so many activities are controlled by higher levels; the CDC's location in Atlanta, which hinders its ability to manage the national political dimension of public policy; and the complex nature of the issue, which made communication to the public, press, and politicians difficult.[119]

The CDC's deficiencies in the fight to continue the SCBW are not unique to that public agency. Rather, they are characteristic of most large, bureaucratic institutions, both public and private. Agile and adept public sector policymaking requires that systems be in place for providing requisite information, such as media and legislative tracking; for resolving internal disputes; and for addressing potential threats to credibility and political capital quickly. Public institutions should assess their capacity in each of these three areas and move to create or bolster such systems as needed.

## NOTES

The views expressed in this chapter reflect those of the author and not those of the Centers for Disease Control and Prevention. The author gratefully acknowledges the assistance of current and former CDC employees in providing information, insight, and editorial direction. Among those whose help was invaluable are James W. Buehler, James W. Curran, Marta Gwinn, Melissa B. Shepherd, and Ronald O. Valdiserri.

1. Feinberg L. D.C. to test newborns, clinic patients for AIDS; blood of 30,000 to be screened for virus. Washington Post: D5; August 18, 1988.

2. CDC. Human immunodeficiency virus infection in the United States: A review of current knowledge. MMWR 36 (Supplement, S-6); 1987.

3. CDC. AIDS and human immunodeficiency virus infection in the United States: 1988 update. MMWR 38 (Supplement, S-4); 1988.

4. CDC. Issues related to the HIV survey in childbearing women (draft memorandum). July 29, 1994.

5. Brandlery K. Women's AIDS Network blasts CDC study suspension. Bay Area Reporter: 3; June 12, 1995.

6. CDC. HIV seroprevalence survey in childbearing women: Testing neonatal dried blood specimens on filter paper for HIV antibody (draft): 6; September 8, 1988.

7. Office for Protection from Research Risks, Division of Human Subject Protections, National Institutes of Health. Ethical and legal aspects of HIV seroprevalence studies that utilize blood specimens from newborn infants in PKU and hypothyroid screening programs (memorandum): 4; February 16, 1988.

8. CDC. HIV seroprevalence survey in childbearing women: Testing neonatal dried blood specimens for maternal HIV antibody (draft): 3; June 1, 1992.

9. Institute of Medicine. Newborn screening for HIV infection. In: HIV screening of pregnant women and newborns, ed. LN Hardy, 25–31(Washington, D.C.: National Academy Press, 1991).

10. Office for Protection from Research Risks, Division of Human Subject Protections, National Institutes of Health. Evaluation of human subject protections in research conducted by the Centers for Disease Control and Prevention and the Agency for Toxic Substances and Disease Registry (memorandum): 7–8; July 28, 1995.

11. Cimons M. Test all newborns for AIDS, scientific council advises. Los Angeles Times: A5; February 9, 1989.

12. Gaims HM. Letter to the Editor. Los Angeles Times: Opinion, 4; March 5, 1989.

13. Institute of Medicine. Executive summary. In: HIV screening of pregnant women and newborns, ed. LN Hardy, 1–7 (Washington, D.C.: National Academy Press, 1991).

14. CDC. Guidelines for prophylaxis against *Pneumocystis carinii* pneumonia for persons infected with human immunodeficiency virus. MMWR 38 (Supplement, S-5): 1–9; 1989.

15. CDC. Guidelines for prophylaxis against *Pneumocystis carinii* pneumonia for children infected with human immunodeficiency virus. MMWR 40 (RR-2): 1–13; 1991.

16. CDC. HIV seroprevalence survey in childbearing women; June 1, 1992.

17. Simonds RJ, Oxtoby MJ, Caldwell MB, Gwinn JL, Rogers MF. Pneumocystis carinii pneumonia among US children with perinatally acquired HIV infection. JAMA 270:470–73; 1993.

18. CDC. Revised guidelines for prophylaxis against *Pneumocystis carinii* pneumonia for children infected with or perinatally exposed to human immunodeficiency virus. MMWR 44 (RR-4): 1–11; 1995.

19. Simonds RJ et al. Prophylaxis against *Pneumocystis carinii* pneumonia among children with perinatally acquired human immunodeficiency virus infection in the United States. New Engl J of Med 332: 786–90; 1995.

20. CDC. Recommendations for assisting in the prevention of the perinatal transmission of human T-lymphotropic virus type III/lymphadenopathy-associated virus and acquired immunodeficiency syndrome. MMWR 34:721–26, 731–32; 1985.

21. WHO. HIV transmission and breast-feeding. World Health Organization Bulletin; September 1992.

22. Connor EM et al. Reduction of maternal-infant transmission of human immunodeficiency virus type 1 with zidovudine treatment: Pediatric AIDS clinical trials group protocol 076 study group. New Engl J of Med 331:1173–80; 1994.

23. CDC. Recommendations of the U.S. Public Health Service Task Force on the use of zidovudine to reduce perinatal transmission of human immunodeficiency virus. MMWR 43 (RR-11): 1–20; 1994.

24. CDC. U.S. Public Health Service recommendations for human immunodefi-

ciency virus counseling and voluntary testing for pregnant women. MMWR 44 (RR-7): 1–15; 1995.

25. CDC. HIV seroprevalence survey in childbearing women; June 1, 1992.

26. CDC. Public health guidelines for counseling and antibody testing to prevent HIV infection and AIDS. MMWR 36:509–15; 1987.

27. Rogers MF et al. Use of the polymerase chain reaction for early detection of the proviral sequences of human immunodeficiency virus in infants born to seropositive mothers: New York City Collaborative Study of Maternal HIV Transmission and Montefiore Medical Center HIV Perinatal Transmission Study Group. New Engl J of Med 320:1649–54; 1989.

28. Garrett L. Test could detect AIDS in newborns. Newsday: 7; June 22, 1989.

29. Associated Press. A test for AIDS infection in newborns. New York Times: B10: June 22, 1989, and Los Angeles Times: Metro 2, 3; June 22, 1989.

30. Shilts, R. And the band played on: Politics, people and the AIDS epidemic (New York: St. Martin's Press, 1987), 587.

31. Ibid., 588.

32. Democratic Governors' Association and Democrats for the 80s. Roper Center, Public Opinion Online. National adult poll of likely voters, accession number 0073309; July 17–22, 1987.

33. U.S. News & World Report/CNN. Roper Center, Public Opinion Online. National adult poll, accession number 0112579, 0112584; March 31–April 1, 1987.

34. Gordon S. Black Corporation. Roper Center, Public Opinion Online. National adult poll, accession number 0179288; July 6–10, 1988.

35. Los Angeles Times. Roper Center, Public Opinion Online. National adult poll, accession number 0078269; July 8–13, 1989.

36. Associated Press. Roper Center, Public Opinion Online. National adult poll, accession number 0011806; May 5–13, 1989.

37. American Association of Blood Banks. Roper Center, Public Opinion Online. National adult poll, accession number 0258557; March 29–April 16, 1993.

38. Altman L. AIDS deaths drop 48% in New York. New York Times: A1; February 3, 1998.

39. Chu SY, Buehler JW, Oxtoby MJ, Kilbourne BW. Impact of the human immunodeficiency virus epidemiology on mortality in children, United States. Pediatrics 87:806–10; 1987.

40. Goldman JJ. 40% increase in AIDS babies in N.Y. seen. Los Angeles Times: Part 1, 4; January 14, 1988.

41. Randal JE. Tracking infection in babies. Newsday: 9; March 8, 1988.

42. Feinberg L. U.S. to test third of new babies for AIDS; no consent sought; blood samples won't be identified by names. Washington Post: A1; August 23, 1988.

43. Lambert B. AIDS survey shows course of infection. New York Times: B1; July 15, 1988.

44. Woodard C. Mothers to get results of AIDS test on babies. Newsday: 6; June 15, 1989.

45. Kolbert E. Mothers to be told if babies have AIDS virus. New York Times: D16; June 16, 1989.

46. Editorial staff. Test all newborns for AIDS. Newsday: 31; April 21, 1989.

47. Navarro M. Testing newborns for AIDS virus raises issue of mothers' privacy. New York Times: A1; August 8, 1993.

48. Editorial staff. AIDS babies pay the price. New York Times: A22; August 13, 1993.

49. Navarro M. Bill on testing newborns for H.I.V. is opposed. New York Times: B5; February 10, 1994.

50. Riley J. FOCUS ON: Mandatory AIDS tests; pain of knowing; doctor, clinician disagree on testing moms, newborns. Newsday: 15; August 25, 1993.

51. Goldberg N, Perez-Rivas M. AIDS test disclosure. Newsday: 23; February 10, 1994.

52. Woodward C. Panel: Easy on mom; says urge, don't force AIDS test at birth. Newsday: 28; December 1, 1993.

53. Gorman C. Moms, kids and AIDS: Can testing and treatment before and after birth help thousands of youngsters threatened by HIV? Time: 60; July 4, 1994.

54. CDC. Recommendations of the U.S. Public Health Service Task Force on the use of zidovudine, 1-20.

55. Connor EM et al. Reduction of maternal-infant transmission of human immunodeficiency virus type 1 with zidovudine treatment: Pediatric AIDS clinical trials group protocol 076 study group. New Engl J of Med 331:1173-80; 1994.

56. Kolata G. Discovery that AIDS can be prevented in babies raises debate on mandatory testing. New York Times: B14; November 3, 1994.

57. Neergaard L. CDC ends AIDS tests on newborns; surprise move fights pending AIDS bill. Associated Press; May 11, 1995.

58. Ibid.

59. Neergaard L. AIDS baby debate heats as lawmakers propose mandatory testing. Associated Press; May 13, 1995.

60. Bunis D. Lawmakers to push for AIDS testing. Newsday: A16; May 14, 1995.

61. Purnick J. When AIDS testing collides with confidentiality. New York Times: B4; May 18, 1995.

62. Editorial staff. If only babies voted. New York Daily News: A18; May 15, 1995.

63. Editorial staff. How politics undermines good science. Los Angeles Times: A20; May 15, 1995.

64. Dwyer J. Feds cave on AIDS testing. Newsday: A10; May 17, 1995.

65. Associated Press. First Lady urges expectant mothers to get HIV tests. May 23, 1995.

66. CDC. U.S. Public Health Service recommendations for human immunodeficiency virus counseling, 1-15.

67. CDC. Implementation and evaluation of USPHS recommendations for HIV counseling and testing of pregnant women, prevention of perinatal HIV transmission, and links to care for HIV-infected women and children: Summary report. May 31, 1996.

68. Curran JW. Personal communication. January 1, 1998.

69. Committee for the Study of the Future of Public Health, Institute of Medicine. The future of public health (Washington, D.C.: National Academy Press, 1988), 114–15.

70. Ibid., 141–42.

71. Curran JW. Personal communication. March 25, 1998.

72. Stein TJ. The social welfare of women and children with HIV and AIDS: Legal protections, policy and programs (New York: Oxford University Press, 1998), 115.

73. Ibid., 73.

74. Kirkpatrick DJ, Taylor CG, eds. Pediatric clinician's guide to AIDS and HIV infection in Georgia (Atlanta: Public Health Committee of the Medical Association of Georgia; Division of Public Health, Georgia Department of Human Resources; Pediatric Infectious Disease Clinic, Grady Health System, 1996).

75. Staff. Legislative master list: AIDS virus/HTLV-III. American Academy of Pediatrics, Washington, D.C., August 2, 1995.

76. Staff. State AIDS reports (Washington, D.C.: AIDS Policy Center, Intergovernmental Health Policy Project, George Washington University, 1990, 1994).

77. Linsky M. Impact: How the press affects federal policymaking (New York: W. W. Norton, 1986), 107–12.

78. Ibid., 105.

79. Ibid., 114–15.

80. Hentoff N. Personal communication. February 5, 1998.

81. Dwyer J. Personal communication. Febraury 2, 1998.

82. Shilts. And the band played on, 588–89.

83. Epstein R. If their baby has virus. (Louisville, Ky.) Courier-Journal: 1A; February 24, 1990.

84. Randal JE. Tracking infection in babies. Newsday: 9; March 8, 1988.

85. Feinberg. U.S. to test third of new babies for AIDS, A1.

86. Epstein. If their baby has virus, 1A.

87. Office for Protection from Research Risks, Division of Human Subject Protections, National Institutes of Health. Evaluation of human subject protections, ii.

88. Ibid., 11.

89. Coughlin SS, Soskolne CL, Goodman KW. Case studies in public health ethics (Washington, D.C.: American Public Health Association, 1997).

90. Mann J. Medicine and public health, ethics and human rights. Hastings Center Report: 6–13; May–June 1997.

91. Curran JW. Personal communication. January 1, 1998.

92. Schlesinger JR. Systems analysis and the political process. J of Law and Econ 11:281–98; 1968.

93. Lindblom CE. The science of "muddling through." Public Administration Rev 19:79–88; 1959.

94. Unknown CDC author. Document concerning the second year of the Survey of Childbearing Women. 1989.

95. Gwinn ML. E-mail: Revival of the perinatal working group. July 13, 1994.

96. Curran JW. Personal communication. January 1, 1998.

97. Gwinn ML. E-mail: NY state perinatal guidelines. September 21, 1993.

98. Curran JW. Personal communication. January 1, 1998.

99. Turnock BJ, Kelly CJ. Mandatory premarital testing for human immunodeficiency virus. JAMA 281:3415–18; 1988.

100. Peterson LR, White CR, and the Premarital Screening Group. Premarital screening for antibodies to human immunodeficiency virus type 1 in the United States. Am J of Public Health 80:1087–90; 1990.

101. Kuhn TS. The structure of scientific revolutions, 3rd ed (Chicago: University of Chicago Press, 1996).

102. George AL. Presidential management styles and models. In: Presidential decisionmaking in foreign policy: The effective use of information and advice, 145–68 (Boulder, Colo.: Westview Press, 1980).

103. George AL. The case for multiple advocacy in making foreign policy. Am Pol Sci Rev 66:751–85; 1972.

104. Porter RB. Three organizational models: Adhocracy, centralized management, and multiple advocacy. In: Presidential decision making, 229–52 (New York: Cambridge University Press, 1980).

105. Ibid., 242.

106. George. The case for multiple advocacy, 759.

107. Porter. Three organizational models, 241–42.

108. Ibid., 243–45.

109. Committee for the Study of the Future of Public Health, Institute of Medicine. The future of public health, 108.

110. Ibid., 107.

111. Ibid., 119.

112. Moore MH. Creating public value: Strategic management in government (Cambridge: Harvard University Press, 1995), 148.

113. Linsky. Impact, 11.

114. Nye JS Jr. The media and declining confidence in government. Harvard International J of Press/Politics 2(3): 4–10; Summer 1997.

115. Curran JW. Personal communication. March 25, 1998.

116. Roper WL. Letter to Representative David E. Bonior. September 16, 1992, CDC controlled correspondence files.

117. Committee for the Study of the Future of Public Health, Institute of Medicine. The future of public health, 115.

118. Curran JW. Personal communication. March 25, 1998.

119. Curran JW. Personal communication. January 1, 1998.

# CHAPTER 13

# Emerging Infections
# and the CDC Response

## RUTH L. BERKELMAN AND PHYLLIS FREEMAN

By the early 1980s many health authorities in the United States viewed the war on infectious diseases as having been largely won. Smallpox had been eradicated, tuberculosis had been all but eliminated, and other diseases such as measles, diphtheria, and polio had been brought under control through childhood immunization programs. The belief that the battle against infectious diseases had been won led to a breakdown in the health infrastructure that had been created at both the national and local levels to prevent and respond to infectious disease outbreaks. Within the nation's leading public health institution, the Centers for Disease Control, research and funding for the control of most infectious diseases had begun to decrease.

A number of public health authorities and health scientists, including Nobel Prize laureate Joshua Lederberg, argued that infectious diseases had not disappeared and remained a substantive threat to humanity. Moreover, changing social and environmental conditions were laying the basis for both the reemergence of existing infectious diseases and the emergence of new diseases. In this chapter Ruth Berkelman, who was deputy director of the CDC's National Center for Infectious Diseases from 1992 to 1997, and Phyllis Freeman, who served on the NCID's Board of Scientific Counselors, describe the efforts by the National Center for Infectious Diseases to draw attention to this impending public health crisis and to place infectious diseases back on the public health agenda. In doing so they provide important insights into both the political culture of the CDC and the legislative process through which emerging illnesses do or do not get recognized and funded.

It is useful to compare the success of the NCID and its Board of Scientific Counselors in working with the media and wider governmental institutions to place emerging infections on the public health agenda with the difficulties that the CDC encountered in responding to the whirlwind of controversy surrounding its Survey of Childbearing Women, described by Lydia Ogden (Chapter 12). The comparison warns us that it is difficult to generalize about the institutional culture of large governmental agencies with multiple divisions and interests. Studies that attempt to explore the ways in which such organizations respond to emerging public health issues need to look carefully at the specific institutional structures, personalities, and dynamics involved in responding to any given public health issue.

---

In the early 1990s, scientists within and outside the federal government undertook an effort to revitalize the Centers for Disease Control and Prevention's capacity to address threats posed by infectious diseases. Why was the effort so needed—and so difficult? Why had the capacity to face infectious disease challenges eroded? Some of the answers to these questions lie with the prevailing belief of many policymakers that the war against infectious diseases was over—and that we had won. But other answers relate to the broader challenge faced by all those seeking to move emerging health issues onto the public health agenda.

This chapter explores the events that led up to the response of the CDC and its National Center for Infectious Diseases to the growing threat of emerging and reemerging infections during the 1990s. It describes how the NCID created its Board of Scientific Counselors (NCID/BSC), which together with NCID scientists called for an overhaul of the government's role in the early detection and response to infectious diseases. The NCID developed a plan entitled *Addressing Emerging Infectious Disease Threats: A Prevention Strategy for the United States* (1994) as its road map for responding to emerging infections. The chapter examines decisions and actions taken by the NCID/BSC in developing and promoting the plan through 1998, at which time it was updated; at the same time, concerns with bioterrorism were escalating and were creating additional momentum to address the challenge of infectious diseases.

The chapter is divided into two parts that offer the complementary perspectives of two individuals who were intimately involved in these activities. The first part, "An Overview," was written by Ruth Berkelman, who was deputy directory of the National Center for Infectious Diseases from 1992 to 1997. It outlines the problem faced by the NCID in the late 1980s and the actions that were taken to transform the nation's capacity to deal with infectious disease threats. The second part, "A Story behind the Story of the Federal Response to Emerging Infectious Diseases," was written by Phyllis Freeman. Freeman was former counsel to the Oversight and Investigations Subcommittee of the Energy and Commerce Committee in the U.S. House of Representatives from 1982 to 1986 and a member of the NCID's Board of Scientific Counselors from 1992 to 1995. This part looks more closely at the internal workings of the Board of Scientific Counselors and its efforts to work with NCID scientists in developing and advancing the goals of the CDC's plan for emerging infectious diseases. From both perspectives, the history of the CDC's response to emerging infections can be seen as an example of institutional learning and adaptation. Nonetheless, any reflection on how the response of the NCID "evolved" should serve more as a cautionary tale about the status of formal public health leadership in the United States than as one of an easily replicable achievement. The favorable outcome was never assured; nor do these accounts suggest that similar challenges to the public's health will be met with comparable resolve.

**PART I / AN OVERVIEW**

## The Alleged Conquest of Infectious Diseases

As early as the mid-twentieth century, scientists were beginning to proclaim triumph over infectious diseases. From Charles-Edward Winslow's 1943 proclamation following the introduction of penicillin—which "forever banished the plagues and pestilences"[1]—to Frank Macfarlane Burnet's 1962 statement that infectious disease had virtually been eliminated "as a significant factor in social life,"[2] the effectiveness of antibiotics and vaccines led to unbounded optimism in the following decades. Mortality from infectious diseases fell rapidly in the United States throughout the beginning and middle of the twentieth century, and a new paradigm of public health emerged. It was thought to be only a matter of time before the developing countries would enjoy the same success against infectious diseases. The eradication of smallpox in 1976 served to emphasize further the technical triumphs against microbial agents. It was no surprise that in 1970 the Communicable Disease Center, arguably the United States' most valuable resource in controlling infectious diseases, changed its name to the Center for Disease Control to reflect a broadened mission that included noninfectious disease issues.[3]

The theory of epidemiological transition, first proposed in 1971, projected an increase in the importance of chronic and degenerative diseases accompanied by a continued decline in mortality from infectious diseases.[4] In 1974, Marc Lalonde in Canada wrote that infectious diseases had "largely been brought under control" and stated a need for increased emphasis on healthy behaviors and a focus on chronic disease prevention.[5] In the United States, the Public Health Service examined the leading causes of death and developed Health Objectives for the Nation. These objectives gave little consideration to the potential need to address new infectious diseases. If anything, Legionnaires' disease and toxic shock syndrome seemed to reinforce the notion that, even if a new disease appeared, the CDC would be able to address it.

As part of the CDC's "reorientation toward lifestyle and environmental issues," the Center for Disease Control became the Centers for Disease Control and Prevention in 1980.[6] New centers continued to be established, including the Center for Chronic Disease Prevention and Health Promotion in 1989 and the National Center for Injury Control and Prevention in 1992. In contrast, the visibility of and emphasis on the National Center for Infectious Diseases generally decreased.[7]

**The Reality of Infectious Diseases**

Despite the earlier predictions to the contrary, infectious diseases remained a serious threat to health in the United States and the world as we entered the twenty-first century. From 1980 to 1992, the death rate due to infectious diseases rose more than 50 percent in the United States, increasing from the fifth leading cause of death in 1980 to the third leading cause of death in 1992. HIV infection, pneumonia, and septicemia all contributed to the increase.[8] The United States also witnessed a resurgence of measles and tuberculosis, diseases thought conquered, while simultaneously facing the emergence of foodborne diseases like *E. coli* 0157:H7 and *Salmonella* in the 1980s. Stuart Levy and others noted that antibiotics were rapidly losing their effectiveness for many conditions in the United States and elsewhere.[9]

The AIDS epidemic surprised public health professionals and contradicted the new paradigm of the diminished importance of infectious diseases. The emergence of this disease as an infectious one may have contributed to the difficulty in its being accepted as a serious public health threat early in the 1980s. AIDS became treated as a unique phenomenon, and in many locales of the United States it was kept administratively and scientifically separate from all other infectious diseases (e.g., in many state health departments, an AIDS unit reported directly to the health commissioner, whereas the communicable disease unit reported separately to either the commissioner or another lower-level official).

Worldwide, the "old killers" like malaria, tuberculosis, and cholera had returned with a vengeance in many areas. Malaria had been fueled by decreased infrastructure for control together with resistance to antimalarial agents. The HIV epidemic had spurred tuberculosis, with multidrug resistance causing additional illness and death; and with contamination of water supplies, cholera regained a strong foothold in South America in the early 1990s after almost a century's absence. In Africa, the emergence of HIV had begun to wipe out many of the gains in life expectancy from vaccine control programs. Scientists and public health experts involved in infectious diseases were also taking note of other new pathogens, some newly recognized, some newly emergent.[10] As the last decade of the twentieth century began, infectious diseases continued to be documented as the leading cause of death worldwide.

Far before most public health leaders took note, a few individuals prominent in the field of infectious diseases had recognized the danger of complacency. Burnet's 1972 edition of the *Natural History of Infectious Diseases,* co-written with David O. White, explicitly recognized the need to maintain control programs or face a "swift return" of earlier plagues.[11] Richard Krause's *The Restless Tide,* published in 1981, called for continued vigilance in the area of infectious diseases.[12] Joshua Lederberg of Rockefeller University also repeatedly warned in the lay and professional press of the threat of both newly emerging and reemerging infectious diseases. These calls for vigilance, although from prominent individuals, went largely unnoticed and unheeded, and the trend toward declining public health outlays for addressing most infectious diseases continued throughout the 1980s.[13]

In 1989, Stephen Morse of Rockefeller University, with the assistance of John LaMontagne and Ann Schleuderberg of the National Institute of Allergy and Infectious Diseases, National Institutes of Health, organized a conference on emerging viruses, which, along with coaxing by Lederberg, spurred the formation of an Institute of Medicine Committee on Emerging Infectious Diseases.[14] Lederberg, Robert Shope of Yale University, Stanley Oaks of the IOM, and others provided leadership to the committee, with Lederberg and Shope officially serving as co-chairs. They received little federal guidance, and specifically, the CDC had minimal input.

The IOM committee challenged the thesis that infectious diseases should be relegated to the past, and in a report issued in 1992, the committee expounded on the increasing vulnerability of the population to emerging and reemerging infectious diseases. Increasing population and urbanization with concomitant crowding; changes in behavior such as sexual activity and use of illegal drugs; changes in technology and industry (e.g., food industry); changes in land use and ecologic conditions (e.g., dam building); rapid increases in international travel and migration; and continued microbial adaptation and change, including the increasing problem of antimicrobial resistance, were cited in the report as factors contributing to the threat of emerging and reemerging infectious diseases. The committee also argued that erosion of the public health infrastructure compromised the U.S. ability to conduct disease surveillance and to detect and respond effectively to outbreaks. Complacency, war, and civil strife had the potential to result in rapid increases in diseases previously controlled. The thesis was clearly presented and persuasive among scientific colleagues, with the report encompassing global issues. Federal agencies, particularly the CDC, were exhorted to respond to the threat and to address a panoply of recommendations.

The IOM published its report on emerging infections as the NCID was struggling with increasing needs without a concomitant increase in resources. With the exception of HIV/AIDS and a few programs such as Lyme disease and measles, which had received targeted appropriations by Congress, the CDC provided few additional resources to the National Center for Infectious Diseases. The NCID was experiencing reductions in personnel devoted to infectious diseases other than AIDS, with further reductions projected (fig. 13.1).[15] CDC laboratory equipment in many areas was not up to date, and services such as laboratory training for professionals in the states had been curtailed in many areas. The majority of the laboratory facilities had been planned and built in the 1950s and 1960s, and most had not been replaced or adequately renovated, and plans to do so lay idle.[16] Problems ranging from crowding to deterioration of the air-handling systems in many laboratories resulted in significant concerns about the safety of the workers.

By the early 1990s and even before, the CDC had lost considerable laboratory expertise in a number of disease areas such as botulism and plague. When the outbreak of plague occurred in India in 1994, expertise was extremely limited both in the United States and worldwide, with the CDC having only one microbiologist trained in plague diagnosis. Indeed, as part of a plan for further personnel reductions, the plague activity had been targeted for potential elimination earlier the same year.

Possibly even more dangerous to the public health, the CDC and the state health departments had not acquired needed expertise in new areas such as cryptosporidiosis and *E. coli* 0157:H7; even as the CDC draft plan was being reviewed, the lack of preparedness was becoming starkly evident. In 1993, Milwaukee experienced the largest outbreak of waterborne disease ever reported in the United States. More than 400,000 individuals in Milwaukee became ill over a short period of time with cryptospordiosis. Over 40,000 sought medical attention, and 4,400 were hospitalized as a result of this epi-

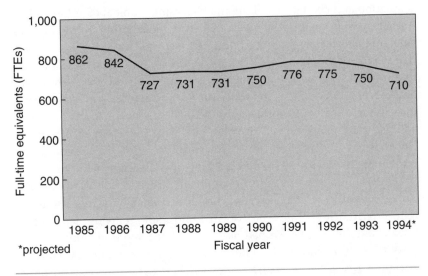

**Fig. 13.1.** Number of personnel (full-time equivalents) assigned to the National Center for Infectious Diseases, CDC, 1985–94. Figure does not include personnel devoted to HIV/AIDS.

demic. Although previous outbreaks of cryptosporidiosis had occurred in the preceding decade, insufficient attention had been paid to developing a diagnostic test that could be widely used, to familiarizing physicians with the emergence of this parasitic pathogen and the disease that it caused, and to developing an understanding of the pathogen and of means to prevent contamination of public water supplies.

The public health response to this and other problems was delayed, both in establishing needed surveillance across the United States and in gathering and disseminating knowledge about prevention, detection, and control of these diseases to health care workers, microbiologists, and the public health community throughout the United States.[17]

### Institutional Response, the BSC, and the CDC Plan

As director of the National Center for Infectious Diseases from 1986 to 1991, Frederick Murphy was aware that the needs of the NCID were countercurrent to the emphasis and growth areas planned for the CDC (primarily the prevention of chronic diseases and the delivery of individual clinical preventive services). Murphy faced crises in both funding and personnel. The core funding for the NCID, adjusted using the NIH Biomedical and Research Development Price Index (BRDPI), was steadily decreasing from 1982 from approximately $35 million to less than $30 million in 1986.[18] During the same period the number of personnel positions allocated to the NCID for infectious disease activities other than HIV/AIDS dropped from 886 positions in 1982 to 842 full-time equivalents (FTEs) in 1986 and to 727 FTEs in 1987 (fig. 13.1).

The erosion of the NCID was clearly evident to Murphy, and he was concerned that the public health needs to address infectious diseases were not being met and would not be met under the status quo. From his experience on the Animal and Plant Health Inspection Service (APHIS) Board of the U.S. Department of Agriculture and his discussions with Jim Glosser, then administrator of APHIS, Murphy was aware of the potential support and leverage that could be gained by an interested and engaged group of external colleagues.[19] He became intent on forming an external advisory group, a board of scientific counselors, for the NCID, to be composed of esteemed colleagues from outside the agency. This idea was new to the agency; few advisory boards for other parts of the organization had been established, and Murphy's request to establish a board met with considerable internal opposition from senior leadership. However, James Mason, director of the CDC, was receptive to the idea, and the charter went forward, with the first meeting held in 1989.

The board was composed of a dozen or more highly respected individuals in the area of infectious diseases. Murphy made the initial nominations, and all were approved by the CDC director and the secretary of the Department of Health and Human Services. These individuals came from academia, industry, public health laboratory and epidemiology organizations representing state health departments, and other professional organizations, such as the American Society of Microbiology (table 13.1). Individuals from the NIH and the Food and Drug Administration served as ex officio members, and liaison membership was also extended to public health organizations in Mexico and Canada.

Even after Murphy's resignation in 1991, the board continued to function. A board meeting in December 1992 built on the fall release of the IOM report on emerging infections. The board recommended that the NCID develop a plan for the CDC that would map the actions needed to be taken to address the threat of emerging infectious diseases. The process moved quickly: within a month, the deputy director (this author) had assembled a team of respected scientists within the NCID and produced a draft plan.[20] In the same month, a subgroup of the NCID board met in Washington, D.C., to review the initial draft with the director of the NCID, James Hughes; its deputy director; and the individuals detailed to draft the CDC plan. Two months later, the NCID held a meeting in Atlanta with individuals across the CDC and public health and infectious disease experts external to the agency. Several NCID board members attended together with Lederberg and other members of the IOM committee; other individual leaders representing major medical, public health, and infectious disease organizations were also present.

The draft was heavily critiqued, and one of the scientists who had worked on the initial draft, Ralph Bryan, was assigned to the deputy director to undertake extensive revisions of the draft. In addition to gaining outside input, the deputy director vetted the draft extensively internally, both with the branches and divisions of the NCID (all interested persons were welcomed) and with senior staff in other centers and offices at the CDC. New drafts were circulated widely both internally and externally to medical and public health professionals, microbiologists, and other interested individuals, and detailed written comments were received from more than two hundred individuals within and outside the agency.

The final draft was completed in January 1994, and the CDC published the plan, *Addressing Emerging Infectious Disease Threats: A Prevention Strategy for the United States,* in April 1994, with a preface signed by the newly appointed director of the CDC, David Satcher.

The CDC plan presented a framework based on disease detection (i.e., surveillance) and response, applied research, prevention, and infrastructure. It articulated goals, objectives, and explicit activities. A list of high-priority activities focused attention on what could and should be done immediately. These included strengthening surveillance at the state and other levels, establishing emerging infections programs to investigate emerging problems more extensively, reestablishing an extramural grants program in applied research, and strengthening public and health professional education related to emerging infections. But as the CDC report was published, the personnel shortage was critical, and the forecast for relief was gloomy.[21] When hantavirus broke out in the southwestern United States in 1993, for example, scientists made repeated appeals for additional personnel so that senior investigators in the Division of Viral and Rickettsial Diseases would not have to perform such tasks as cleaning and feeding the laboratory animals. Later the same year, and in accordance with Executive Order 12839, the administration set a personnel reduction goal of 250,000 by 1999, and target FTE reductions were given to the individual agencies; the National Center for Infectious Diseases was directed to draft plans to reduce NCID by as many as 95 FTEs at the time the CDC plan on emerging infections was being published.

**Table 13.1.** The NCID's Board of Scientific Counselors, Roster, November 1991

Walter Bowie, D.V.M., Dean Emeritus, School of Veterinary Medicine, Tuskegee University, Tuskegee, Alabama

Gail Cassell, Ph.D., Professor and Chairman, Department of Microbiology, University of Alabama at Birmingham, Birmingham, Alabama

Jeffrey Davis, M.D., State Epidemiologist, State of Wisconsin Department of Health and Social Services, Madison, Wisconsin

Herbert DuPont, M.D., Director, Center for Infectious Diseases, The University of Texas Medical School at Houston, Houston, Texas

Bernard Fields, Chairman, Department of Microbiology and Molecular Genetics, Harvard Medical School, Boston, Massachusetts

Phyllis Freeman, J.D., Associate Professor, University of Massachusetts Law Center, Boston, Massachusetts

Scott Halstead, Acting Director, Health Sciences Division, Rockefeller Foundation, New York, New York

Harlyn Halvorson, Ph.D., President and Director, Marine Biological Laboratory, Woods Hole, Massachusetts

William Hand, M.D., Director, Division of Infectious Diseases, Emory University School of Medicine, Atlanta, Georgia

George Hill, Ph.D., Professor and Director, Division of Biomedical Sciences, Meharry Medical College, Nashville, Tennessee

Donald Hopkins, M.D., Senior Consultant, Chicago, Illinois (Former Deputy Director, CDC)

Carlos Lopez, M.D., Private Practitioner, Atlanta, Georgia

John Maupin Jr., D.D.S., Morehouse School of Medicine, Atlanta, Georgia

Michael Osterholm, Ph.D., State Epidemiologist, Minnesota Department of Health, Minneapolis, Minnesota

Philip Russell, M.D., Professor of International Health, The Johns Hopkins University, Baltimore, Maryland

*Ex officio members:*

John Bennett and John La Montagne, National Institute of Allergy and Infectious Diseases, National Institutes of Health

Mary Ann Danello, Ph.D., Food and Drug Administration

*Liaison representatives:*

Joseph Losos, Director General, Laboratory Center for Disease Control, Ottawa, Canada

Ernesto Calderon, Center for Infectious Diseases, National Institute of Public Health, Mexico

## Gaining Support for the CDC Plan

In effect, *Addressing Emerging Infectious Disease Threats* provided a road map for rebuilding the nation's infectious disease prevention infrastructure. But it was not at all clear that the path the plan charted would be followed. Transforming its recommendations into funded initiatives would require a sustained effort to mobilize political and public support for the plan. NCID staff and BSC members therefore began in the early 1990s to lay the groundwork for the plan by increasing public awareness of the threat of emerging infections and the need for new public health investment.

Even before publication of the CDC plan, the growing nucleus of individuals within and outside the agency interested in the challenge of emerging infectious diseases worked to increase awareness of the issues. In the late 1980s, Murphy had initiated a program focus for the NCID entitled "New and Emerging Infectious Diseases." Anthony Fauci, director of the NIAID, and John LaMontagne convened a task force in 1991 to examine future challenges and opportunities facing microbiology and infectious diseases sciences, and one of the six themes chosen to cover this subject in a report published in 1992 was "Emerging Infectious Diseases." In 1992, Bennie Osburn and colleagues at the School of Veterinary Medicine at the University of California organized an international conference on the infectious diseases of animals, choosing as its focus "New and Emerging Diseases."[22] By 1993, sessions at professional meetings were being organized around the theme of emerging infectious diseases, and conferences were being organized by the National Academy of Sciences and by the Harvard School of Public Health.

The thinking and activities of the board members and particularly its chair, Gail Cassell, were critical to forward movement of the agency's efforts in emerging infectious diseases. In particular, Cassell's dual role as chair of the NCID board and president of the American Society for Microbiology was critical. She and others at the ASM, particularly Janet Shoemaker, director of public and scientific affairs, were instrumental in gaining congressional support for emergency funding to address the need for investigation of hantavirus as well as sharing the larger issue of emerging infectious diseases with policymakers. Cassell's leadership is well illustrated also by her many oral and written communications, such as the article entitled "New and Emerging Infections in the Face of a Funding Crisis," published in May 1994 in *ASM News,* just a month after publication of the CDC plan.

The NCID developed fact sheets, slide sets, and other materials for distribution, materials that might be useful to NCID board members and others interested in the issue of emerging infectious diseases. Satcher and Hughes sent letters regarding the CDC's planned response to all state health commissioners, state epidemiologists, state directors of public health laboratories, chairs of departments of infectious disease in medical schools, deans of schools of public health, executive directors of various professional organizations, and others identified as having a professional interest in emerging infectious diseases. Enclosed with the letters were the executive summary of the IOM report, the CDC prevention strategy that articulated the CDC's proposed response, and other materials also developed by the NCID. The CDC also provided materials to congressional staff and to others upon request. NCID scientists, BSC members, and others were available for congressional briefings sponsored by such organizations as the ASM, the American Public Health Association, and the George Washington Health Policy Forum. These activities were designed to enhance awareness of both public health and infectious disease in the professional community and among health policymakers. The activities were also intended to emphasize the need to address emerging infectious diseases as well as to offer specific guidance for meeting the challenge.[23]

## The Role of the Media

NCID scientists and staff and BSC members were aware of the important role the media played in mobilizing public awareness of public health issues. They thus made special efforts to work with news reporters and science writers and to train NCID scientists to work with the media.

Several lay scientific writers also became interested in the thesis of emerging infectious diseases in the early 1990s and initially engaged in discussions with a small nucleus of interested scientists.[24] Richard Preston's article "Crisis in the Hot Zone" appeared in 1992 in the *New Yorker* at the same time as the release of the IOM report on emerging infectious diseases.[25] His initial inspiration for the article and subsequent book *The Hot Zone* came from Stephen Morse of Rockefeller University. Laurie Garrett, a *Newsday* writer, wrote frequently of the issue for lay readers and spent a year with faculty at the Harvard School of Public Health; her book *The Coming Plague* subsequently won a Pulitzer Prize. NCID scientists took steps to inform Richard Preston and Laurie Garrett, both of whom were publicly speaking on the issue of emerging infectious diseases, of the CDC's plan on emerging infectious diseases, and both writers were openly supportive in the media of the CDC's efforts to enhance its response capabilities.

Pulitzer Prize–winning reporter Mike Toner also wrote frequently of emerging issues like antibiotic resistance as well as the public health needs in infectious diseases. One major article in 1995 was devoted to the poor state of the CDC's laboratories and the need for congressional appropriations to be restored for the construction of new laboratories. Toner quoted three NCID board members, including Cassell and Michael Osterholm, Minnesota state epidemiologist, and two CDC officials, including Satcher and Joe McDade, associate director for laboratory science of the NCID, with all individuals basically providing the same message, that the laboratories were in "horrendous" condition and that the new laboratories were desperately needed.[26]

By 1994, the NCID had trained a number of scientists to respond to media requests, and the frequency of such requests began to increase.[27] With the occurrence of outbreaks from *E. coli* 0157:H7 to Ebola virus to group A streptococcal infections ("the flesh-eating bacteria"), media interest ran high. Emerging infections became a cover story of *Time, Newsweek,* and *US News & World Report* in 1994–95, and these stories featured the CDC and the work accomplished by various professionals within the NCID. The issue became a focus for stories on national evening news, CNN, "Nightline," "Good Morning America," and the National Press Club in the same years. Such stories were useful in educating the public not only about emerging infectious diseases but also of the important role of public health in protecting the population's health.

### Securing a Place on the National Public Health Agenda

The success of the NCID/BSC efforts to place emerging infections on the public health agenda began to be apparent in 1994. David Satcher, the new CDC director, took up the cause in his keynote speech to the American Society for Microbiology. Not coincidentally, BSC chair Gail Cassell served as president of ASM. Also in 1994, at the request of the Department of State and under the auspices of the Committee on International Science, Engineering and Technology (CISET), National Science and Technology Council, Satcher chaired a federal working group on emerging and reemerging infectious diseases. Seventeen federal agencies were involved, and the report, *Infectious Diseases—A Global Health Threat,* was published in September 1995. At the National Conference on International Health in June 1996, for which emerging infections was the theme, Vice President Al Gore was the keynote speaker and announced a presidential decision directive on global infectious diseases. The address was symbolic that emerging infectious diseases were clearly a part of the executive branch public health agenda.

Congress first appropriated resources ($6.7 million) to begin implementation of the plan in 1994 (fiscal year 1995) in the absence of a presidential budget request (fig. 13.2).[28] In subsequent years, the president's budget proposed increased appropriations, and Congress responded in a bipartisan manner. In fiscal year 1996, an additional $10.7 million was provided to the annual budget, and in fiscal year 1997, $25.7 million, for a cumulative annual increase of $44.1 million to the NCID budget, approximately a 50 percent increase to the non-HIV budget in the NCID in fewer than three years following publication of the plan.[29]

In terms of improved financial resources, the success of the effort is easy to measure. Some amelioration of the personnel resources can also be documented in subsequent years, and gaps that were clearly identified in specific disease areas were, if not closed, at least narrowed. The high-priority activities listed in the 1994 plan were implemented by 1997.[30] In addition, with a high number of laboratory personnel having recently retired without replacements, the CDC initiated a public health laboratory fellowship program in 1995 in collaboration with the Association of Public Health Laboratory Directors. This program has already infused some newly trained microbiologists into public health laboratories. Congress also appropriated funding for construction of a new laboratory building that opened in 2000. In 1998, an updated plan, *Preventing Emerging Infections: A Strategy for the 21st Century,* together with a growing perception of an increased risk of bioterror-

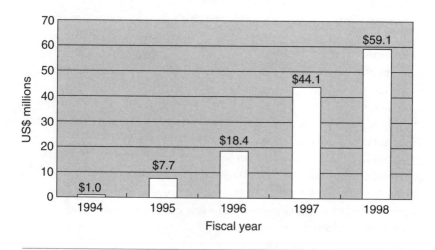

Fig. 13.2. Congressional appropriations for emerging infections in fiscal year 1994 and for CDC implementation of the 1994 plan on addressing emerging infectious disease threats in fiscal years 1995–98.

ism, provided further impetus to improve the CDC's capacity to address infectious diseases.[31]

The NCID provided resources to selected state health departments to improve their infrastructure for surveillance and response. The effectiveness of such resources can be measured only indirectly, since no model standards of what is "adequate" capacity for addressing infectious diseases have been set. However, early incidents related to foodborne diseases illustrate success in filling some critical needs and led to the capacity to detect outbreaks earlier and to increase the safety of the food supply.[32] The programs facilitated gathering information needed to ascertain quickly whether any such disease was occurring in the United States. The programs also provided a framework for the subsequent development and success of Food-Net, a collaborative effort by the CDC, USDA, and the Food and Drug Administration as increased attention focused on overall safety of the food supply.

Internationally, initial efforts to enhance detection and response capability were focused on training efforts in selected diseases. James LeDuc, an NCID scientist, was key to many of the international efforts and was detailed to the World Health Organization in the early 1990s. David Heymann, in turn, developed an organizational unit at the World Health Organization focused on global outbreak detection and response,[33] as well as other aspects of addressing the challenges posed by emerging infectious diseases.

### The Success of the CDC Plan

In summary, after the publication of the CDC plan, scientists were successful in gaining recognition of the threat of emerging infectious diseases and in initiating efforts to address these threats at a time when the attention of many public health policymakers was diverted to other health issues. The experts in the field of infectious diseases both within and outside the CDC united in their efforts, and the NCID board played a critical role in establishing this unity and serving as a liaison between the public and private sector. These scientists effectively coalesced to provide one clear message to the Department of Health and Human Services, Congress, the public, and their colleagues, using both professional channels and the media. These joint efforts, predominantly by scientists, were successful in increasing the U.S. capacity to address emerging infectious disease threats even in the absence of a recognized "community of suffering."

## PART II / A STORY BEHIND THE STORY OF THE FEDERAL RESPONSE TO EMERGING INFECTIOUS DISEASES

Why have many of the activities described in Part I of this chapter taken place? It is important to dispel any notion that they occurred as part of the natural course of events. Gestation of the CDC plan and its delivery into the political arena were carefully directed. Progress toward achieving the goals set forth in the plan ultimately required members of the NCID board to transfer the threat of "emerging infections" from a previously quiet realm of scientists, largely ignored by those making budget priority decisions, into the quirky but crucial domain of national budget politics. This part of the chapter tells the story of how this transfer was managed.

### The Institute of Medicine Report of 1992

It seems highly unlikely that the CDC plan would ever have been conceived had Joshua Lederberg and Stephen Morse at Rockefeller University and their colleagues elsewhere not persisted in drawing attention to their well-informed concerns in the 1980s. It was their efforts that generated early institutional responses: the NIH co-sponsored conference in 1989 that generated coverage of the issues by the *New York Times* and featured Lederberg's preoccupation with the mounting microbial threats to humanity, the appointment of a study panel by the Institute of Medicine in 1991, and its 1992 report, *Emerging Infections: Microbial Threats to Health in the United States.*

A central recommendation in this report was the call for a national effort to strengthen surveillance, suggesting that the CDC could be responsible for strategy development. Until the end of 1992 the threat of the spread of infectious disease had been recognized, analyzed, and worried over almost exclusively by scientists working outside government agencies in academic institutions.[34]

Yet a strategy featuring better surveillance was virtually meaningless to any audience beyond the experts. This point is crucial because rousing a substantial governmental response often depends on informed public interest and pursuit of a remedy that can be readily appreciated. Systems for early detection of old or new infectious diseases (surveillance) never did provoke a groundswell of public support, even though they are essential for preventing outbreaks and epidemics. Nor has surveillance stirred potent constituent activism. No "political hero" ready and able to intervene with colleagues at top levels in the executive branch or with members of Congress has adopted surveillance systems as a cause. Threats to public health, whose agents can

be confirmed only in laboratories, are unlikely candidates for inciting constituent support beyond the experts. Yet it is precisely when pathogens are
under early scrutiny in laboratories and still invisible to everyone else that
public health strategies can be most effective in preventing dangerous consequences for humans. Stepping up support for all those activities that form
the basis of a comprehensive program capable of addressing infectious disease threats (e.g., surveillance, laboratory support, epidemiology) would
become integral to the vision and culture of the CDC/DHHS regular budget
process over the next three years. But there was little prospect of new support in either the executive or legislative branches of federal government in
1992 when the IOM released its report.

Consequently, little optimism for significant program development to
meet the microbial threats beyond HIV endured from Frederick Murphy's
era into the era of his successor, James Hughes. When NCID staff convened
their regular meeting with the Board of Scientific Counselors in December
1992, it was immediately clear that the NCID's leadership and members of
the BSC shared the motivating perception that the IOM report offered a tremendous opportunity—and one that would not come knocking soon again.

However, no significant progress in detecting infections early or protecting the public from them could be made without a significant infusion of
new funds into the NCID. These would have to come from shifting dollars
already appropriated for some other purpose within CDC or from a new
congressional appropriation specifically for this purpose. Although other
centers within the CDC had succeeded in promoting program and budget
initiatives for additional funding within recent memory, the NCID had not.
In fact, the prospect of gaining support in this way appeared so dim to some
CDC staff that the NCID emerging infections campaign was sometimes
greeted within the NCID and other CDC centers with cynicism and questions about whether promoting such an initiative would simply waste effort.

But attempts to shift or "reprogram" money from elsewhere in the CDC to
the NCID would also require support of the CDC director and probably of
Congress.[35] At least a new appropriation offered an appealing tactical advantage of avoiding immediate competition among the CDC centers for
resources already appropriated—and eagerly awaited by any unit already
designated to benefit. In contrast to the relatively well funded program areas
of HIV/AIDS and immunization, the NCID was sometimes referred to as the
"unfunded infectious disease center" responsible for "everything else."
"Everything else" amounted to more than one hundred diseases with greater global disease burden in 1990 than even HIV and tuberculosis combined.[36] Within the NCID, budget materials separated the center's budget

in two distinct categories: "AIDS" and "non AIDS." By 1992 the NCID funding for "everything else" (called "non AIDS"—hardly an inspiring label) amounted to approximately $50 million, far less than the hundreds of millions dedicated to HIV/AIDS.

By the time the CDC's plan for responding to emerging infections was being prepared, winning a congressional appropriation for work to be carried out by any unit of the federal government without top-level executive branch sponsorship required even more prodigious effort from those able to take the proposal directly to influential members of Congress from a base outside government. To understand why this was so, it is necessary to trace changes in the federal budgetary process during the 1980s and early 1990s.

## Budget Politics

Starting with the election of President Ronald Reagan (and of a Republican majority in the Senate) in 1980, the process for gaining approval at each step in the formal chain of command for new appropriations from Congress grew in complexity and diminished in likelihood for attainment. No "initiative" for new funds originating within the CDC could advance for consideration by the cabinet agency (the DHHS) without winning the approval of the CDC director, who was a political appointee of the president.[37] It had not been easy previously, even when the "tax and spend" Democrats were the majority party in both branches of Congress and controlled the presidency. However, both before and after the new era of fiscal conservatism swept in with President Reagan, any item for new funding could be proposed by politically invisible agency personnel (such as NCID staff) or instigated at a higher level within the executive branch based on either expert or political interest or both. This was not possible during the Reagan era, making it more difficult for the NCID staff to advance its cause.

Similarly, before President Reagan assumed control over the federal budget process, budget preparation had a bottom-up character.[38] Units within agencies explained any reasons for expanding activities and estimated their budgetary needs to superiors, who accumulated these requests, approved some, and reduced or refused others. The budgets sent up to the Office of Management and Budget by cabinet agency heads reflected all these decisions.[39] In contrast, President Reagan's OMB initiated a top-down approach by imposing on agencies a cap, or maximum, on the request that could be sent forward for inclusion in the president's budget; the intent was to discourage any increases, including those for new initiatives.

The Reagan administration's determination to control the budget ever more tightly resulted in more restricted forms of communication between the administration and Congress. And the administration diminished the transparency of the budget process to make it more difficult for Congress and for constituent groups to advocate so knowledgeably for budget requests made but denied within executive branch agencies. Many federal officials who had regularly disclosed budget information became reluctant to do so for fear they would lose their jobs or suffer other adverse consequences. All these changes obscured the history of professional judgment represented by program initiatives and funding requests made at each step before the president's budget was finalized for presentation to Congress. Thus, starting in the early 1980s, interested members of Congress, citizens' groups, and professional lobbyists had to work harder to glean the history of internal debates over program and budget priorities within the executive branch.

In 1981 the new president called for the most dramatic reductions in health spending in memory—a cut of 25 percent in authorizations for health programs seeking appropriations, including all those implemented through Public Health Service agencies, including the CDC and the NIH.[40] As budget battles raged through the 1980s, the Reagan and Bush administrations maintained greater pressure to constrain budget requests and the flow of budget information.[41]

Prospects for health funding did not improve in the early 1990s. Although health reform was a central piece of Bill Clinton's political agenda when he assumed office in 1992, new spending on health programs faced a difficult legislative path, thanks in part to the divide that existed between a Republican-controlled Senate and a Democrat-controlled House of Representatives. Prospects for new funding decreased further in 1994 when the Republican Party won a majority in the House of Representatives and Speaker Newt Gingrich's "Republican Revolution" was in full force. With the Speaker's widely publicized determination to eliminate agencies and to achieve dramatic budget cuts, success by the usual route of up-the-chain-of-command approvals within the executive branch for requesting new funding of Congress—with or without support of the CDC director—would have been enormously labor-intensive and difficult.

A few NCID board members who were already familiar with budget politics felt from the start that without hope of support from the CDC leadership, Hughes and his colleagues could have no realistic expectation of funding to implement the plan through "normal channels"—even with the new IOM report in hand. NCID leadership and the politically experienced members of the board understood that no progress could be made unless the NCID changed its traditional approach and unless some members of the board moved beyond scientific discussions to intensely tactical ones. No new prevention strategy could reach fruition without translating the scientific insights into programs and without a new level of advocacy for them. The former was clearly a role suitable to NCID scientists; the latter would have to be assumed by members of the board and others outside federal employment, including those who had shaped the IOM report.

## From Science to Public Interest Advocacy

There was never any question that the commitment shared among NCID scientists and board members was well grounded scientifically. Primary motivation arose from a shared determination that it would be professionally irresponsible in the extreme to allow budget politics to quiet the cautionary voices of the experts. To accept the budget constraints as inevitable would be to obscure from public view the knowledge that without increasing investment in detection and response, infectious diseases, new and old, would undoubtedly cause a great deal of preventable harm and expense.

As early as the December 1992 meeting of the NCID's Board of Scientific Counselors, key decisions were made, some immediate, others more evolutionary in nature. Among these were the resolve to write a plan of action as the basis for support, to think big, and to pursue funding more aggressively. The new efforts would depend on carefully orchestrated and monitored teamwork among the "insiders" (NCID staff) and "outsiders" (board members and others). The NCID would pursue collaboration with the CDC's other centers and across numerous scientific and medical institutions essential for implementation of a viable program. Introduction of the contents of such a science-based plan to the media, the general public, and policymakers would proceed simultaneously. All documents used to inform audiences beyond the scientific community would be drawn from the CDC plan and the Institute of Medicine report. This contributed admirably in keeping all those involved in the project "on message" no matter how many derivative materials they drafted for use in advocacy.

"Thinking big" meant packaging together many smaller problems for presentation as one inclusive concept. The wrapper for marketing the concept to the lay public and politicians was adopted from scientific discussions about "emerging" and "reemerging infections." A typical requirement for political marketability is the creation of the perception that what needs action—and money—is both new and very important. If the problem is not seen as new, some may infer that the problem must not have been important before—or, therefore, now. If it is not big or at least growing, it is likely not to become politically significant ever. With this in mind, despite the persistence over decades of many of the problems in need of attention (e.g., *Salmo-*

*nella* and *Shigella*), the adjectives "new," "emerging" and (newly) "reemerging" took precedence in the vocabularies of the NCID scientists and board members.

By grouping the long list of disease-causing agents under one banner, "the problem" to be addressed grew from a series of single diseases to a more imposing aggregation imbued with newness. The choice of terms already in use, moreover, might build on existing support without stretching the science or offending scientific sensibilities. The decision to package a single priority called "emerging and reemerging infectious diseases" represented a departure from previous NCID strategy whereby, as recently as January 1992, "emerging and reemerging infections" had appeared on a list as one of ten NCID priorities (NCID internal document distributed to the NCID board). The decision to fashion the campaign around many infections grouped together under one label was taken against the conventional political advice of some health lobbyists and congressional staff to advocate for funding disease by disease, given previous successes with constituency-based advocacy.

Another choice, although not one ever brought to a public forum such as a meeting of the board, was for members who were so inclined to pursue every avenue possible for wider recognition of the problem and new funds to address it. These options were discussed at informal gatherings of members of the board who volunteered to attend, often late at night over brandy in the hotel where out-of-town members lodged. In these sessions the group discussed multiple strategies to elicit more enthusiasm from CDC leadership, to jump up the executive branch chain of command by appealing directly to the newly assembled Clinton transition team, and to seek congressional support directly. Such activities were not typical within the institutional culture of the CDC.

### Teamwork between "Insiders" and "Outsiders"

The division of tasks between the two teams—the NCID "insiders" and the board "outsiders"—evolved rather naturally, including a division of labor that protected NCID scientists from stepping over any line of questionable propriety for federally employed professionals. NCID scientists shouldered the massive effort of drafting the plan and all the related materials (summaries of components of the CDC plan, slide sets for teaching in medical and public health schools). Similarly, the inside team informed and consulted every center in the CDC, starting early in 1993, to build consensus for CDC-wide endorsement.

Well before emerging infections had become the NCID/BSC's cause célèbre, the board had criticized the CDC for its institutional timidity and insularity. From the perspective of Harlyn Halvorson, a past president of the ASM and a scientific leader for many decades, this sort of criticism dated from the 1970s when groups such as the ASM had started to press the CDC to build external constituencies for the benefit of a wider range of scientific input as well as for institutional support. These sorts of criticisms drew no fire from the NCID emerging infections team members, who seemed to agree with the assessment. Just as "HIV/AIDS" and "non AIDS" budgets managed within the NCID were seen as distinct, it seemed that two institutional cultures had developed within the NCID since the rise of a substantial HIV/AIDS program. Because those working with HIV/AIDS interacted regularly with advocacy groups, infected individuals, politicians, and the press, they considered advocacy to be an inevitable aspect of their professional lives. Those laboring in other areas often did not.

### The "Insiders"

In this context, and as few NCID board members were experts on the technical capacity of the CDC to work on every potential infection, the NCID scientists took the initiative to acquaint the board with the strengths and growing gaps in scientific expertise. NCID board members were grimly impressed by the data demonstrating the decrease in NCID scientists with experience appropriate to the rapid identification of important pathogens to inform strategies for protecting the public.[42]

Somewhat later, Joe McDade, the associate director for laboratory sciences, and Ruth Berkelman, deputy director of the NCID, exposed board members to the condition of the laboratories. A tour through these uniformly overcrowded and sometimes ill-equipped facilities prompted a letter from the NCID board to the CDC's new director, David Satcher, calling for a formal review of the facilities and of their safety. Such education of the NCID board about staffing and about laboratory capacity at the CDC and in the states played a crucial role. It helped to motivate some board members to take up the challenge of increasing funding and others at least to tolerate unaccustomed advocacy in the midst of scientific deliberations.

With a new boldness, the NCID organized meetings with broad representation from professional groups to review drafts of the CDC's plan on for responding to emerging infections. NCID staff intently solicited reviews from every participant and were rewarded with hundreds of detailed critiques and suggestions. The NCID reviewed every submission, culling substantive contributions. The CDC's plan as published demonstrated the NCID's ability to involve and make good use of the best thinking of hundreds of scientific collaborators in the United States and beyond. In the same period, senior NCID scientists authored (with Osterholm as coauthor) an article in *Science* that appeared just as the completed plan was released by the CDC in April 1994. Its title, "Infectious Disease Surveillance: A Crumbling Foundation," and content presented the public health threat more starkly than any previous NCID contribution to the literature. When NCID staff were requested by congressional offices to provide routine briefings on NCID activities, some used that opportunity to teach congressional aides about the public health significance of emerging and reemerging infectious diseases.

As Berkelman notes in Part I of this chapter, NCID staff also took care to respond to inquiries by the press more fully than had been customary, often tutoring reporters on the science to diminish distortion while encouraging public interest. With previously uncharacteristic media savvy, journalistic interest was quietly coaxed to the topic. In 1994-95, stories in decidedly more visible media (*Newsweek, Time,* and *U.S. News & World Report*) offered the U.S. public a broader view of infectious disease. Players within Congress who would become essential to the appropriations strategy took special note of these reports.[43]

### The Board's "Outsider" Role

The activities assumed by various members of the board who volunteered included contacting the Clinton transition team and mounting a congressional appropriations strategy—along with continuing efforts to engage the CDC's directors and acting directors (three between December 1992 and 1996) as allies. By 1993 all elements of the campaign waged outside the CDC were in full swing: raising the profile of "emerging infections" at professional meetings sponsored by many groups such as the American Society for Microbiology and the Infectious Diseases Society of America (IDSA) as well as at those sponsored by the NCID; increasing the flow of information for consumption by the general public and policymakers; contacting the Clinton transition team; and, finally, the mounting of an appropriations strategy. Only a higher threshold of frustration among infectious disease experts launched the last two—as it is more difficult to engage academic scientists and other constituencies represented on the NCID board in overtly political acts than in sharing their expertise with peers at professional meetings or with the public via media outlets. Such advocacy to improve the overall capacity to combat infectious disease, rather than to promote funding for specific programs, often disease by disease, was unprecedented.[44]

Early on, one small ad hoc group of NCID board colleagues met informally to discuss how to approach the transition team. As President Clinton had been elected in November 1992 but had not yet taken office, his team would be selecting defining issues for the new presidency while scrambling to change the outgoing administration's budget for submission to Congress by President Clinton early in his first term. Several in this ad hoc board group knew individuals on the team, so the question of who to contact was easy. Far more perplexing was what should be said. How could these board members capture the pressured team's imagination—not just those assigned to health issues but the broader range of Clinton advisers?

At a hurried lunchtime discussion a few members of the ad hoc board group adopted an analogy offered by Osterholm for communicating the essence of disease surveillance—that of disease surveillance as the smoke detectors of public health. They elaborated the image to explain that some fires had already broken out (e.g., hantavirus, Ebola, *E. coli* 0157:H7), others were smoldering inconspicuously but might flare (e.g., foodborne pathogens lurking in lovely, imported produce), and still others might grow into epidemics—as HIV/AIDS had. The toll in human suffering and deaths might be significantly diminished if controls could be activated by a system of smoke detectors—a disease surveillance system as detailed in the CDC plan. The human and medical care costs might be contained, but only if early detection of infectious disease threats could propel prompt development of diagnostics, therapeutics, and prophylactics and strategic use of all. At NCID the experts were conscientiously preparing the smoke detectors for installation, but the system would be of no consequence for protecting the public unless money was appropriated for its installation.

Sitting together when the board was not in session, an ad hoc group coached one of its own through the contents of what would become a memo to the Clinton transition team, combining the smoke detector analogy with scientific documentation of the threats and the nature of response for which new funds would be necessary. The delivery of this document was the first of many contacts by board members with administration and legislative branch operatives, contacts that were prohibited or at least strongly discouraged for NCID staff. Although every U.S. citizen retains a right to petition Congress in his or her own name, one can represent a federal agency's position only when invited to do so by one's superiors. It is customary for CDC center directors and their designees to be invited to brief higher-level executive branch officials or congressional aides or members when technical experts are needed to elaborate on an issue already selected by the administration—or to respond to congressional requests for information. But NCID staff knew they were not to communicate with members of Congress or their aides without explicit invitation from Congress or by their CDC superiors. NCID staff responded to congressional requests for information as opportunities appeared, but it was up to board members to reach out when no request had been forthcoming.

Discussions about how to mount a credible campaign for new congressional funding without administration acquiescence began as the memo to the Clinton transition team was in process. Because members of Congress respond most readily to potential voters (or to large contributors, which to this author's knowledge included no NCID board member), it was essential to build a legislative appropriations strategy around one or more of the health groups with which board members and NCID staff were professionally involved. As the board boasted a former president of the American Society for Microbiology (Halvorson) and its incoming president (Cassell), the ASM once again was to become a key focus of subsequent activities. Board members were active in a variety of other groups. But along with the ASM, it was a relatively obscure organization, the Council of State and Territorial Epidemiologists (CSTE), that became part of the crucial connection to Congress.

By virtue of their considerable personal attributes and their ability to speak for organizations, Cassell (then chair of the University of Alabama's Department of Microbiology, incoming president of ASM, co-chair of NIH director Harold Varmus's review of NIH intramural programs, and chair of the NCID board) and Osterholm (then state epidemiologist of Minnesota and president of the CSTE) assumed visible leadership roles in the campaign. Throughout the months during which the CDC plan was under preparation and thereafter, Cassell would speak in many professional and academic settings, and to congressional staff members, on behalf of the ASM. By the 1990s the ASM could speak for approximately forty thousand scientist members, and its history of effective budget advocacy for the NCID's counterpart at the NIH, the National Institute for Allergy and Infectious Disease (NIAID), was well known in congressional offices. Osterholm, as a state epidemiologist in Minnesota and president of the CSTE, could speak credibly for the states, even though his organization lacked both the large membership and political visibility of the ASM.

Osterholm was already an experienced hand in health advocacy, but Cassell's scientific career had never before involved her in translating her expert opinion into a campaign for congressional support. Her willingness as NCID board chair to assume such an unaccustomed role may have been enhanced by the readily available wisdom of politically seasoned colleagues including Halvorson and Philip Russell. As a past president of the ASM and a scientific leader for many decades, Halvorson had often engaged in this sort of public affairs advocacy. He had been chairing the ASM's Public and Scientific Affairs Board when Murphy established the NCID Board of Scientific Counselors. Russell, a retired general and former commanding officer of the U.S. Army's infectious disease program centered at Fort Detrick, reinforced the ad hoc group of board members' sense of urgency and brought a long-term perspective on government science and public health from his long association with military, civilian (NIH and CDC), and international (World Health Organization) biomedical programs of research, surveillance, product development, and field operations. With coaching and encouragement Cassell embarked on a new element of her scientific career.

Well known to the small ad hoc NCID board group that came together at night in the hotel to scheme was the distinctly divergent approach characteristic of NIH Institutes in pursuit of appropriations in contrast to that of the CDC and its centers. Cassell took this collective wisdom with her directly into her role as lead advocate in working with the ASM to attract a new level of congressional appropriations for implementing the CDC Plan. Because in retrospect this seemingly obvious step also amounted to a major departure from CDC ethos, differences between the NIH style adopted by board activists and the predominant approach favored by CDC directors since the 1970s become central to understanding how the CDC plan attracted new appropriations.

### The NIH, the CDC, and the Congressional Budget Process

Although many of the same formal impediments to direct contact with Congress had existed for NIH staff as for those in CDC centers before 1980 and after, several important differences help to explain distinctions in institutional culture and, perhaps, in funding patterns. In fiscal year 1993 (the federal budget year ending September 31, 1993), when activism around the CDC's emerging infections plan began in earnest, the NIH budget totaled $10.3 billion, the CDC budget just $1.7 billion. With a campus just outside Washington, D.C., and many readily available constituencies of academic biomedical researchers who depend on NIH extramural programs to support their work, the NIH has enjoyed a far more lucrative relationship with the U.S. Congress than has the CDC.

The legislation authorizing NIH activities is very broad, so most NIH interactions with Congress focus on the appropriations process without the need for congressional approval (authorization) for new programs. Each NIH institute has a separate budget, and each institute director testifies yearly about the institute's appropriation request, using the contents of the president's budget as the focus for the formal presentation. The NIH has not been shy about encouraging its constituents—be they medical school deans, professional organizations including the Association of American Medical Colleges, the ASM, or groups of disease sufferers—to make their views known in congressional and senatorial districts around the country where biomedical research institutions are located or where affected citizens reside, as well as to key members of budget and appropriations committees. (The NIH now boasts a Web site displaying its grants year by year by state— a system easily correlated with congressional districts: http://grants1.nih. gov/grants/award/state/state03/htm.)

Through the 1980s and 1990s it was not uncommon for congressional aides to be well briefed on the financing needs felt within a particular institute even when the NIH director had denied some of the same requests for funding in preparing the overall NIH budget request for the administration. Nor was it uncommon, either before the election of President Reagan in 1980 or after, for members of congressional appropriations committees to ask institute directors for their "professional judgment" about how much funding would be needed for certain areas of research.

The "professional judgment budget" became such a well-known contrivance for contrasting the funding requests in the president's budget with those actually desired within any particular institute that it acquired a nickname, the "PJ budget." Several members of the NCID board understood how NIH institute directors and their staffs managed to participate in the behind-the-scenes appropriations process: congressional aides ask agency staff about the "professional judgment budget" and use this information to prepare questions for members of the House and Senate appropriations committees to ask NIH institute directors during public hearings.

The comfort level among the colleagues based in the NIH, congressional offices, and the AAMC and other biomedical sciences organizations for such delicate interactions has been attained in part by geographical proximity and frequency of personal contact but also by another phenomenon. It has not been uncommon for NIH administrators to move into jobs in organizations like the AAMC,[45] or for biomedical scientists to work for a time in the Congress,[46] or for congressional staff to be attracted to jobs at NIH.[47] Over the years such individuals have created a culture in which they work side by side for shared goals. With this professional foundation for interacting with Congress, and with a Congress ever attentive to medical advances benefiting institutions and disease sufferers in every part of the United States, NIH staff and directors have felt more protected than at risk when helping appropriations committees understand their budgets.

The situation at the CDC has been quite different. The CDC, headquartered in Atlanta, is geographically isolated from the Washington culture rich in political experience. The CDC's authorizing statute is less broad, meaning the CDC must often seek congressional authorization (formal approval for a new program or set of activities) before seeking budget increases through the appropriations process. Thus, the CDC must court both congressional authorization and appropriations committees—and their many members—whereas the NIH can focus more of its considerable staff resources on appropriations. Nor is most of the CDC's budget organized to correspond to the administrative units, centers, and programs. Rather, funds are directed to the agency through series of budget lines that the CDC director's office distributes among several centers. In stark contrast to the situation at the NIH, few beyond the CDC director's office and a very small number of congressional staff and independent health organization activists have ever really understood how much of each budget line had been or would be devoted to which diseases or programs and in which of the CDC's centers. As there are no congressionally appropriated center budgets, CDC center directors (with the exception of the National Institute of Occupational Safety and Health [NIOSH]) are not offered the same opportunity to testify before the appropriations committee as is each NIH institute director. Thus, it is the CDC director and whomever he or she may designate to accompany him or her to appropriations committee hearings who present the president's budget and answer questions of appropriations members. There is no parade of CDC center and program directors available to answer appropriations committee members' questions. Even if there were, without statutorily distinct centers with their own budget lines, more ambiguity exists at the CDC than at the NIH about who would control allocation of increased funds. Generations of CDC leaders have allowed the overlapping scientific domains of CDC centers and programs to add to the budget confusion, possibly because this has centralized budget power in the hands of the CDC's director.[48]

Although there has always been some measure of behind-the-scenes preparation among congressional aides for the CDC appropriations hearings involving contact with CDC staff, the communications have not tended to reach as deeply into the scientific ranks at the CDC as they have at the NIH. With the several exceptions, those in the office of the CDC director managed much of the communication with Congress. Even as successive CDC directors declined to meet with the biomedical sciences leaders as described by Halvorson, they did begin to work with another coalition of health groups promoting CDC priorities they endorsed. In retrospect this had the disadvantage of targeting budget increases only to those areas selected by

CDC directors and producing much smaller budget gains overall than those enjoyed by NIH, but it did not tend to create many alternative or potentially competitive bases of power within the CDC.

From 1993 to 1996 the board that former NCID director Murphy had established in 1989 was able to carry out its work in ways that resembled the more successful NIH approach than the traditional CDC one. NCID board members, led by their chair, turned to the ASM and the broader coalition of biomedical research groups as the focal link to Congress. It was Shoemaker, director of public affairs of the ASM with more than fifteen years of working closely with congressional appropriations committees to enhance research funding at the NIH, who made many of the calls and accompanied Cassell to the many meetings on Capitol Hill. She recalls that appropriations staff knew of Cassell primarily from her prestigious role as co-chair of the NIH director's commission to review intramural research. Thus Cassell's stature and credibility in congressional eyes may have derived more from her long association with the NIH than from her role as chair of the NCID board. Once having gained an audience among the crucial players in Congress, Cassell and Osterholm (and NCID scientists when specifically invited) used drafts of the CDC plan (and by April 1994 the official version), and the burgeoning media attention, to teach Congress exactly why new funds were needed and for what they would be used. Public health groups including the CSTE and the American Public Health Association and the CDC coalition helped, but it was the combination of the ASM's opening the doors and NCID board leaders with the planning document and press clips in hand that created NIH-style budget momentum for the CDC emerging diseases initiative.

The board's enthusiasm for promoting the emerging infections message outside the CDC as well as within provided a new level of legitimacy and support for NCID staff efforts. Had it raised controversy to recruit NCID scientists away from their regular duties to work extensively on the CDC's plan, or to organize international meetings on emerging infections, the NCID board was there to justify such extraordinary efforts and arrangements. All those who gave generously of their time and talents were rewarded by active support of their peers at the NCID and on the board.

Ad hoc advocacy proceeded from the December 1992 board meeting onward. Although the memo to Clinton's transition team early in 1993 failed to mobilize high-level executive support, two other elements of the advocacy bore fruit, one outside all normal CDC operations and the other by forging an alliance with a new CDC director. First, the major efforts of NCID staff and the board to increase the visibility of the issues at professional meetings, in the media, and with health groups did advance funding to implement the CDC's plan. Congress first appropriated new resources in 1994 based on the plan ($6.7 million) to begin implementation of it. In subsequent years, the president's budget proposed increased appropriations, and by 1998 Congress responded with $59.1 million.

## The Continuing Challenge of Political Latency

For those seeking meaningful institutional responses to public health threats, it is important to question what the events recounted here portend for the future. Committed professionals in and outside government collaborated effectively, projected their concerns to a variety of audiences beyond the scientific community, and induced development of a CDC plan and implementation of some of its elements. Formal channels for budget advocacy were pried open and offered a less labor-intensive and controversial alternative to professional insurgency.

Capacity for early detection and strategies to protect the health of the public from infections are no longer on the wane, as there is greater appreciation that infectious diseases continue to pose significant threats despite seemingly authoritative declarations relegating such plagues to history. There never arose among the board any voice of opposition to the formal agenda for enhancing the NCID's capacity–or to the many volunteer efforts that remained purposefully off that agenda. And never did a substantive challenge arise to the credibility of the scientific judgments underpinning the advocacy. Surely these are signs of success. Participants felt successful, even triumphant, at times.

Yet the same events are troubling when viewed through a wider-angle lens. "Emerging infections" is not likely to be a rare example of a looming public health crisis that will be identified by scientists while remaining invisible to the public and to most policymakers. For example, increasing contamination of groundwater, which is of much concern to experts, appears not to have attracted a national response commensurate with the threat it represents for human health. That means society will continue to depend on public health institutions to mount effective responses. Even when these organizations sustain the technical acumen to detect a problem early (e.g., laboratory capacity) and the talent to plan an effective strategy, they will often be unable to advance protection of the public's health without increased resources. Generating those new resources depends on support from the broader public, or at least from highly motivated health leaders. If senior public health officials (state or federal) and their allies feel no responsibility to find the needed funds, or are actively discouraged from seeking them by political appointees more attuned to cutting budgets than protecting against even predictable problems that may occur in another politician's term, many of these problems will fester and erupt on someone else's watch. The incentives to devote personal time and professional assets to such a cause when significant rewards are doubtful will continue to discourage all but the exceptional among public health leaders.

Had the dynamics among those pursuing a response to emerging infections not been so positive, how many of the characters in this tale would have persisted in their efforts? Even had they done so despite less auspicious circumstances, there are all too few directors of state or federal public health agencies who use their boards creatively, indicating when it would be essential to reach out politically, contacting those in key executive or legislative positions with their assessments of challenges for which government action will be an essential ingredient. Nor are there many such boards or advisory bodies whose members assume significant advocacy responsibilities with or without prompting. Some public health agencies have no board or comparable body. Unfortunately public health departments—be they federal, state, or municipal—are not among the more powerful of our societal institutions. They are not even among the more powerful of health institutions because they lack the rapt attention that patients and politicians accord medical care—the need for which is clear enough once someone has the misfortune to fall ill. Nor are public health agencies high in the pecking order among public institutions. Budget requests from those dealing with economic affairs often receive more generous political attention.

Even if public health agencies were more prominent, our democratic processes do not seem well tuned to support institutional responses to yet unseen threats to public health. Protecting the public from health threats that have yet to become visible will require startling boldness and persistence by scientists and public health leaders if the sorts of impediments confronted in this one case are to be surmounted in the many that will challenge us in the future.

### NOTES

1. Winslow, *Conquest of Epidemic Disease.*

2. Burnet, *Natural History of Infectious Disease,* 3rd ed.

3. The CDC had begun to expand to noninfectious activities including urban rat control and evaluation of family planning in the 1960s to accommodate the needs of state health departments; it furthered its reputation in Washington, D.C., as a "can-do" organization in regard to public health activities (David Sencer, pers. comm.).

4. Tyler and Herold, "Public health and population."

5. Lalonde, *New Perspective on the Health of Canadians.*

6. Etheridge, *Sentinel for Health.*

7. Responsibilities for infectious diseases are currently divided into three separate areas at the CDC, each reporting to the director of the CDC. The National Immunization Program and the National Center for HIV, STD, and TB Prevention are responsible for the prevention and control of many of the vaccine-preventable diseases and for HIV, sexually transmitted diseases, and tuberculosis prevention and control programs, respectively. The third organizational unit, the NCID, is responsible for the prevention and control programs, including the laboratories, of the remaining (more than one hundred) known infectious pathogens.

8. Pinner et al., "Trends in infectious disease mortality in the United States."

9. Levy, *The Antibiotic Paradox.* As early as 1948, Betty Hobbs at the University of London Hospital called attention to the issue of antibiotic resistance and the problems of hospital-acquired infections with resistant bacterial strains, but few were listening even in the 1960s and 1970s as others voiced concerns (Harlyn Halvorson, pers. comm., June 24, 2000).

10. Murphy, "New, emerging, and reemerging infectious diseases"; Morse, "Examining the origins of emerging viruses."

11. Burnet and White, *Natural History of Infectious Disease,* 4th ed.

12. Krause, *The Restless Tide.*

13. Burnet and Lederberg were both Nobel laureates; Krause was the director of the NIAID.

14. The Institute of Medicine was established in 1970 and is part of the National Academies, a group of nonprofit organizations created as honorary societies for top scientists. The National Academies were established by Congress to provide advice to government on scientific issues that affect policy decisions.

15. Unpublished data, CDC.

16. A biosafety level 4 laboratory, for work on such pathogens as Ebola virus for which there is no cure, was built in the early 1980s, and approximately 10% of the scientists working in the laboratories used this laboratory.

17. Berkelman, "Emerging infectious diseases in the United States"; Berkelman et al., "Infectious disease surveillance"; Berkelman, Pinner, and Hughes, "Addressing emerging microbial threats in the United States."

18. Unpublished data, CDC.

19. Fred Murphy, pers. comm.

20. Ralph Bryan, Robert Gaynes, C. J. Peters, and Robert Pinner were detailed for approximately four weeks; Judith Aguilar and Meredith Hickson also provided support.

21. To document the problems facing local and state public health agencies, Michael Osterholm, a member of the NCID board and of the governing body of the Council of State and Territorial Epidemiologists, examined the capacity of state public health agencies to address emerging infectious diseases threats. He surveyed all fifty states and found few state or federal resources available to states for the surveillance of infectious diseases not specifically targeted to receive congressional appropriations

(i.e., AIDS, tuberculosis, sexually transmitted diseases, Lyme disease, and vaccine-preventable diseases); he also documented that twelve states had no professional staff assigned full-time to the surveillance of foodborne infectious diseases (Osterholm, Birkhead, and Meriwether, "Impediments to public health surveillance in the 1990s").

22. Murphy, "New, emerging, and reemerging infectious diseases."

23. The American Society for Microbiology was a particularly vocal advocate to congressional staff and others throughout the plan's development and following its publication; the CDC coalition, led by the American Public Health Association and to which the ASM belonged, also included addressing emerging infectious diseases on its agenda by the mid-1990s.

24. Henig, *A Dancing Matrix;* Preston, *The Hot Zone;* Garrett, *The Coming Plague.*

25. Preston's future bestseller, *The Hot Zone* (1994), was based on this article.

26. Toner, "Virus hub."

27. Howard, "Perspective."

28. Approximately $1 million was appropriated by Congress to the CDC in fiscal year 1994 for emerging infectious diseases; these appropriations were not directly related to the plan, and the authors are unaware of the specifics related to this appropriation. The appropriation was not part of an overall strategy.

29. In 1994, the CDC had estimated that full implementation of the plan would cost approximately $125 million annually.

30. Berkelman, Pinner, and Hughes, "Addressing emerging microbial threats in the United States"; Hughes, "Addressing the challenges of emerging infectious diseases."

31. CDC, *Preventing Emerging Infectious Diseases.*

32. For example, Colorado used its newly gained resources provided through the CDC to develop the technical capacity to perform molecular subtyping on isolates of *E. coli* 0157:H7 that are submitted to their laboratory from health care providers throughout the state. In 1997, within months of developing this capacity, Colorado detected a small outbreak of *E. coli* 0157:H7. Through use of this new technology and enhanced collaborations between the CDC and USDA, contaminated beef was clearly implicated and led to a recall of 25 million pounds of ground beef, averting a potential nationwide epidemic. In another instance, the initiation of Emerging Infections Programs in four state health departments was well timed to offer a framework for a rapid U.S. response following recognition of new variant Creutzfeldt-Jacob disease in the United Kingdom. New variant Creutzfeldt-Jacob disease has been associated with the emergence of bovine spongiform encephalopathy in cattle, the so-called mad-cow disease.

33. Heymann and Rodier, "Global surveillance of communicable diseases."

34. An important exception was the former director of the NCID, Fred Murphy. He had established the NCID board because he did recognize the importance of the infectious disease threat but had been unable to attract sufficient agency support to make what he felt to be creditable progress. By 1992 he had resigned for lack of governmental backing when he tried to exert leadership in this area. The NCID board was carry-

ing out his legacy. It was Murphy who appointed some of those who became most active in the campaign that won new support for this agenda.

35. Funds appropriated to the CDC by Congress are designated by "line item," each of which has a stated purpose. Since there were insufficient funds in the "Infectious Diseases" line item to implement the emerging infections plan, funds would have had to have been redirected from another line item. Such reprogramming requires approval of the appropriations committees.

36. Murray and Lopez, *Global Burden of Disease.*

37. Political appointees in high-level federal jobs are not within the Civil Service and therefore do not enjoy its employment protections. Political appointees are expected to support and advocate administration policies. They serve at the pleasure of the president. They are often asked to resign when a new president takes office, or they are dismissed and replaced by members of a new president's team to assure loyalty to the new administration.

38. Within the DHHS any initiative had to survive intense examination and competition among all requests for new funds proposed by sister agencies (including, for example, the much larger NIH). If a request for new funds survived DHHS review, it then went before the even more skeptical Office of Management and Budget for scrutiny before possible incorporation in the president's budget. For a request that did win a spot in the president's budget, all the officially designated executive branch political appointees tracked and usually advocated for it throughout the appropriations process. Once the president submits a budget to Congress, the formal process requires it to go first to the House of Representatives for consideration, later to the Senate. The House Budget Committee determines what portion of the total federal budget it will allocate to each House appropriations committee. Each House appropriations committee then makes specific funding decisions within its jurisdiction, item by item, up to the amount the House Budget Committee has indicated as the maximum for that area. This process is repeated in the Senate before the House and Senate appropriations committees negotiate their differences. Negotiations over differing funding levels allocated by the House and Senate committees are conducted by "conference committees" made up of a small number of members of each appropriations committee. In these "conferences" designated members from each party on the House appropriations committees meet with their counterparts from the Senate appropriations committee with parallel jurisdiction. They often meet day and night until negotiations are complete. Then the appropriations bills are voted on by the House and by the Senate and sent to the president. Support could wane or grow as budget negotiations proceeded, depending on any budget caps agreed upon between the president and Congress for the entire federal budget, the allocations made to each budget area (such as defense, transportation, energy, or health), and the competition among agencies and their programs (continuing and new) within the particular appropriations committees making the final decisions about the federal budget—specifically, how much to allocate for health and how much in line items to direct to each agency and program. These decisions are made by the House and Senate Budget Committees in advance of

the appropriations process, which designates specific amounts to each agency budget, line by line.

39. Decisions by higher-level officials to reduce or deny budget requests were passed back to the originators of the requests in a communication called a "pass back." This history of requests and pass backs was summarized in a spreadsheet. This and other budget documents were routinely used by advocates for the NIH and much less so for those of the CDC.

40. President Reagan called for a 25% reduction in authorizations for health programs funded through the appropriations process. Congressional authorization is a step that precedes the budget and appropriations steps discussed in note 38. Most federal programs must be "authorized" before becoming eligible for appropriation of federal funds. In this first step, Congress grants authorization for an agency to carry out the functions for which funds may or may not be appropriated subsequently. As part of the authorization process, an authorization committee may recommend a specific amount for consideration by appropriations committees by including it in the authorization language of the bill. The 25% reduction applied to all "discretionary" health programs, meaning all except those funded outside the congressional authorization and appropriations process. Some major health expenditures, such as those for Medicaid, are funded as "entitlements," which means the funding is not allocated through the appropriations process described earlier.

41. President Reagan did not succeed in obtaining the 25% cut in health programs, although he did reach his targets in other budget areas, only because one House authorization committee with health jurisdiction, the Committee on Energy and Commerce, organized opposition by publicizing the likely consequences of such cuts. In response to being educated about the likely toll of the cuts, a sufficient number of Republicans broke rank with the president to protect health programs. Similar pressures were applied each year. So long as the Democrats held the majority in the House, organized opposition to administration-led efforts to reduce health program budgets persisted.

42. At a regularly scheduled meeting in 1993 NCID leaders invited Michael Osterholm, the state epidemiologist from Minnesota who was also a member of the board, to report from the states' perspective about alarming limitations in capacity for detection and response to disease outbreaks. He reported to both NCID staff and its board his findings based on survey data from 1992. We might not otherwise have encountered this information before its publication, first in 1994 in *Science* (Berkelman et al., "Infectious disease surveillance") and in *ASM News* (Cassell, "New and emerging infections") and in more detail in 1996 (Osterholm, Birkhead, and Meriwether, "Impediments to public health surveillance").

43. Janet Shoemaker, director of Public Affairs for the ASM, recalls that these articles played a significant role in bringing the issue to the attention of congressional staff and members and in their understanding that the public health threat was both more ominous and larger than the episodic outbreaks that they assumed could be addressed within existing funding levels.

44. Although surveillance is an essential component of public health infrastructure, it did not even attract appropriate attention as part of the health care reform debates—carried on simultaneously with this effort throughout 1993-94. Thus, NCID leaders and NCID board activists were acutely aware of just how difficult it would be to "sell" surveillance to the administration and Congress. This experience spawned more frustrated reflections on the health reform process from the perspective of the limited thinking in proposals about protecting health, including the failure to understand and promote surveillance (Cassell, "New and emerging infections"; Freeman and Robbins, "Health reform without public health").

45. For example, both John Sherman and Tom Kennedy, leaders at the AAMC for many years, had worked previously within the NIH (pers. comm. with Sam Korper, former director of legislative affairs, NIH).

46. The ASM operates a year-long program for ASM fellows through which academic scientists have an opportunity to learn how federal science policy is made by working in congressional offices. Some continue to work with Congress after the fellowship.

47. For example, Susan Quantius, a highly regarded member of the staff, left Chairman Porter's House Appropriations employ in the late 1990s to work for the Association of American Universities and later the NIH (Janet Shoemaker, ASM, pers. comm.).

48. The congressional appropriations committees have a mechanism for directing funds to particular programs–they "earmark" them for specific purposes by language in the statute or the accompanying report. The NIH has discouraged earmarking because it limits flexibility in use of budget. The number of congressional earmarks in the CDC budget has increased markedly in the past decade, and many within the CDC and the science community are concerned about the trend for the same reasons it has been resisted at the NIH. The earmarks may be more common for the CDC because that budget is more confusing than that of the NIH. Another factor may be the more distant relations between CDC scientists and congressional aides than those between the aides and NIH staff and advocates. Many who have worked on congressional appropriations staffs or in science advocacy organizations continue to report that they understand the CDC budget less clearly than that of the NIH, despite the fact that the NIH budget dwarfs that of the CDC. Many also report that they understand much less about how the CDC uses its line item funds, those with earmarks and those without, than their counterparts working with the NIH.

REFERENCES

Armstrong GL, Conn LA, Pinner RW. Trends in infectious disease mortality in the United States during the 20th century. *JAMA* 1999; 281:61–66.

Begley S, Cohen A, Dickey C. A special report: Commandos of viral combat. *Newsweek* May 22, 1995. (Also in same issue, with no author: Killer virus: beyond the Ebola scare: What else is out there?).

Berkelman RL. Emerging infectious diseases in the United States, 1993. *J Infect Dis* 1994; 170:272–77.

Berkelman RL, Bryan RT, Osterholm MT, LeDuc JW, Hughes JM. Infectious disease surveillance: A crumbling foundation. Policy Forum. *Science* 1994; 264:368–70.

Berkelman RL, Pinner RW, Hughes JM. Addressing emerging microbial threats in the United States. *JAMA* 1996; 275:315–17.

Brownley, S. The hot zone. *U.S. News & World Report,* March 27, 1995.

Burnet, FM. *Natural History of Infectious Disease.* 3rd ed. Cambridge: Cambridge University Press, 1962.

Burnet FM, White DO. *Natural History of Infectious Disease.* 4th ed. Cambridge: Cambridge University Press, 1972.

Cassell GH. New and emerging infections in the face of a funding crisis: Strong programs and adequate resources are imperative for the nation's health. *ASM News* 1994: 60;251–54.

CDC. *See* Centers for Disease Control and Prevention.

Centers for Disease Control and Prevention. *Addressing Emerging Infectious Disease Threats: A Prevention Strategy for the United States.* Atlanta: CDC, 1994.

———. *Preventing Emerging Infectious Diseases: A Strategy for the 21st Century.* Atlanta: CDC, 1998.

The disease detectives. *National Geographic,* January 1991.

Etheridge E. *Sentinel for Health: A History of the Centers for Disease Control.* Berkeley: University of California Press, 1992.

Foreman CH. *Plagues, Products, and Politics.* Washington, D.C.: Brookings Institution, 1994.

Freeman P, Robbins A. Health reform without public health: A formula for failure. *Am J Pub Health Pol* 1994; 15:3:261–82.

Garrett L. *The Coming Plague: Newly Emerging Diseases in a World out of Balance.* New York: Penguin Books, 1994.

Henig RM. *A Dancing Matrix: Voyages along the Viral Frontier.* New York: Alfred A. Knopf, 1993.

Heymann D, Rodier GR. Global surveillance of communicable diseases. *Emerg Inf Dis* 1998; 4:362–65.

Howard, RJ. Perspective: Media coverage of emerging and reemerging diseases behind the headlines. *Statist Med* 2001; 20:1357–61.

Hughes, JH. Addressing the challenges of emerging infectious diseases: Implementation of the strategy of the Centers for Disease Control and Prevention. In: *Emerging Infections,* edited by Scheld WM, Armstrong D, Hughes JM, 261–69. Washington, D.C.: ASM Press, 1998.

Institute of Medicine. *Emerging Infections: Microbial Threats to Health in the United States.* Edited by Lederberg J, Shope RE, Oaks SC. Washington, D.C.: National Academy Press, 1992.

Krause RM. *The Restless Tide: the Persistent Challenge of the Microbial World.* Washington, D.C.: National Foundation for Infectious Diseases, 1981.

Lalonde M. *A New Perspective on the Health of Canadians: A Working Document.* Ottawa: Queen's Printer, April 1974.

LeDuc JW. WHO program on emerging virus diseases. *Arch Virol Suppl* 1996; 11:13–20.

––––––. World Health Organization strategy for emerging infectious diseases. *JAMA* 1996; 275(4): 318–20.

Lemonick M. Revenge of the killer microbes. *Time,* September 12, 1994.

Levy SB. *The Antibiotic Paradox: How Miracle Drugs Are Destroying the Miracle.* New York: Plenum Press, 1992.

Morse S. Examining the origins of emerging viruses. In: *Emerging Viruses,* edited by Morse S, 10–28. New York: Oxford University Press, 1993.

Murphy FW. New, emerging, and reemerging infectious diseases. *Adv Virus Res* 1994; 43:1–52.

Murray CJL, Lopez AD. *The Global Burden of Disease.* Vol. 1. WHO 1996.

National Center for Infectious Diseases/Centers for Disease Control and Prevention. Board of Scientific Counselors, NCID, November 15–16, 1990, Meeting Recommendations and Comments.

Osterholm MT, Birkhead GS, Meriwether RA. Impediments to public health surveillance in the 1990s: The lack of resources and the need for priorities. *J Public Health Manag Pract* 1996; 2(4): 11–15.

Pinner RW, Teutsch SM, Simonsen L, Klug LA, Graber JM, Clarke MJ, Berkelman RL. Trends in infectious disease mortality in the United States. *JAMA* 1996; 275:189–93.

Preston R. *The Hot Zone.* New York: Random House, 1994.

Tales from the hot zone. *US. News & World Report,* March 27, 1995.

Toner M. Virus hub: A crowded CDC: Aging, crammed facilities pose a safety concern. *Atlanta Journal/Constitution,* May 13, 1995, E1.

Tyler CW Jr., Herold JM. Public health and population. In: *Public Health and Preventive Medicine,* 13th ed., edited by Last JM, Wallace RB, 41–53. Norwalk, Conn.: Appleton and Lange, 1992.

U.S. Public Health Service. Healthy people: The surgeon general's report on health promotion and disease prevention. 1979.

Winslow CA. *The Conquest of Epidemic Disease.* Princeton, N.J.: Princeton University Press, 1943.

# CHAPTER 14

# Hepatitis C and the News Media

## *Lessons from AIDS*

LAWRENCE D. MASS

All the contributions to this volume demonstrate that the media play a major role in determining whether an emerging illness becomes a part of the public health agenda. Communities of suffering often make a special effort to involve the media in their cause. Public health institutions are also aware of the need to inform the media of their policies regarding emerging illnesses.

Dr. Larry Mass is a New York physician who writes freelance newspaper and journal articles on health problems, particularly those that affect the gay community. Dr. Mass is acknowledged to have written the first newspaper article on AIDS in the United States. In this chapter, Mass describes his efforts to publish articles on hepatitis C (HCV), which he saw as a major emerging public health issue that was being ignored. He relates the frustration he experienced as various editors refused to publish his material and compares this experience with his earlier efforts to publicize the emerging HIV/AIDS problem in the 1980s.

It could be argued that the news media, as well as public health institutions, were slow to respond to both the emergence of HIV/AIDS (as chronicled in Randy Shilts's *And the Band Played On: Politics, People, and the AIDS Epidemic*) and the more hidden epidemic of HCV. In both cases, Mass was on the front line of activists pushing for public recognition. For Mass writing as an activist physician, difficulty in garnering media attention for HCV reflected the same biases and conservatism on the part of the media leadership that he had encountered in his attempts to chronicle the emergence of AIDS. On the other hand, it is difficult to assess the extent to which the responses of various editors to Mass's articles reflected their disinterest in the problem of HCV versus their evaluation of Mass's work. Mass argues that the media were unwilling to print what they saw as alarmist pieces. For members of the news media, knowing how to respond to an emerging illness is not always clear. Reporters want to be on top of stories concerning potential threats to the public health. At the same time, they are hesitant to be too far out in front on an emerging illness lest they be viewed as raising alarms prematurely and be accused of being irresponsible fear mongers. In dealing with emerging illnesses, knowing what to cover can be particularly challenging: there is uncertainty within the medical establishment concerning the extent of a particular problem, as well as disagreement over the etiology of an illness or its case definition. Media representatives often find themselves caught between communities of suffering, which are seeking attention for their illness, and public health officials, who are acting as gatekeepers to public health resources. For activists like Mass who see the need for greater alarm, this media reticence can be frustrating.

Whatever the source of the editors' resistance to Mass's articles, it is clear that few articles on the emerging threat of HCV appeared in major U.S. newspapers or journals prior to 1997 and that Mass's campaign coincided with the emergence of media coverage of the problem. It was not until the convening of an NIH panel on HCV in March 1997 that a number of newspapers and journals such as the *San Francisco Chronicle, Boston Globe, Minneapolis Star Tribune,* and *U.S. News & World Report* began publishing pieces on HCV. From 1997 until 2002, some 458 articles were published on HCV and prevention. Still, it may not have been until *Newsweek* ran a cover story, "Hepatitis C: The Insidious Spread of a Killer Virus," in April 2002 that the disease attracted the attention of the general public in this country. Thus, Mass appears to be justified in suggesting that the news media were slow to respond to the emerging epidemic of HCV.

---

In any consideration of the lessons of history, perhaps the most widely quoted observation is that of the philosopher Santayana: "Those who cannot remember the past are condemned to repeat it." Tangentially, there is another observation about the process and lessons of history that people often quote: "The personal is political." What follows is an effort to elucidate the bigger picture of "emerging illnesses and the media" with reference to the story of my own personal experience of trying to cover the emergence of a disease that has been characterized by leading addiction authority Dr. Mary Jeanne Kreek as "the emergent and preeminent public health problem of the twenty-first century, surpassing that of HIV": hepatitis C.[1]

This personal history begins where my involvement in the AIDS epidemic began, at Greenwich House West (GHW) Methadone Maintenance Treatment Program (MMTP) in New York City, where I served as medical director from 1980 to 1998. There, in 1981, I got the first telephone calls about a mysterious new disease in gay men in New York City medical center intensive care units. Those calls were from my friend and colleague Dr. Joyce Wallace, whose later initiatives on behalf of street workers (prostitutes), although widely recognized and applauded, remain chronically underfunded. The subsequent unfolding of the AIDS epidemic is now relatively well documented history.[2]

*Introductory note on hepatitis C.* Although the numbers of new cases of hepatitis C in this country have dropped—from an average of 230,000 new cases per year in the mid-1980s to 36,000 in 1996—and although spontaneous remission (disappearance of detectable virus) is observed in up to 25 percent of patients, nearly 85 percent of those who become infected develop chronic infection, and up to 60 percent of those develop chronic liver disease. Typically, HCV may cause so low grade an infection that it remains subclinical and undetected for years, often decades, before it is finally diagnosed. At that point, major, often end-stage liver disease has already taken place. In patients with transfusion-associated HCV, 10-20 percent have evidence of cirrhosis within five years after onset. In the United States, conservative estimates are that there are now more than 4 million persons infected with HCV and 180 million worldwide (approximately five times as many as those infected with HIV). At present, HCV is responsible for 8,000–10,000 deaths per year in the United States, but that figure is expected to triple in the next ten to twenty years. Liver failure due to HCV is now the number one cause of liver transplantation in the United States. Antiviral treatments include Intron and Rebetron (combining interferon with ribavirin). Up to 40 percent remission or cure rates with these treatments are being reported.[3]

My awareness of HCV did not unfold in precisely the same way the AIDS epidemic did, though the similarities are impressive. For years, throughout the 1980s and into the 1990s, new admissions to methadone maintenance treatment often—in fact, usually—showed liver enzyme abnormalities that were unexplained. Because HCV wasn't identified until around 1990 and testing didn't become available until 1992, we didn't realize that that was what we were dealing with. Like everyone else, I was reading the occasional pieces on HCV that began appearing in the mainstream press, along with those in the medical journals. Acknowledging that there was still much we didn't know, they were forthright and informative. When it was realized that the disease had a much higher chronicity and complications rate than other forms of hepatitis and that treatment was experimental and associated with debilitating side effects and had limited success, this information was likewise presented to the public in articles that were responsible, if never very disturbing. When all was said and done, you could walk away from most such articles, the few there were, reassuring yourself that this was a disease that was unlikely to actually touch your own life, one you therefore didn't need to worry much about. It was at this point, in the mid-1990s, that I became increasingly uncomfortable.

Because of the indigency of most methadone maintenance clinic patients, including ours, testing for HCV has not been routine, despite an estimated occurrence in this population of as high as 90 percent. I am not a trained epidemiologist, and I was aware that this disease had been around for some time—recognized as chronic nonA–nonB hepatitis. However, something within me reasoned that if this overwhelming majority of our patients were infected with this bloodborne illness—which eventually, however slowly, progressed in significant percentages to major and eventually fatal complications, for which the only successful treatment (for advanced disease) is liver transplantation, an extremely expensive, high-risk medical procedure that might itself need to be repeated on the same patient after several years—then we must be dealing with a very serious epidemic, even if it was qualified by epidemiologists as "an epidemic of recognition."

At the time I began my own efforts to write about hepatitis C in 1997, routine testing for HCV among injecting-drug users was not considered justified for three reasons: first, because the positivity rate was so high that, in the presence of otherwise unexplained liver enzyme elevations, it could be presumed; second, because treatment often involved substantial morbidity and its promise was so uncertain; and third, because of the long incubation period of the disease. As Dr. Edwin Salsitz of Beth Israel Medical Center has often observed in his discussions of HCV, many of these patients are at greater, more immediate risk of morbidity and mortality from smoking. Because most MMTP patients are referred out for primary care, it seemed to make the most economical and conservative sense to advise patients of their likelihood of exposure and leave the follow-up, including initial HCV testing, to their primary care service. What we did not know is that with combination treatment the sustained remission incidence can be as high as 40 percent, the prognosis being better for those with certain genotypes. Although this does not necessarily mean that everyone who is HCV positive should therefore be treated—as with cancer of the prostate, the likelihood of slow progression may influence the patient to wait—it does make earlier treatment, even in the absence of liver enzyme elevations, a more recommendable option.

But the reason I was becoming so uncomfortable wasn't solely because this was an emerging public health crisis. My discomfort was more acute than that. It had to do with my realization that the way HCV was unfolding before the public via the media was strikingly similar to what had happened with AIDS; that the truth of the epidemic's seriousness was being blunted to the point of obfuscation; and that some of the major players involved—the Centers for Disease Control and Prevention, the *New York Times*—were the same and seemed to be making the same mistakes, at least in kind.

Doing the kind of independent, community-oriented work I do as a journalist and writer, I've never made much money, nothing that could begin to support me. Three years into the AIDS epidemic, I was hospitalized with a major depression/burnout and withdrew from the scene, leaving AIDS coverage to such giants as Randy Shilts and Laurie Garrett. Working full-time as a physician, I have not had the resources of energy and financial freedom to reenter the fray. But the situation with HCV seemed so drastic that, even at the risk of another burnout, I had no other choice. For me, speaking out about hepatitis C became a moral imperative.

As with the AIDS epidemic, my first efforts to speak out about HCV were via the gay press, since this was a population I was very worried might be at risk, notwithstanding the reassurances of public health officials and guidelines that mostly but not always reassured us that sexual transmission was virtually negligible. The greatest priority, of course, was to reach those at greatest risk—injecting-drug addicts, a constituency that has no political representation or press but which has become accustomed, via AIDS, to getting its health care information from the gay, lesbian, bisexual, and transgender (GLBT) community (organizations, press). So I wrote what was not by any means the first media piece on HCV but the first to raise the question of sexual transmissibility and prevalence in the gay community. Entitled "Is Hepatitis C a Sexually Transmitted Disease?" the piece appeared in *Lesbian and Gay New York* (*LGNY*).[4] *LGNY* editor Paul Schindler followed up with his own investigations.

These first pieces were being published in the midst of what was being called locally "the Sex Panic Wars." Reacting defensively to two highly controversial and widely discussed exposés of unsafe sex and drug use and reescalating rates of HIV in the gay community and the implications thereof—Gabriel Rotello's *Sexual Ecology* and Michelangelo Signorile's *Life Outside*—a group called Sex Panic had formed. It claimed that Rotello, Signorile, and of course legendarily critical ACT-UP founder Larry Kramer were erotophobic and trying to return us to conservative values and restraints that were viewed as inimical to the cultures of gay and sexual liberation. As the editor of a new collection called *We Must Love One Another or Die: The Life and Legacies of Larry Kramer,* I wrote in defense of Rotello, Signorile, and Kramer, alongside the pieces about HCV and another subject: the other sexually transmitted diseases (STDs), such as syphilis, gonorrhea, herpes, and the hepatitides, those that had been virtually forgotten in the tireless debates about the relative safety of oral sex for HIV transmission. What struck me most was the extent to which this was all precisely a redux, a repeat, of what were, de facto if not by actual name, the Sex Panic Wars in 1981—the same arguments and a lot of the same players.

Despite my championing of their work and perspectives, and notwithstanding the pertinence of my points about HCV and the other STDs, none of these individuals acknowledged any of this work or these issues in the ensuing flood of interviews, roundtables, letters to the editor, and articles involving them during that period, apart from a flattering mention of my early AIDS coverage in *Sexual Ecology*. What I saw as my inability to get more explicit support, however implicit such support would seem from my own work and my friendships with them, was such that I called them on it. That is, I spoke to them, exchanged e-mails, etc. In response, Signorile advised that I try to be "bolder" in my outreach and confrontations. Rotello seconded this advice. Later, when Larry Kramer told me he was working on an op-ed piece on the Sex Panic Wars for the *New York Times* and wanted to know what I thought of it all, I conveyed my perspective of this being just one big repeat of the 1981 debates, a perspective that subsequently emerged as Larry's own in the *New York Times* editorial.

Next, I went to *Out* magazine, which had the largest circulation of any gay publication. Although I had never written for it, I naively assumed that my contacts and closeness, or what I thought of as such, with Signorile, Rotello, and Kramer and my credentials as a cofounder of Gay Men's Health Crisis (GMHC) and as the first to write about AIDS, were such that my proposal to do a serious piece on all this—on hepatitis C, the other/forgotten STDs, and Sex Panic—would be taken seriously. It was not. The editor at the time did look at the proposal, but I was subsequently informed by a staffer that "these issues" would be covered by the magazine's own people. They never were. The *Advocate* proved more accessible. Instead of having me do a piece, a staffer interviewed me. In the course of this small, inconspicuous piece, my concerns reached what was to be their widest audience in the gay press.[5]

What was so unsettling was how closely this inaccessibility to the gay press paralleled my experience in the early days of the AIDS epidemic. In early 1982, I wrote a piece called "An Epidemic Q & A," containing basic questions and answers about AIDS, that most of the gay press, including the *Advocate*, declined to publish at a time when it was not otherwise publishing any other articles/information about the epidemic. Meanwhile, in early 1982 I wrote what would have been the first feature article on AIDS to be published in the mainstream press (if the *Village Voice* could ever have been designated as such) in the New York City area. Sluggishly and reluctantly commissioned by the *Voice*, it was entitled "The Most Important New Public Health Problem in the United States," which is what Dr. James Curran, head of the newly formed CDC task force on the epidemic, was calling it. The commission was at the initiative of Arthur Bell, who, however, was understandably so preoccupied with his own writing about his descent into blindness from diabetes that his support was never more than tepid. By contrast there was no support whatsoever from Richard Goldstein, who otherwise seemed to enjoy picking my brain (I was regularly invited to visit with him at his home and discuss issues) but with whom my friendship had become strained in the aftermath of my writing for the *Native* having already scooped the biggest story of the era. After sitting on the article for weeks, senior editor Karen Durbin (whom Goldstein has succeeded in that position) finally returned my calls and explained, "It's not a *Voice* piece." "But it's crucial that this basic information get out," I protested. "Can't we edit it?" At the other end of the phone there was silence. Months passed before the *Voice* would finally publish a feature, a personal story by Stephen Harvey that conveyed some of the bewilderment and disruption in the life of a gay man with this new disease that had yet to be named but which told little of the facts and seriousness of the epidemic.[6] In those days, of course, we didn't know what we do now. People were uncertain, mistakes were made, and I didn't have much of a track record as a writer and journalist. But now, here it was, 1997, with the example of AIDS and my own record established, and the same thing was happening again with a comparable emerging public health crisis, hepatitis C.

Having done everything I had the wherewithal to do to get the gay press and the para-gay *Village Voice* to deal with hep C, I now turned my efforts to the mainstream press. Coincidentally, at this time, I had met a *New York Times* staffer at my gym—John Wilson, an associate editor in the "Science Times" division. Over several dinners, I told John of my concerns, especially about hepatitis C, and of the necessity of having more and better information reach a broader audience. John was an assistant to Linda Villarosa, an openly lesbian editor of the "Science Times" section. He suggested I speak with her. Carefully following his suggestions, I sent her several communications, including my bio, my four books, and the fact that my first feature article on the AIDS epidemic, "Cancer in the Gay Community," was among the opening displays of the Newseum, the new museum of journalism and the media in Arlington, Virginia. Notwithstanding what, again, I would have thought were significant credentials, and with all due respect to the legendary workloads that *Times* editors and writers have to manage, and notwithstanding one communication that indicated some awareness of me and my work, I was treated defensively. Instead of being invited to discuss my topic and subjects in a reasonable manner and setting, I was asked to pitch a proposal (in writing), which I did. The subject was HCV, and the piece was to be called "The Emergent and Preeminent Public Health Problem of the Twenty-First Century." The pitch was more informational than stylistic, and to the extent that it made any effort at being stylistic, it was written in what I called "the dry, reportorial style of the *Times*." Some weeks later, after repeated pleas for follow-up, I finally got a reply from Villarosa. "Your tone is so dry. Is there anything you really, really care about?" Insensitive as this was, I answered that I did indeed care about HCV but wondered if she might not be looking for something more personal, like the story of what it was like to be a middle-aged doctor who was also a writer always desperate to find time for his writing, who had done a lot of unpaid independent and community work and who just lost much of his job in the wake of managed care. I never heard back from her. Later, when I met her and her daughter, Kali (named after the Hindu goddess), at a party, she said, "We'll talk." We never did. A month or so later, the *Times* ran a piece on HCV by Denise Grady that was typically defensive in approach. Though bigger than a news item, it was not a front-line feature. On the contrary, it was more devoted to dispelling alarm than raising concern. Entitled "How Widespread Is the Threat?" it was mostly about how this was not even a real epidemic, since the numbers of new cases had dropped.[7] The implication was that this is not something for most real folks to be very concerned about, since it was primarily confined to drug addicts.

The encountering of this brick wall of complacency had a Kafkaesque quality, as we say about a situation that seems surrealistically out of time. I turned to Larry Kramer again, who at one point took Villarosa's side. "I know what she means," he said. "Your tone can be very dry." Though an academic tone, in my case reflecting my training and orientation as a physician, could indeed be a problem, it wasn't my tone, any more than it was Larry's tone of stridency, that prevented us from getting a better, fairer hearing—he in the early days of AIDS, and me, now, for HCV in its early period of unfolding before the public. The problem was the complacency and conservatism of the media and their bureaucrats. Meanwhile, Kramer continued to be my principal resource and support. He next advised me to call Adam Moss, editor of the Sunday *New York Times* magazine, which had terminally suffocated Kramer's passionate voice in the one piece he did for it on AIDS drugs, drug companies, and price gouging. The *Times* magazine had yet to do anything on HCV. Clearly, Kramer appreciated my passion and sent his own strong note of support to Moss. So I called Moss and was advised to send in a proposal, which I did. A week later, I got an e-mail from Moss's assistant, Katherine Haughton, politely explaining that they didn't think such a piece would be right for the magazine. So I sent an e-mail back to her, to Moss, and also to Linda Villarosa, John Wilson, and Larry Kramer saying, very simply and clearly, what I still believe to be true and pertinent: the *New York Times* may be our newspaper of record, but it has some pretty glaring failures in its history. In the e-mail, the two great examples of relegation and foot-dragging I cited were the early reports of mass murders of Jews in what became the Holocaust of World War II and the early unfolding of AIDS.

As is now thoroughly documented in every major history and dramatization of AIDS (e.g., Shilts's *And the Band Played On* and Kramer's *The Normal Heart*), the *New York Times* was notorious in its downplaying of the epidemic during the first two years of reporting. There were front-page stories on a flu epidemic among Lippizaner stallions, but AIDS had yet to rank a front-page article. In that early period, when Larry Kramer was literally screaming bloody murder, trying to raise the alarm to fourth level, I wrote a piece in the gay press called "The Case against Panic." In retrospect, Larry was right on target, and I was not. As a result of Larry's very unacademic passion and persistence, we eventually got GMHC, ACT-UP, major press coverage, and, ultimately, whole new classes of antiviral and experimental drugs, to say nothing of countless legal considerations and an array of other support.[8] Though nowhere near as vociferous as Larry, I'm proud to say that I was otherwise allied with him in most viewpoints and objectives, especially our shared anger at the lack of first-quality, front-line mainstream coverage, especially in the *New York Times*. When I directly complained to Lawrence Altman about the paucity and relegation of AIDS coverage in the *Times* at the second international AIDS conference in Atlanta in 1985, Altman said: "That's because we're not an advocacy journal." For the record, Altman never returned my calls to him about HCV.

Just as the *Times* so catastrophically downplayed the seriousness of the AIDS epidemic in those earliest years of reporting, I said in that e-mail to Moss, Villarosa, and the others, so is it doing now with hepatitis C. I never heard anything further from any of them, except Larry Kramer.

When I related these developments to Larry Kramer—the one colleague who, however preoccupied with AIDS and his own life and work, and however critical he might be about my approach, fully understood and supported my position—he suggested I try *New York* magazine, where he would put in a strong word of recommendation with Maer Roshan, features editor there. I also knew Roshan. Eventually, the piece was commissioned, and it was accepted and published in March 1999. But so much important time had passed that any feelings of achievement were considerably blunted. Another reason my enthusiasm was tempered is that, apart from a cover call-out ("Is hepatitis C the new AIDS?") that I had to push very hard for, the piece did not get the front-page, banner headline attention that was the whole point.[9] Apart from a cover story in *U.S. News & World Report* in 1998,[10] that kind of front-page coverage hadn't happened in a major publication.

There is one other big media consideration for HCV, which I'd hoped to secure in efforts to win out over cover stories on fashion empires and gourmet wars. Larry Kramer may have had a Herculean impact on the broadcasting of the news about AIDS, but probably the biggest jumps in public consciousness took place, first, when it was learned that Rock Hudson was dying of AIDS and, second, years later, when Magic Johnson was diagnosed with the disease. Despite considerable efforts at sleuthing, no comparable superstar name emerged with hepatitis C, though that was just a matter of time. In the absence of such a "hook," it was unlikely that a cover story on HCV would appear in *New York* or the *New York Times*.[11]

A footnote to the *New York* piece, for which I interviewed Dr. Douglas Dieterich, a leading physician-advocate for improved HCV testing, treatment, and media coverage: a year later Dieterich himself came out of the HCV closet, acknowledging that he had undergone treatment for the disease (which he had acquired as a health care provider) during the previous year and was now in sustained remission.

The other big piece of this puzzle is the CDC and its directives. From the earliest days of the AIDS epidemic, superior leadership was provided by Dr. James Curran, who took the bold step, very early, of telling gay community physicians how serious this epidemic was. I remember him looking us in the eyes as he said: "We are going to be living and working with this epidemic for the rest of our lives." Beyond that, as I've already noted, it was Curran who first captured the seriousness of the epidemic in phraseology for the public, in calling it, officially, "The Most Important New Public Health Problem in the United States." Even if we couldn't get the media to devote commensurate attention to this epidemic, public health officials and agencies seemed to be in the right place about it.

When it came to finer points of behavioral advice, however, the caution and reticence of the CDC were sympathetic but, as history has shown, excessive. With extreme sensitivity to the fragility of the gay community, a minority without civil rights protections, and under considerable pressure from us, and in the wake of the swine flu disaster, CDC epidemiologists and officials erred on the side of caution in not issuing stronger, more explicit behavioral recommendations early on. What they failed to do was to explicate something closer to what Larry Kramer was so crudely trying to say when the character Dr. Brookner in his play *The Normal Heart* calls upon gay men to "just stop having sex." In the absence of better, more conclusive evidence that AIDS was being caused by a single primary organism, the CDC issued guidelines that were too general: to limit the number of partners with whom you have sex.[12] Now, in the case of HCV, there aren't even recommendations for routine testing of those with a history of multiple sexual partners and high-risk sex, even involving persons known to be hepatitis C positive. Of course, there are differences between HCV and HIV. The former is not very transmissible sexually, not nearly so much as hepatitis B or HIV. Notwithstanding this important point, however, I would propose that the same excessive reticence with which the CDC failed to issue stronger, more presumptive guidelines about sexual behavior in the earliest period of AIDS is reflected in the current CDC guidelines and recommendations for HCV, those issued in the October 16, 1998, *MMWR.*[13] Since we don't know for sure, and in the best possible interests of protecting public health, even at the risk of being wrong and spending too much money on unnecessary testing, shouldn't we be erring on the side of having stronger recommendations for routine testing and affirming the possibility of sexual and nasal transmission and precautions? To recapitulate, I'd say that what's most striking about the unfolding of the "epidemic of recognition" known as hepatitis C is the extent to which the mistakes that were made with AIDS are being made again, and with regard to two of the public health's biggest institutions: the *New York Times* and the CDC.[14]

"What are the critical elements that determine whether a particular ailment or disability is recognized as a public health issue or whether it remains hidden, the experience of isolated sufferers," asked the editors of this volume. At a symposium on HCV in May 1998, epidemiologist Dr. Sharon Stancliffe expressed her concern that in light of the disease's containment within high-risk populations, we might create fear and panic by sounding the alarm at a broad public level. On the surface Dr. Stancfliffe was right. In the United States, AIDS never did enter the heterosexual mainstream to the extent that was feared and predicted by some, but the reach and depth of its devastation, which Dr. Stancliffe of course acknowledges, are such that exaggeration was all but impossible. This is likewise the case, I would propose, with HCV.

That *New York Times* piece by Denise Grady was entitled "Hepatitis C: How Widespread Is the Threat?" In retrospect, that approach, of emphasizing its marginalization, gained us very little in dealing with AIDS. With AIDS, it was important not only to target the affected populations for education and support groups and fund-raising but also to maintain general public awareness and concern. Especially since so many of the marginalized are isolated as such, their only sources of information are those of the general public. Drug addicts have very little in the way of information networks and hotlines and support groups. At Beth Israel, the country's leading provider of methadone maintenance programs and services, we don't even have routine HCV testing for a population that is believed to be up to 90 percent positive, to say nothing of support groups, which the CDC should be commended for urging. With HCV, we need to do what we did with AIDS: raise the alarm and with the broadest possible public. The disease needs to be boldly and honestly presented. Headline and front line, it needs to come out of the closet. That's how AIDS activists were able to have such an impact not only on AIDS but on medical research and health care altogether in this country. The approach cannot and must not be to try to further marginalize those affected in the interests of not alarming or presuming upon the public.

Following the publication of my piece on hepatitis C in *New York* magazine, I continued my efforts to reach out to the media, especially in light of a significant development, a kind of proof of the whole issue of media obfuscation that even I had missed out on in the *New York* piece. On October 15, 1998, the U.S. Committee on Government Reform and Oversight issued a report to the Committee of the Whole House on the State of the Union entitled: "Hepatitis C: Silent Epidemic, Mute Public Health Response." Here are some of the conclusions of that report:

[1.] Called "the silent epidemic," the spread of Hepatitis C Virus . . . has evoked a Federal public health response almost as mute. . . .

[2.] Plans to "look back" for people infected through blood have sputtered, and little has been accomplished. Disease reporting and surveillance is uneven. Research into HCV is uncoordinated. . . .

[3.] Unless confronted more boldly, more directly, and more loudly by the Department of Health and Human Services . . . , the threat posed by hepatitis C will only grow more ominous. . . .

[4.] The time for pondering the appropriate, pro-active public health response to hepatitis C is past. The time for aggressive implementation is at hand.[15]

Using this report as a touchstone, I approached several publications to consider doing editorials, in the absence of more substantial features and in light of this epidemic's inevitable political fallout, a phenomenon that was already happening. When you have an editorial about something, it underscores its seriousness and importance. Already, hepatitis C was the number one cause for liver transplant requests, a demand that is expected to increase as much as eightfold in the next fifteen years. Already, disability claims by veterans had reached crisis proportions, prompting surveys by Veterans Administration hospitals that proved what was most feared: that escalating percentages (in some studies as high as 50%) of the country's population of 3.5 million veterans were testing positive for HCV.[16] A bill was presented to Congress that would give Vietnam-era veterans with HCV service-connected disability. In the aftermath of his aggressive promotion of this vastly expensive enterprise, Ken Kaiser was coincidentally not rehired as undersecretary of health for the Veterans Administration in Washington.

The first periodical I approached about doing an editorial on this emerging catastrophe primarily afflicting the underserved and disenfranchised was the *Nation*. There, I managed to persuade senior editor Richard Lingeman to commission me to do an editorial, which I submitted a week or so later. Though it was somewhat longer than what he had asked for, I submitted it with segments blocked off for potential cutting. The following week I was informed that the piece had been killed. There was no explanation and no offer to revise it, no reassurance that the subject would be otherwise dealt with, and to my knowledge it has not been.

Next, I approached Mary Suh, head of the op-ed page of the *New York Times*, and I encountered a different kind of resistance. Suh felt that the problem was that we had no "hook." Had a megacelebrity recently been revealed to have the disease, or were the release of a major new study imminent—something more than the political fallout and media obfuscation and general urgency I was so concerned about—then, she implied, we might have something to go with. Otherwise, she just couldn't see it. The fact that the *Times* had yet to have an op-ed editorial or opinion piece about this major public health crisis that was so seriously afflicting and killing so many New Yorkers didn't phase her. That wasn't enough. Timing, finding a "hook," that's what was missing—the way just such a hook seemed to be missing and prevented the *New York Times* from doing feature coverage, to say nothing of an editorial, in the early period of AIDS.

In some desperation, I now inquired again with Richard Goldstein at the *Village Voice,* whom I'd criticized in print for his being a principal player in the *Voice's* obfuscation of the early unfolding of AIDS. Politely, he offered to let me submit my ideas to one of the staffers, who might then utilize them for his or her own report. In other words, he was unwilling to consider anything by me on its own terms. I e-mailed Richard back, saying that, notwithstanding any bad feeling there may have been between us, I was shocked at his behavior in light of the seriousness of the health issues we were talking about and my own widely recognized work vis-à-vis AIDS coverage, in real contrast to his and the *Voice's* during the early period of AIDS. Whatever your feelings about me, I urged, the *Voice* needs to do something major on this major New York City, national, and international public health crisis. One of the *Voice's* science writers, Mark Schoofs, otherwise deeply involved in covering the holocaust of AIDS in Africa, for which he was awarded a Pulitzer Prize (1999), seemed an ideal choice. To date, the *Voice* has yet to do anything more than the occasional item on hepatitis C (e.g., a story on New York City police officers being denied disability benefits for HCV). An interesting aside to all this is that the medical director of addiction treatment programs at Greenwich House in 1998–99 and subsequently at Beth Israel, Dr. Randy Seewald, is married to Jeff Simmons, a managing editor of the *Voice.*

In July, I took my case to the Gay Men's Health Summit Conference in Boulder, where I did a presentation called "Hepatitis C and the Gay Community," which discussed at some length the implications of the certainty that hepatitis C is emerging as an STD in that community. At least in part because of gay community internecine politics—the Gay Health Summit Conference was organized and headed by Eric Rofes, a Sex Panic figurehead considered an unmitigated enemy of gay sex and lifestyle critics Rotello, Signorile, and Kramer—the gay press's coverage of the event was muted.

My final, rather desperate and pathetic effort to garner greater media attention for HCV took place at a cocktail party. I had been asked to be on the host committee of a benefit for prominent media persons with the National Lesbian and Gay Journalists Association (NLGJA). There was some controversy about the event, which was honoring former mayor Ed Koch, who had been indicted by gay politicos as a closeted homosexual and who had mirrored the *New York Times* in impeding early responses to AIDS, as well as Norman Pearlstine, editor and chief of *Time* magazine, who had likewise come under fire for *Time*'s having overlooked gays and lesbians in its summary coverage of people, events, and communities of significance in its end-of-the-millennium feature issue. As I later explained to Michelangelo Signorile, who criticized the event for political incorrectness in a piece for the *New York Observer,* I had an ulterior motive for participating: I had hoped to corner Mike Wallace, another of the evening's honorees, to see if I might not get a mini-hearing for my plight re hepatitis C. Well, I did get that intimate moment with Mike Wallace. I knew I would have to choose my words carefully and would have to be terse. Very quickly and crisply, I explained that I was a physician and that I had been the first journalist anywhere to write about AIDS, etc., but that there was now another, comparably serious public health crisis that was still largely hidden, the way AIDS was early on. Used to being "hit on," he immediately went into gear, cutting me off and telling me to put whatever it was in the form of a proposal and send it not to "60 Minutes" but to an alternative news show that CBS had developed with which he had no direct linkage. He made it crystal clear that there was to be no further discussion of it at that moment. Nor was there any offer to use him as a reference or contact person.

In retrospect, I have tried to cross-examine myself as to whether my appeal for greater attention to HCV is not overwrought. After all, there are other illness that are chronically with us, that continue to wreak devastation on a great scale, including some of those specifically under discussion here–for example, tuberculosis. The question that needs to be asked is whether hepatitis C should be thought of as any different in character and importance and priority than hepatitis B or tuberculosis. Because the identification and characterization of hepatitis C are so recent and public and health care provider awareness of it is still so limited; because of the much greater incidence, chronicity, and seriousness of complications (notwithstanding its typically prolonged incubation period of as long as two or three decades), together with it's the illness's greater mortality; because of the co-morbidity, expense, and unavailability of treatment (since the great majority of those infected are in underserved communities); and because of the currently severe limitations of resources for public and patient education, it is my contention that in this crucial period of characterizing HCV and educating the public about it, the most appropriate illness for analogy with it is not tuberculosis or hepatitis B so much as AIDS. Once a commensurate level of public health priority and public awareness is achieved, HCV may begin to take its place alongside the other major public health concerns with which we continue to do battle but which no longer require the kind of banner headline attention they did in earlier phases of characterization and emergence before the public–e.g., tuberculosis, smoking, AIDS.

Several years ago, in the heyday of the wars in Bosnia, Susan Sontag, who had spent some time there as a humanitarian ambassador, wrote an essay about her experience that was published in the *New York Review of Books*. One of the things we thought we learned from the Holocaust, she said, was that with today's advanced media, with the ubiquity of television, camcorders, instant communications anywhere and everywhere, what happened in World War II, specifically genocide, could probably never happen again. Bosnia, she concluded, had fully demonstrated that to be not true.[17] Even though she was making a point that had in fact already been made by others vis-à-vis the Killing Fields massacres in Cambodia, she was dramatically and unassailably correct. In this context, I am not surprised that we have learned so little from the AIDS epidemic, that we are making the same mistakes again, and so soon. We might add a line to Santayana's observation that those who do not remember the past are condemned to repeat it: But even when they do remember the past, they're likely to repeat it if they think they can get away with it.

NOTES

1. Mary Jeanne Kreek, presentation on hepatitis C at the American Methadone Treatment Association Conference in New York City, September 1998.

2. See the following histories of AIDS and the media: Sandra Panem, *The AIDS Bureaucracy: Why Society Failed to Meet the AIDS Crisis and How We Might Improve Our Response* (Cambridge: Harvard University Press, 1988), 123. Edward Alwood, *Straight News: Gays, Lesbians and the Media* (New York: Columbia University Press, 1996), 211–212. John-Manuel Andriote, *Victory Deferred: How AIDS Changed Gay Life in America* (Chicago: University of Chicago Press, 1999), 49–51. "Cancer in the Gay Community," by Lawrence D. Mass, M.D. (*New York Native*, July 27, 1981, cover story), the first feature article on the AIDS epidemic, was among the opening displays of the Newseum, the museum of journalism and the media in Arlington, Va. (see Andriote, *Victory Deferred*, p. 2 of illustrations).

3. Miriam J. Alter et al., "Recommendations for Prevention and Control of Hepatitis C Virus (HCV) Infection and HCV-Related Chronic Disease," *MMWR*, October 16,

1998, 1–38. "Experts Warn of a Tripling of Deaths from Hepatitits C by 2017," *New York Times,* March 27,1997, sec. A, p. 21.

4. Lawrence D. Mass, M.D., "Is Hepatitis C an STD?" *Lesbian and Gay New York,* May 25, 1997, 1.

5. John Gallagher,"Forgotten but Not Gone: Other Sexually Transmitted Diseases," *Advocate,* July 8, 1997, 35.

6. Panem, *The AIDS Bureaucracy;* Alwood, *Straight News,* 218.

7. Denise Grady, "Hepatitis C: How Widespread Is the Threat?" *New York Times,* December 15, 1998, Science Times, F1, F9.

8. See Rodger McFarlane, "We Must Love One Another or Die: Larry Kramer, AIDS Activism and Monumental Social Change," in Lawrence D. Mass, ed., *We Must Love One Another or Die: The Life and Legacies of Larry Kramer* (New York: St. Martin's Press, 1999), 271–281.

9. Lawrence D. Mass, M.D., "The Body Politic: C-Sick," *New York,* March 29, 1999, 20–21.

10. Nancy Shut, "Hepatitis C: A Silent Killer," in "Hepatitis C: The Next Epidemic," *U.S. News & World Report* cover story, June 22, 1998, 60–66.

11. On August 6, 2001, the *New York Times* published its first and thus far only front-page story on hepatitis C: "A Health Danger from a Needle Becomes a Scourge behind Bars: Prison Authorities Seek a Response to High Hepatitis C Rates," by David Rohde, A1, B9.

12. Larry Kramer, *The Normal Heart* (New York: New American Library, 1985); Randy Shilts, *And the Band Played On: Politics, People, and the AIDS Epidemic* (New York: St. Martin's Press, 1987); see also Lawrence D. Mass, "Five Interviews from the First Two Years of the AIDS Epidemic (1981–1983)," in *Homosexuality and Sexuality: Dialogues of the Sexual Revolution,* vol. 1 (New York: Haworth/Harrington Park Press, 1990), 113–161. See also Lawrence D. Mass, foreword to *The Golden Boy,* by James Kenneth Melson (New York: Haworth/Harrington Park Press, 1992), vii–xvi.

13. Alter et al., "Recommendations for Prevention and Control of Hepatitis C," 18–33.

14. "Hepatitis C and the Chemically Dependent Patient: A Symposium," sponsored by the New York State Office of Alcoholism and Substance Abuse Services, the U.S. Center for Substance Abuse Treatment, Mount Sinai Hospital, and New York University Medical Center, March 1, 1999. See Chapter 11 in this volume.

15. U.S. Government Committee on Government Reform and Oversight, submitted report to the Committee of the Whole House on the State of the Union, "Hepatitis C: Silent Epidemic, Mute Public Health Response," October 8, 1998, originally available on the Internet at www.house.gov/reform/hr/reports/hepatitis.htm. For further information, contact the Committee on Government Reform and Oversight

16. Ibid.—see Koop reference on p. 10 of this report.

17. Susan Sontag,"Godot Comes to Sarajevo," *New York Review of Books,* October 21, 1993, 52–60.

Deborah Barrett, Ph.D., is a sociologist and psychotherapist in Chapel Hill, North Carolina. Her research includes the study of international population discourses and national population policies as well as the sociology of chronic illness.

Ruth L. Berkelman, M.D., is a professor in the Department of Epidemiology, Rollins School of Public Health, Emory University. Her research interests include emerging infectious diseases and bioterrorism, disease surveillance, and public health policy.

Peter J. Brown, Ph.D., is a professor of anthropology and international health at Emory University. He is also director of the Center for Health, Culture, and Society. His research interests include gender and health, especially in minority populations; obesity; the social epidemiology of suicide; the social organization of an Alzheimer's Special Care Unit; and cultural factors in the prevention of chronic disease.

Steven Epstein, Ph.D., is an associate professor of sociology at the University of California, San Diego. He is also affiliated with UCSD's interdisciplinary Science Studies Program. His research interests include the study of controversy, expertise, and knowledge production in the biomedical arena.

Phyllis Freeman, J.D., is coeditor of the *Journal of Public Health Policy* and a professor in the College of Public and Community Service and John W. McCormack Graduate School of Policy Studies, University of Massachusetts, Boston.

Howard Frumkin, M.D., M.P.H., is an internist, occupational medicine physician, and epidemiologist who is a professor and chair of environmental and occupational health at Emory University's Rollins School of Public Health and professor of medicine at Emory Medical School.

**Diane E. Goldstein** is an associate professor of folklore at Memorial University of Newfoundland. Her areas of research include belief studies, folk medicine, folk religion, supernatural traditions, applied folklore, the ethnography of speaking, and narrative.

**Peter J. Krause, M.D.,** is a professor of pediatrics at the University of Connecticut School of Medicine. His research is in tick-borne diseases, with a particular focus on babesiosis, Lyme disease, and human granulocytic ehrlichiosis.

**Howard I. Kushner, Ph.D.,** is the Nat C. Robertson Distinguished Professor of Science and Society at Emory University. Kushner has a joint appointment in the Rollins School of Public Health and in the Graduate Institute for the Liberal Arts and is associate director of the Center for Health, Culture, and Society. His current research includes a collaborative investigation of Kawasaki disease and of risk and protective factors in addictive behaviors.

**Lawrence D. Mass, M.D.,** is an attending physician in the Department of Medicine of Beth Israel Medical Center in New York City, where he is a unit direct in the Division of Addiction Treatment Services/MMTP. He is a co-founder of Gay Men's Health Crisis and was the first journalist to write about AIDS in the United States.

**Michelle Murphy, Ph.D.,** is an assistant professor in the History Department and the Institute of Women and Gender Studies at the University of Toronto. She is interested in the history of technoscience, health, and social movements in the twentieth century.

**Lydia Ogden, M.A., M.P.P.,** is the associate director for policy and planning of the Global AIDS Program (GAP) at the Centers for Disease Control and Prevention. Her responsibilities include legislation, budget development, interface with the Department of Health and Human Services and the administration more broadly, and strategic planning.

**Randall M. Packard, Ph.D.,** is the director of the Institute of the History of Medicine at the Johns Hopkins University and has a joint appointment in the Bloomberg School of Public Health. He works on the social history of disease and the history of international health. His current research is on the global decline and resurgence of malaria.

**Sandy Smith-Nonini, Ph.D.**, is an assistant professor in the Department of Sociology and Anthropology at Elon University. Her research interests are the anthropology of health and medicine, cultural studies of the state, economic globalization, and Latin America.

**Ellen Griffith Spears** is a doctoral candidate in American studies and A. Worley Brown Southern Studies Fellow at Emory University. She worked at the Atlanta-based Southern Regional Council from 1991 to 2002, serving as managing editor of the council's journal, *Southern Changes,* and associate director from 2000 to 2002. She is currently working on a social history of environmental health in Anniston, Alabama.

**Andrew Spielman, Sc.D.**, is a professor of tropical public health at the School of Public Health and Center for International Development, Harvard University. His research interests are in the dynamics of transmission of vector-borne infection, with a particular focus on West Nile virus, malaria, and babesiosis.

**Colin Talley, Ph.D.**, is an associate research scholar at the Center on Medicine as a Profession, Columbia College of Physicians and Surgeons. He is currently studying the relationship between the medical profession and the tobacco industry.

**Sam R. Telford III, Sc.D.**, is an associate professor in the Department of Biomedical Sciences at the Tufts University School of Veterinary Sciences. His research interests are in the ecology and epidemiology of vector-borne infections, particularly Lyme disease, ehrlichiosis, babesiosis, tick-borne encephalitis, and tularemia, and in wildlife parasitology.

**Christian Warren, Ph.D.**, directs the Rare Book Room and Historical Collections at the New York Academy of Medicine and coordinates the activities of the Academy Fellows' Section on Historical Medicine. Warren is a historian of nineteenth- and twentieth-century America, and his research focuses on the social dynamics of health, class, race, and the natural and built environments.